Interactions of Nanomaterials with Emerging Environmental Contaminants

ACS SYMPOSIUM SERIES **1150**

Interactions of Nanomaterials with Emerging Environmental Contaminants

Ruey-an Doong, Editor
National Tsing Hua University
Hsinchu, Taiwan

Virender K. Sharma, Editor
Florida Institute of Technology
Melbourne, Florida, United States

Hyunook Kim, Editor
University of Seoul
Seoul, Republic of Korea

Sponsored by the
ACS Division of Environmental Chemistry

American Chemical Society, Washington, DC

Distributed in print by Oxford University Press

Library of Congress Cataloging-in-Publication Data

Interactions of nanomaterials with emerging environmental contaminants / Ruey-an Doong, editor, National Tsing Hua University, Hsinchu, Taiwan, Virender K. Sharma, editor, Florida Institute of Technology, Melbourne, Florida, United States, Hyunook Kim, editor, University of Seoul, Seoul, Republic of Korea ; sponsored by the ACS Division of Environmental Chemistry.

pages cm -- (ACS symposium series ; 1150)

Includes bibliographical references and index.

ISBN 978-0-8412-2916-7 (alk. paper)

1. Pollution control equipment--Materials--Congresses. 2. Catalysts--Congresses. 3. Nanostructured materials--Industrial applications--Congresses. 4. Nanostructured materials--Environmental aspects--Congresses. 5. Pollution--Congresses. I. Doong, Ruey-an. II. Sharma, Virender K. III. Kim, Hyunook. IV. American Chemical Society. Division of Environmental Chemistry.

TD192.3.I58 2013

620.1'159--dc23

2013039591

The paper used in this publication meets the minimum requirements of American National Standard for Information Sciences—Permanence of Paper for Printed Library Materials, ANSI Z39.48n1984.

Distributed in print by Oxford University Press

PRINTED IN THE UNITED STATES OF AMERICA

Foreword

The ACS Symposium Series was first published in 1974 to provide a mechanism for publishing symposia quickly in book form. The purpose of the series is to publish timely, comprehensive books developed from the ACS sponsored symposia based on current scientific research. Occasionally, books are developed from symposia sponsored by other organizations when the topic is of keen interest to the chemistry audience.

Before agreeing to publish a book, the proposed table of contents is reviewed for appropriate and comprehensive coverage and for interest to the audience. Some papers may be excluded to better focus the book; others may be added to provide comprehensiveness. When appropriate, overview or introductory chapters are added. Drafts of chapters are peer-reviewed prior to final acceptance or rejection, and manuscripts are prepared in camera-ready format.

As a rule, only original research papers and original review papers are included in the volumes. Verbatim reproductions of previous published papers are not accepted.

ACS Books Department

Contents

Indexes

Preface

Emerging environmental contaminants are newly identified or previously unrecognized pollutants, which primarily include human and veterinary pharmaceuticals and personal-care products, surfactants, plasticizers, flame retardants, metals and metalloids, various industrial additives, pesticides, and pesticide metabolites. The use of novel nanomaterials with unique characteristics has been demonstrated to increase the removal efficiency of emerging pollutants, which provides a promising strategy to control the distribution of environmental contaminants. In addition, nanomaterials also can serve as ideal platforms for precise and accurate detection and sensing of emerging contaminants in the environment and biological fluids; this is because of their novel characteristics on optical and electrochemical properties. The interactive research of nanomaterials with emerging environmental contaminants will improve our understanding of the implication and application of nanomaterials in the environment.

This book is derived from the symposium "*Interactions of Nanomaterials with Emerging Environmental Contaminants*" at the 244th ACS National Meeting in Philadelphia during the fall of 2012 sponsored by the American Chemical Society (ACS) Division of Environmental Chemistry. Many topics addressing issues of interaction of emerging environmental pollutants with nanomaterials (including physical, photochemical, and biological interactions) were presented in this symposium, and they constitute the main content of this book.

This book contains 12 peer-reviewed chapters that cover various aspects of interaction of various nanomaterials with environmental contaminants. These chapters can be organized into two major sections: (I) interaction of nanomaterials with biomolecules for biosensing and detection of contaminants (Chapter 1-4) and (II) interaction of nanomaterials with contaminants to enhance the removal efficiency and rate of emerging pollutants (Chapter 5-12). Chapter 1 by Ren, Zang, Qie, and Baker provides an overview of the adjuvant effect of emerging nanomaterials. The nanoparticles can serve as drug delivery to deliver antigens to species targets. In contrast, ambient nanoparticles exhibit adverse effect on human and ecological health. To address the accurate and precise detection of environmental contaminants in the environment (as well as in the human body), Chapter 2 and Chapter 3 have developed photoluminescent gold nanomaterials — including gold nanoparticles and gold nanodots — for the monitoring and detection of metal ions, proteins, and bacteria in an aquatic ecosystem. In Chapter 2, Unnikrishnan and Huang have synthesized luminescent core-shell Au nanodots using different kinds of capping ligands (including alkanethiols, proteins, DNA, thiol derived carbohydrates, and aptamers) as the shell layer for detection and determination of mercury ions, proteins, and *E. coli*. Chang and his team (Chapter 3) have developed several functional Au nanoparticles and Au nanodot-based sensors that allow the sensitive and selective detection of mercury, lead, and copper ions through analytes induced changes in colors, absorption, and fluorescence. In addition, a biosensing system (AlGaN/GaN high electron mobility transistors immobilized with antibodies) is developed by Wang and his team (Chapter 4) to effectively detect a short peptide only containing 20 amino acids, which opens a door for investigation into nanomaterials with biomolecules.

The applications of nanomaterials for various reactions (including adsorption, photocatalytic degradation, and reductive dechlorination) are also addressed. Huang, Padhye, and Wang (Chapter 5) describe the generation of N-nitrosamines from transformation of amines catalyzed by activated carbons. This interaction is important because activated carbon and amine are often used in water treatment plants. In Chapter 6, Du and Jing explore the dynamic adsorption process of propranolol at the TiO_2/water interface on the molecular level. In addition to adsorption behavior, Hung and his colleagues (Chapter 7) combine TiO_2 and carbon nanotubes to adsorb and look at the photocatalytic decomposition of bisphenol A, which is an endocrine disrupting chemical widely existing in the environment.

Iron-based nanomaterials are effective catalysts for the removal of emerging pollutants in the environment. Ren, Han, Al Anazi, Nadagouda, and Dionysiou (Chapter 8) discuss the application of different iron-based nanomaterials for environmental remediation — including water treatment, groundwater remediation, and soil decontamination. Ferrate, ferrites, and TiO_2-composite magnetic iron oxides are used for the removal of contaminants of emerging pollutants. In addition to iron oxide catalysts, zerovalent metals are also common nanomaterials widely used for reduction of emerging pollutants. Su, Tso, Peng, and Shih (Chapter 9) have used nanoscale zerovalent iron (nZVI) and bimetallic Pd/Fe as the dual functional tools for adsorption and reduction of aromatic contaminants (including decabrominated diphenyl ether, hexachlorobenzene, penrachlorophenol, and Congo red). The combination of biological or sequential Fenton treatment on the mineralization of some emerging contaminants is also evaluated in this chapter. In Chapter 10, McPherson, Goltz, and Agrawal summarize the role of polyelectrolyte stabilization and catalytic metal modification in the enhanced performance of nZVI. The addition of polyelectrolyte stabilizers to nZVI decreases particle agglomeration and reduces particle size — resulting in the increase in reactivity and transport in porous media. The modification of nZVI with Pd and Ni for the enhancement of reactivity is also summarized and discussed. In addition, three field studies using Pd-nZVI for the remediation of chlorinated compounds in groundwater are introduced. Sharma, Siskova, and Zboril (Chapter 11) review the recent development of nanoscale zerovalent iron and magnetic bimetallic Fe/Ag nanoparticles with core-shell structures. They also show the usefulness of these nanomaterials on nutrient removal, transformation of halogenated aromatic contaminants, and antimicrobial activity. In addition to zerovalent iron, Lee and Doong (Chapter 12) have demonstrated the feasibility of using another environmentally friendly metal to remove chlorinated compounds. In this chapter, the authors review and discuss the reductive dechlorination of chlorinated hydrocarbons and emerging pollutants by zerovalent silicon and bimetallic Fe/Si and Ni/Si. More importantly, they demonstrate the synergistic effect of nickel ions and polyethylene glycol on the dechlorination rate.

The understanding of interaction of emerging contaminant with various nanomaterials is essential to exploring the applications of nanotechnology in the environment. We hope that this collection will benefit graduate students who are engaged in research and development in the advancement of nanotechnology and environmental science and technology. We wish to thank Anne Brenner and Timothy Marney of the editorial department of ACS for their assistance in preparing this volume and for keeping us on schedule.

Ruey-an Doong
Department of Biomedical Engineering and Environmental Sciences
National Tsing Hua University
101, Sec. 2, Kuang Fu Road
Hsinchu, 30013, Taiwan
radoong@mx.nthu.edu.tw (e-mail)

Virender K. Sharma
Chemistry Department
Florida Institute of Technology
150 West University Boulevard
Melbourne, Florida 32901, United States
vsharma@fit.edu (e-mail)

Hyunook Kim
Department of Environmental Engineering
University of Seoul
Seoul, Republic of Korea
h_kim@uos.ac.kr (e-mail)

Interaction of Nanomaterials with Biomolecules for Biosensing

Chapter 1

The Adjuvant Effect of Emerging Nanomaterials: A Double-Edged Sword

Hong Ren,[1,2] Quanxuan Zhang,[*,3] Liangyi Qie,[4] and Gregory L. Baker[3,†]

[1]Athinoula A. Martinos Center, Department of Radiology, Massachusetts General Hospital, Harvard Medical School, Charlestown, Massachusetts 02129, U.S.A.
[2]Department of Chemistry and Chemical Biology, Harvard University, Cambridge, Massachusetts 02138, U.S.A.
[3]Department of Chemistry, Michigan State University, East Lansing, Michigan 48824, U.S.A.
[4]Department of Geriatrics, Key Laboratory of Cardiovascular Proteomics of Shandong Province, Qilu Hospital of Shandong University, Jinan 250012, China
[*]E-mail: zhangqua@msu.edu
[†]This manuscript is in memory of my advisor and friend, Dr. Gregory L. Baker who passed away unexpectedly while this paper was being written.

Nanoparticles have growing applications in industry, consumer products, biology and medicine. One of those applications involves the interaction between nanoparticles and immune components. The nanoparticles can act as antigen carriers to deliver and release antigens to specific targets, and enhance the immune response against a variety of antigens as adjuvants. The adjuvant effects of nanoparticle size, shape, surface charge, linkage method on the immunological response are also discussed. In contrast, as-prepared nanomaterials and ambient particulate matter (PM) from air pollution exhibit adverse adjuvant effect *in vitro* and *in vivo*, and recent advances to address their potential risk on human health are also included in this review.

Introduction

Nanomaterials are characterized by their sizes which are in the range of several nanometers to several hundreds of nanometers, well below the micrometer range. Because of their nanoscale dimensions and hence large specific surface area, nanomaterials exhibit remarkable physicochemical properties, such as optical property (*1*), catalytic property (*2*), mechanical property (*3*) and drug delivery property (*4*), which are usually not active for their bulk materials. Due to these specific properties, research and development of new nanomaterials have steadily increased, which can be reflected by the increasing number of publications on nanomaterials research, from about 110 publications in 1990 to 4200 publications in 2000 followed by a burst to more than 77900 publications in 2013 (key word: nano, ISI web of knowledge, 05/2013). As a result, nanomaterials, rapidly introduced into electronic devices, construction and composite materials, are more and more present in workplaces as well as consumer products since large-scale producing, handling and processing facilities of nanomaterials are easily available.

Although properties of nanomaterials are impressive from physicochemical viewpoint, they also raise safety concerns of nanomaterials. One major concern is that nanomaterials may lead to potential toxic effect on environment or human health, because nanomaterials readily penetrate cell membranes, travel throughout the body, and deposit in target organs. Therefore, it may trigger injurious responses (*5*, *6*). In recent years, nanotoxicology has become one of the major research focuses in nanoscience. While the number of publications dealing with nanotoxicity study was about 1350 in 2000, this increased to 5500 in 2010, then it rapidly jumped to more than 8200 in 2013 (key words: particle/toxicity, ISI web of knowledge, 05/2013). These clearly indicate that nano toxicology has been widely recognized and gained more and more attention. These studies would help to determine whether and to what extent these properties may present a threat to environment and human beings, and guide applications of nanomaterials in daily life as well (*7*). Among several mechanisms proposed to explain the adverse effect of nanomaterials, generation of reactive oxygen species and oxidative stress has received the most attention (*8*). Dimension, surface chemistry, surface charge and aggregation of nanomaterials are related to their nanotoxicity (*5*). It is difficult to identify the health risk of each new nanomaterials because all material properties need to be taken into account during the toxicity study.

Despite safety concerns that nanomaterials could present adverse effect on human beings, it has been reported that nanomaterials serving as effective vaccine/drug carriers could improve and/or facilitate the extended release of antigens, and hence enhance the immune response level and quality of antigens (*9*, *10*). This has drawn more and more research focuses worldwide in nanomedicine field, especially in vaccine immunology, and nanomaterial has been referred as nano adjuvant to enhance the immunogenicity of specific vaccine or antigen. From ISI web of knowledge (key words: particle/adjuvant, 05/2013), there were only 380 publications on nano-adjuvant research in 2000 and it increased to more than 1350 in 2010, and then rapidly jumped to more than 1800 publications in 2013.

An adjuvant is a substance which is not immunogenic itself, but increases or prolongs immune response if introduced in combination with a vaccine or antigen *in vitro* or *in vivo* (*11*). Ideally, adjuvants should be biodegradable, stable with long shelf life, cheap to produce (*12*). They are introduced to enhance immunogenicity by reducing the administration amount of required vaccine or antigen for protective immunity without inducing immune responses against themselves (*13*). Nano adjuvants take the advantages of nanomaterials to deliver vaccines, drugs, peptides and nucleic acids efficiently to targets, control their release and induce immune response while protecting the integrity of delivered vaccines from enzymatic degradation or from degradation at physiological conditions in body (*9*, *14*, *15*). However, due to the concerns about safety and toxicity, there is only limited vaccine adjuvant approved for human use in the United States, such as aluminum hydroxide, or alum. Despite the wide use of alum, it has comparatively weak immune response and works only with certain diseases. So nanomaterial as adjuvant in vaccine immnunology is one of the major areas currently being studied widely because new, safe and efficient nano adjuvants are highly demanded. Apparently, further efforts in improvement or optimization of nanomaterial-based adjuvants in immunology, such as new nanoparticles, particle sizes, surface properties and loading methods of vaccine etc., are required to help induce, strengthen and prolong the immune response before nano adjuvants are to be used with vaccine in human beings.

Due to their positive adjuvant effect on immunology and their adverse adjuvant effect on human health, nanomaterials could be viewed as a double-edged sword. Therefore, the purpose of this manuscript is to present an overview of the positive adjuvant effect of nanomaterials in immunology where nanomaterials, more specifically, nanoparticles are developed as vaccine or antigen carriers/adjuvants. Two categories of nanoparticles from previous reports (polymeric and inorganic nanoparticles) will be discussed. We also consider the fact that human exposure to nanomaterials in the environment is more and more common, which brings potential adverse effect when allergens/antigens are present in the body. So the adverse adjuvant effect of as-prepared nanomaterials and ambient particulate matter from air pollution will also be discussed at the end of this manuscript.

Adjuvant from Polymeric Nanoparticles

Over the last several decades, many nanoparticles-based delivery systems have been developed and these systems have received extensive interests as potential adjuvants for immunology. Based on the polymer ingredients, the nanoparticles can be divided into biodegradable and non-biodegradable polymeric carriers.

Nanoparticle Adjuvant from Biodegradable Polymers

Polymeric nanoparticles formulated from different biodegradable polymers and copolymers have been widely explored as controlled delivery vehicles of different agents, such as peptides, proteins and nucleic acid vaccines (*16, 17*). They are well established delivery systems and the encapsulation of biomolecules in polymeric nanoparticles protects them from extreme conditions while maintaining their integrity and activity, and herein enhances the immune response of the antigens (*18*).

Polyester is a thermoplastic polymer with labile aliphatic ester bonds in the backbones which can degrade hydrolytically under physiological conditions (*19*). It is one of major nanoparticle materials used in vaccine delivery to determine the antigen immunogenicity. Poly(lactide) (PLA) and poly(lactide-*co*-glycolic acids) (PLGA), the most popular materials for formulation of polymeric nanoparticles, have been approved by FDA for the applications in human beings. The application of these PLGA nanoparticles in antigen delivery enhanced cellular and humoral immune response compared to the free antigen dose (*20, 21*), which confirms their adjuvant effect. Both nano- and micro-particles of PLGA can be used as adjuvants and antigen carriers to facilitate presenting antigens to T cells and increase the immune response (*22, 23*). In 2012, Bershteyn and coworkers synthesized PEGylated phospholipid-enveloped PLGA (50:50) microparticles (MP) conjugated with thiolated protein antigens (ovalbumin, OVA) and studied the adjuvant effect of MP carriers in mice (*22*). The $CD8^+$ T-cells responses on day 7 from different dose combinations of MP, antigens, monophodphoryl lipid A (MPLA, an immunostimulatory molecule) showed very different levels. The $CD8^+$ T-cells response of mice from antigen-MP was significantly greater than soluble antigen immunization alone at both doses (either no detectable or slightly higher responses compared to control experiment). It showed much higher response than alum-immunized (FDA approved) mice which was not detectable compared to control, even at very high alum dose (Figure 1A). Thus these results clearly identify that lipid-enveloped PLGA MP is one efficient adjuvant in immune response in mice. More recently, Moon and coworkers synthesized similar lipid-enveloped PLGA nanoparticles and then conjugated antigen, *vivax* malaria protein (VMP001) to the lipid membrane of particles with MPLA (*23*). The mice were immunized twice on days 0 and 21, and the sera were collected and analyzed on days 35 and 120 to compare the antibody responses promoted by the immunization of free VMP001 and VMP001-nanoparticles. As shown in Figure 1 B-F, the nanoparticle vaccines generated a more balanced Th1/Th2 antibody response. The level of IgG_1 and IgG_{2c} were persistent while IgG_{2b} and IgG_3 responses were lost on day 120. Compared to free VMP001+MPLA, the immunization with nanoparticles showed much higher antibody responses (Figure 1 B-F), though cytokine production by restimulated spenocytes *ex vivo* did not show greater differences in either nanoparticles loaded or not loaded groups.

Particle size is a critical factor that affects the immune response. Generally, small nanoparticles are known to be more effective targeted-delivery system than large particles because small nanoparticles can easily penetrate the biological barriers. So one would expect stronger immune response from small nanoparticle-antigen conjugates. However, several reports related to degradable particles from PLA, PLGA or sulfobutylated poly(vinyl alcohol)-*graft*-PLGA have shown that immune response from model antigens entrapped in particles exhibits no such trend (*24–27*) and the relation between immune response and particle size varies with different polymer materials and antigens. But there is evidence that the size of particle adjuvants may be related to different categories of the immune responses with preferred humoral immune response for large particles and preferred cellular immune response for small particles (*25*, *28*). Overall, polyester nanoparticle vaccines elicited enhanced immunogenicity and prolonged antibody response in mice which proved the adjuvant effect of nanoparticles (*29*).

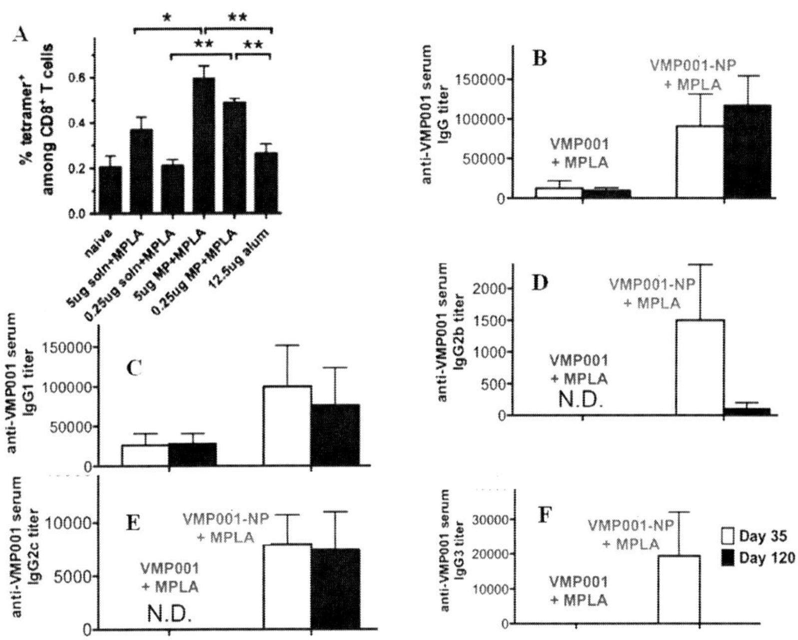

Figure 1. (A) Frequencies of ova-specific CD8$^+$ T-cells in spleens were analyzed by peptide-MHC tetramer staining and flow cytometry. (, $P<0.05$; **, $P<0.05$). (Reproduced with permission from reference (22). Copyright 2012 Elsevier Ltd.). (B-F) C57Bl/6 mice were immunized s.c. on days 0 and 21 with 25 μg of MPLA and 1 μg of VMP001 in either soluble or VMP001-NP formulations, and anti-VMP001 IgG sera were characterized on days 35 and 120 for (B) IgG, (C) IgG$_1$, (D) IgG$_{2b}$, (E) IgG$_{2c}$, and (F) IgG$_3$ titers. (Reproduced with permission from reference (23). Copyright 2012 PLoS One)*

Chitosan (poly(D-glu-cosamine)), the most investigated polysaccharide, can be prepared from natural polymer (chitin) via partial *N*-deacetylation. It is soluble in acidic solutions and prone to chemical or biological functionalization due to the presence of highly reactive amino groups in its structure (*30*). It is widely studied as nano vaccine carriers and adjuvants against a variety of antigens, such as H1N1 hemagglutinin antigen delivered by chitosan-coated poly(ε-caprolactone) (PCL) nanoparticles (*31*), swine influenza DNA vaccine (*32*) and hepatitis B surface antigen (rHBsAg) (*33*, *34*). And the results suggest enhanced immunological properties of these chitosan-based nano adjuvants. Zhao and coworkers reported that spherical chitosan nanoparticles conjugated with swine influenza antigen (plasmid DNA) were prepared by complex coacervation method with high encapsulation efficiency and high antigen stability (*32*), which indicates that the encapsulated DNA was protected from degradation after incorporated with nanoparticles. The presented much higher antibody serum IgG titers (2 to 3 times higher than naked DNA), using ELISA from BALB/c mice which were immunized with antigen-loaded chitosan nanoparticles, naked antigen and blank chitosan nanoparticles, suggest significant adjuvant immune response from chitosan nanoparticles. Even more, the antibody presentation sustained almost same level at prolonged time (up to 8 weeks during the tested time interval) which indicates continuous release profile of the incorporated DNA antigen from chitosan nanoparticles. This may bring the potential to increase DNA vaccination efficiency and is worth of further study for clinical use in the future. A similar adjuvant effect was also reported by Prego and coworkers (*33*, *34*) when ionic gelated chitosan nanoparticles loaded with hepatitis B antigen (HB) were prepared and intramuscularly administrated to mice (two doses). Although the immune response (IgG level) from antigen-loaded nanoparticles existed a slight delay compared to FDA approved alum, the adjuvant immunological effect of nanoparticles was ~9-fold higher than the alum vaccine with extended antigen release profile accounting for the prolonged immune response. The surface charge and composition of the antigen-loaded chitosan nanoparticles also play important roles in modulating the immune response (*34*). When incorporating nanoparticles with less excess or excess of antigen, the resulted antigen-nanoparticle complexes could have positive ζ potential (CSNC+, leaving chitosan as predominant composition on the particle surface when less excess antigen was used) or negative ζ potential (CSNC-, leaving antigen as predominant composition exposure towards the external medium when excess antigen was used). The immunological response of CSNC- (two doses) was much lower than CSNC+, even lower than the control (alum-antigen) during the 27-week study, and it was not able to induce effective immunogenicity against antigen (*33*, *34*). This is consistent with previous reports that exposure to repetitive antigens induced weak immune response (*35*), and cationic surface charge of nano adjuvant enhanced the immune response (*36*). This might be explained by the fact that CSNC+ could facilitate antigen internalization through strong association with outer membrane of dendritic cells (*37*), and therefore, CSNC+ could induce adequate presentation to the immune system and an enhanced activation of APCs with further development of a strong adaptive immune response. Despite that immunization via multiple doses are widely applied in immunology research, single-dose approach of antigen (HB) on chitosan nanoparticles also show high specific and long-lasting IgG antibody levels against antigen (*34*), which suggest nano adjuvant could achieve the goal of vaccination by reducing the injection frequency while maintaining efficient immune protection.

Other biodegradable polymers have also been studied as adjuvants in immunonology, such as micro/nano-particles from poly(ε-caprolactone) (PCL) (*38*), nanoparticles from poly(anhydrides) (*39*) and poly(γ-glutamic acid) (*40*) etc. The results from those studies all suggest nanoparticles formulated from these polymer materials exhibit adjuvant immune response while co-administrated with specific antigens.

Nanoparticle Adjuvant from Non-Biodegradable Polymers

Unlike biodegradable polymers, non-biodegradable polymers cannot be broken down or degraded *in vivo* by hydrolysis and/or bacteria digestion. Hence the study on non-degradable nano adjuvants in immunology attracted much less attention compared to biodegradable polymer due to slow clearance and the risk of chronic toxicity after administration. However, they are considered to have extended immune response and thus improve the immunogenicity due to the prolonged persistence of nanoparticles in tissues (*21*). Especially, a great number of non-biodegradable nanoparticles have been proved to be biocompatible, though not biodegradable. So there are still considerable reports related to their adjuvant studies on immune response.

Polystyrene or latex nanoparticles have been incorporated with OVA to investigate their adjuvant immunogenicity (*41–44*). When antigen OVA was covalently linked to polystyrene nanoparticles (*41*), it displayed much higher MHC class I-restricted T cell immune response and Ab titers than those currently used adjuvants (Alum, Freund's complete adjuvant (CFA), QuilA, monophodphoryl lipid (MPL)) after one or two immunization. And the conjugates of antigen and nanoparticles via covalent linkage showed higher immune response than soluble antigen alone or a simple mixture of antigen and nanoparticles. Especially, the effect of particle size was addressed in the study and the result suggests that 40-nm sized particles exhibited superior antibody and cellular immune response which indicate they are ideal as immune adjuvants in the model. Moreover, in tumor models EG7-OVA (ovalbumin-expressing EG7 tumors), E7-HPV (HPV 16-expressing tumors) and their established tumor, mice immunized with nanoparticle-antigen conjugates cleared the tumor after certain time challenge, whereas mice treated with antigens only all showed grown tumors. Other literatures (*42*, *43*) using latex nanoparticles also demonstrated 1000-10000 fold more efficient immune response than soluble antigen alone via MHC-class I or II molecules presentation.

Another biocompatible but non-biodegradable polymer, poly(*N*-isopropylacrylamide) (PNiPAAm) with thermo-responsive property, has been studied as nano adjuvant to investigate antigen-specific immune responses (*45*, *46*). Shakya et al. synthesized well-controlled PNiPAAm with lower critical solution temperature (LCST) at 32 °C. PNiPAAm conjugated with collagen type II and formed clear and visible white precipitate around the injection site

(>32 °C), which confirms the thermo-responsive property while immunization. And this adjuvant enhanced immunogenicity of collagen type II with induced collagen-related arthritis, and PNiPAAm covalently conjugated with antigen induced much weaker arthritis than that with physically adsorbed antigen, which suggests that one of major mechanisms corresponding to the adjuvant effect of polymeric nanoparticles might be depot effect (slow release of the entrapped antigen). Moreover, this polymer adjuvant also showed enhanced immune response to OVA. These results suggest PNiPAAm may be used as one general adjuvant in immunology and vaccination.

However, several factors might need to be considered in order to use non-biodegradable polymer nanoparticles as immunological adjuvants, such as toxicity, particle aggregation and accumulation in the tissue after immunization, which require further studies on *in vivo* clearance and safety administration.

Although several factors of the conjugates of polymeric nanoparticle and antigen (from either biodegradable or non-biodegradable polymer) can affect the adjuvant immunogenicity, such as particle size and covalent linkage or physical adsorption, there are no clear trends which can be used to predict whether nanoparticles would exhibit negative or positive adjuvant effect. However, it is clear that positive surface charge of nanoparticle and antigen conjugates can form complex with DNA easily, facilitate higher transfection efficiency and herein enhance the adjuvant immune response (*47*, *48*).

Adjuvant from Inorganic Nanoparticles

The development of functional, inorganic nanoparticles has progressed exponentially over the past two decades. As an alternative to polymeric nanoparticle adjuvant, inorganic nano adjuvants have also been widely studied as antigen carriers in vaccine immunology due to the fact that there are a variety of inorganic nanoparticles being prepared (*49*) and utilized in biomedical applications (*50*), such as imaging (*51*), tumor detection (*52*), and drug delivery (*53*). These inorganic particles may also have prolonged adjuvant effect due to the slow clearance from tissues. Particle size and surface functionality or charge can be easily tuned during particle preparation or via simple surface chemical reactions (*50*). This may extend the research varieties of inorganic nano adjuvants. Several inorganic nanoparticles acting as immune adjuvants will be included here in details for understanding their adjuvant activities.

Alum ($Al(OH)_3$) is the first adjuvant and has been approved and used in human beings for long time. Recently, aluminum oxide nanoparticles have been addressed more attention perhaps due to the similar component to alum. Antigen peptomers (head-to-tail linked 18-mer synthetic peptides) were covalently conjugated to surface-derivatized aluminum oxide nanoparticles (355 nm mean diameter) and their systemic immunization study was reported with the peptomer-particle conjugates without co-administration of hydrophilic adjuvant muramyl dipeptide (MDP) resulted in the highest serum antibodies titers (*54*). Both free C4 (the 4th conserved region of HIV-1 gp120) peptides and C4 peptomers with and without MDP exist lower antibody titers. This suggests the peptomer–particle formulation provided adjuvant activity when administered systemically. More recently, Maquieira and coworkers have first covalently coupled non-immunogenic hapten (one small molecular mass compound) to aluminum oxide nanoparticles as carrier and adjuvant for haptens immunization (*55*). Aluminum oxide particles (amorphous 40 and 3000 nm, crystalline 300 nm) were prepared and coupled with hapten covalently. The immune response is related to particle size (Figure 2) and crystallinity with the highest response from crystalline 300nm particles-hapten conjugates with or without Freund's adjuvant. It is consistent to previous reports with higher response from crystalline materials (*56*), and particle size effect is also consistent with previous results (*54*, *57*). The results also suggest the covalent linkage in aluminum oxide nano adjuvant system is essential to induce the enhanced immune response as compared with no response from free hapten or a simple mixture of particles and hapten.

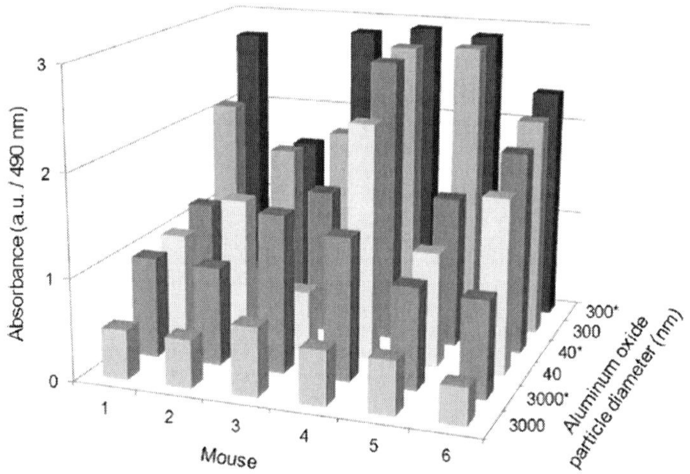

Figure 2. Effect of the size of aluminum oxide particles in the atrazine−hapten conjugates on mouse serum antibody titers, expressed as absorbance values. Asterisk (*) denotes immunization using Freund's adjuvant. (Reproduced with permission from reference (55). Copyright 2012 American Chemical Society.)

Gold nanoparticles have also attracted much attention as antigen carriers and adjuvants. Chen *et al.* conjugated pFMDV (VP1 protein of foot-and-mouth disease) with gold nanoparticle with different sizes (2-50 nm) to form nanoparticle vaccines and immunized mice model to study the size-dependent immunogenicity (*58*). The results showed significant size-dependent immuno-genicity against pFMDV with the highest immune response from 8 and 12 nm particle conjugates. And gold nanoparticle was demonstrated to be an ideal candidate as vaccine carrier due to no detectable antibody-binding activity. The effect of size and shape of gold nanoparticles on immune response was studied by Niikura and coworkers (*59*). Different sized and shaped gold nanoparticles were synthesized (Figure 3 A-D) via seeding growth method and coated with anionic polymer (PSS-MA) to electrostatically attach West Nile virus envelope (WNVE) protein to produce

20 and 40 nm spherical (Sphere40-E, Sphere20-E), rod (Rod-E, 40 × 10 nm), and cubic (Cube-E, 40 × 40 × 40 nm) particle-antigen conjugates (AuNP-Es) for *in vivo* and *in vitro* immunization study. Sphere40-Es induced the highest level of WNVE specific antibodies and Rod-Es induced only 50% of that from Sphere40-E (Figure 3 E). Results from cell uptake experiments of nanoparticles showed the uptake of Rod-Es is more efficient than others, which suggest that antibody production was not dependent on the uptake efficiency of different AuNP-Es. Although the mechanism of shape-dependent WNVE antibody production needs to be further investigated, this report (*59*) will pave the way for future development of nano vaccines. Other researchers also demonstrated gold nanoparticles along with alum (*60*) or with Fc fragment from human IgG (*61*) can improve the cell uptake of antigen and enhance the immunogenicity against respective antigens.

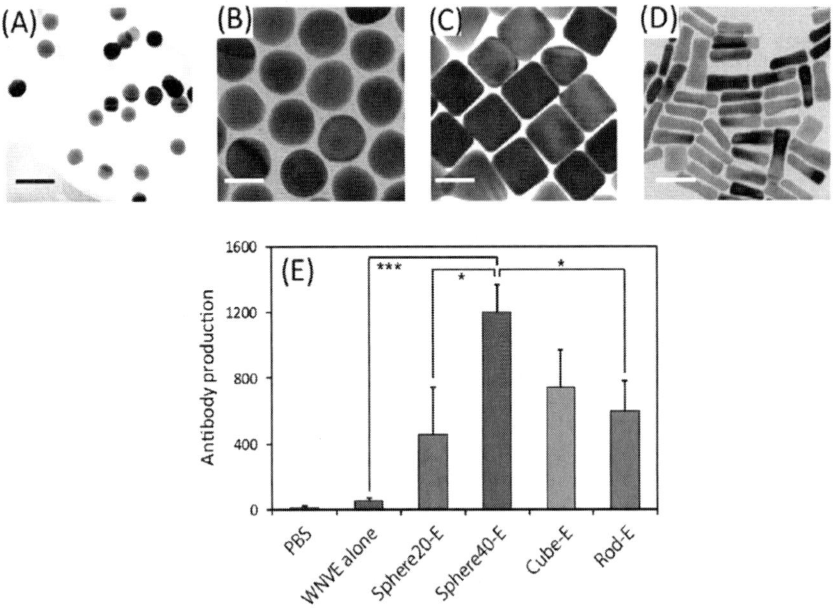

*Figure 3. TEM images of as-prepared (A) Sphere20, (B) Sphere40, (C) Cube, and (D) Rod gold nanoparticles before conjugated with antigen (Scale bar: 40 nm). (E) WNVE-specific IgG ELISA end point titers in mice immunized twice at 3-week intervals with 100 ng WNVE/animal/dose of AuNP-E. Significant differences: *p < 0.05; ***p < 0.001 (mean ± SEM, n = 10). (Reproduced with permission from reference (59). Copyright 2013 American Chemical Society.)*

Mesoporous silica nanoparticle (MSN) is one of the most popular materials used as antigen delivery systems due to its high surface area and porous structure. Wang and coworkers (*62*) prepared three different sized MSNs as vaccine adjuvants and investigated the effect of their pore structure on immune response. After bovine serum albumin (BSA) was entrapped into MSNs, oral immunization suggests MSN-BSA conjugates (430 nm) displayed the highest response compared to those from 130-nm and 1-2-μm MSNs which might attributed to the releasing rate (depot effect) and pore structure for different sized MSNs. MSN-BSA conjugates (430 nm) also showed much higher immune response than free BSA or BSA-CFA adjuvant, which suggests MSNs are acting as effective adjuvants. Co-presentence of Th1 and Th2 responses was proved by the comparison of IgG_1 and IgG_{2a} titer's change after 4 week vaccination. Most recently, OVA model antigen was efficiently conjugated to amino-functionalized MSN (AM-MSN, 90 nm) at 72 mg OVA/g MSN compared to only 29 mg OVA/g non-aminated MSN (*63*) due to the stronger electrostatic interaction between OVA and AM-MSN. Mice immunization using OVA-AM-MSNs at high (10 μg) and low (2 μg) loading of OVA induced enhanced antibody immune response, but cell-mediated immunity was observed when mice were immunized with high loading of OVA on AM-MSNs, not with low loading of OVA on AM-MSNs. This indicates that there might be a minimal threshold amount of antigen loading on AM-MSNs required to induce both cellular and humoral adaptive immunity. No discernible local or systemic morphology change at the injection site was observed after administration of OVA-AM-MSNs. These results indicate the adjuvant activity with OVA antigen and a new nano vaccine method. Non toxic mesoporous silica cylinders (SBA-15) have also been used to investigate the adjuvant immunogenicity against antigens, BSA (*64*) and *Micrurus* snake toxins and *Int1*β (*65*). The results clearly suggest that SBA-15 increased the immunogenicity compared to respective antigens alone or other common adjuvants, and positively modulated the immune response of low responder individuals into high antibody producers. These cylindrical silica nanomaterials stimulated mutually TH1 and TH2 immune responses.

Single-walled carbon nanotubes (SWNT) were covalently conjugated with Wilm's tumor antigen, WT1Pep427, and were used to immunize mice with or without another immunological adjuvant, Titer max (*66*). The SWNT-WT1Pep427 conjugates with Titer max induced specific serum IgG response against antigen, whereas antigen alone, SWNT-antigen alone or antigen-Titer max alone did not induce such response. This clearly suggests carbon nanotubes could serve as antigen carriers and induce immune response against weak MHC class II antigen. Quantom dots (QD, CdSe/ZnS) and Iron oxide nanoparticles (IO) conjugated with merozoite surface protein 1 (rMSP1), a recombinant malaria vaccine antigen, were also used as adjuvants to immunize mice in literatures (*67–69*). QD-rMSP1 induced significantly higher antibody response and parasite inhibitory antibodies against rMSP1 than those from rMSP1 with other conventional adjuvant together. The enhanced immunogenicity might be due to the uptake of QDs by dendritic cells resulting in the activation of dendritic cells and secretion of key cytokines (*67*).

Similar to polymeric nano adjuvants, there are a variety of inorganic nano adjuvants due to the extensive development of synthesis of inorganic nanoparticles. Their adjuvant effect on immune response is related to their size, shape, linkage method between antigen and nanoparticles, their chemical composition and antigen loading amount on nanoparticles. These inorganic nanoparticles are biocompatible due to the low toxicity examined *in vitro*, which suggests they are novel vaccine carriers and adjuvant candidates for the future clinical application.

Both polymeric and inorganic nanoparticles can act as adjuvants, combined with the antigens, usually through a depot effect via slow dissolution after administration. This overcomes the problem of rapid loss of free antigen and allows lower administration dose of antigen while maintaining similar or enhanced immune response to benefit the vaccination development.

Allergic Adjuvant Effect from Nanomaterials and Ambient Particulate Matter

We have discussed a variety of nanoparticles prepared from different materials as adjuvants to enhance immune responses against specific antigens and prevent human diseases. As mentioned above, adjuvants are intended to lower antigen dosage and improve efficiency of the delivered vaccine (*12*); hence they are beneficial to vaccine immunology and could be used in future novel vaccine development. However, like a double-edged sword, nanoparticles have also been reported to exhibit adverse allergic effect raised from allergen/nanoparticle complexes (*70*). Yang and coworkers (*71*) have reported that silica nanoparticles (nano-SiO_2, 10-20 nm) showed adverse effect on lung function of rats with OVA immunization compared to the saline-treated control rats. The OVA-sensitized and saline-sensitized control rats were treated daily with intratracheal instillation of nano-SiO_2 solutions (0.1 mL of 0, 40 and 80 μg/mL) for 30 days, rather than few immediately high dose exposure, which simulates the real environmental exposure. The results suggest that nano-SiO_2 has adverse effect on large airways of the lung as evidenced by the increasing inspiratory and expiratory resistance while increasing the dose of methacholine (MCH) challenge in airway hyper-responsiveness assessment. The pulmonary histological assay of lung indicates the obvious airway remodeling in rats via exposure to 80 μg/mL nano-SiO_2, and OVA-treated rats exhibit much more aggravated inflammation than that from the saline-treated control ones. This may attribute to the unbalance of Th1/Th2 cytokine accelerated by nano-SiO_2 through the increasing IL-4 production in the tissue. We have studied the enhanced allergic effect of engineered silica nanoparticles (SNP, polyethylene glycol coated particle surface, 90 nm hydrodynamic diameter) on airway diseases by assessing the magnitude of OVA-induced histopathological and immunological responses in the lung of mice (*72*). Female BALB/c mice were intranasally sensitized with allergen OVA along with co-exposure to SNP (0, 10, 100 and 400 μg) and secondary OVA challenge. The significantly greater level of OVA-specific serum IgE and IgG1, airway eosinophil inflammation, mucous cell metaplasia, and Th2 and Th17 cytokine gene and protein expression were observed from the SNP/OVA-mice compared to those from OVA-mice or SNP/saline-mice, and these results indicate that airway exposure to engineered SNP during the sensitization of mice to OVA enhances allergic airway disease with a dose-dependent fashion upon secondary OVA challenge. Besides silica nanoparticles, the adverse adjuvant effect of other nanomaterials, such as carbon black nanoparticles (CNP, (*73*)), single-walled carbon nanotubes (SWCNT, (*74*)), silver nanoparticles (AgNP, (*75*)) and latex nanoparticles (LNP, (*76*)) have also been studied in literatures. CNP (14 and 56 nm, (*73*)) was used and intratracheally administrated with bacterial endotoxin (lipopolysaccharide, LPS) in ICR male mice and the results indicate that CNP can aggravate the LPS-related lung inflammation with 14-nm CNP having more severe inflammation than 56-nm CNP. The enhancement is mediated through the local expression of IL-1β and chemokines. Inoue and coworkers (*74*) also studied the impact of repeated pulmonary exposure to SWCNT in ICR male mice model immunized by allergen OVA on allergic inflammation. The adjuvant activity of SWCNT for the enhanced titers of OVA-specific IgG1/IgE was evidenced by the increased lung levels of proinflammatory cytokines and chemokines related to allergy compared with OVA alone. The exacerbation of lung inflammation is possibly through the inappropriate activation of antigen-presenting cells and is associated with the greater oxidative stress in the airway from SWCNT-OVA treatment. However, two kinds of SWCNT with different dimensions (0.8-1.2 nm in diameter and 100-1000 nm in length vs 1.2-2 nm in diameter and 1-15 μm) showed almost identical enhanced allergic inflammation, which indicates the negligible size effect of SWCNT. More recently, the adverse asthmatic effect of AgNP (33 nm) was reported by Cheng and coworkers using proteomic approach in allergen OVA-AgNP sensitized female BALB/c mice model (*75*). The protein profiles of bronchoalveolar lavage fluid (BALF) and plasma suggested that metabolic, cellular, and immune system processes were associated with pulmonary exposure to AgNPs, and significant allergy responses were observed after AgNP exposure in both control and allergic mice as determined by greater level of OVA-specific Ig E. The size effect of adjuvant activity of spherical LNP (25, 50 and 100 nm, carboxylated particle surface) on lung inflammation in mice induced by LPS or OVA allergen was studied by Inoue and coworkers (*76*). Surprisingly, LNP with all sizes in mice model sensitized by OVA-LNP did not enhance significant allergic asthma. However, pulmonary exposure to LNP showed enhanced lung inflammation induced by LPS and the enhancement was mediated through the increased local expression of proinflammatory cytokines and chemokines. The enhancing effect on lung inflammation from LPS combined with LNP smaller than 50 nm in diameter was more prominent than that with larger diameter (100 nm). These results suggest that exposure to LNP, in particular to smaller sized nanoparticle can directly/indirectly facilitate systemic inflammation accompanied by lung inflammation.

As discussed above, nanomaterials exhibit enhanced allergic effect in mice models under the sensitization of allergen, especially in asthma-like airway diseases. However, ambient particulate matter (PM, ranging from nanometers to micrometers) is one of the major air pollutants from fossil combustion, vehicular traffic, consumer products and bulk manufacture and handling of nanomaterials (*5*). Exposure to and inhalation of PM are more common for human in the real world and are believed to be associated with the increasing allergic disorders from cardiorespiratory diseases, such as asthma and rhinitis (*8*, *77*). This is also evidenced by epidemiology studies showing a close relation between residential freeway proximity and allergic disease incidence (*78*). So the adverse adjuvant allergic effect of PM on airway diseases or other diseases has drawn extensive interests and the recent advances will be reviewed and discussed.

Large numbers of *in vitro* and *in vivo* studies upon exposure to PM showed enhanced adjuvant effect of allergic inflammation to antigen/allergen models, especially for asthma-like airway diseases (*79–85*). This clearly indicates the adverse health effect of PM on airway diseases which could exacerbate further serious allergic inflammation, especially on already-sensitized subjects. Li and coworkers (*81*) demonstrated that the collected ultrafine PM (<150 nm) could enhance significant allergic inflammation in mice model and act as adjuvant to promote T_H2 polarization. The results also showed bigger PM (<2500 nm) exhibits lower adjuvant effect of allergic inflammation compared to ultrafine PM (UFP, < 150 nm). The enhanced allergic inflammation can be suppressed partially by adding thiol antioxidant (*N*-acetyl cysteine), which suggests the adjuvant effect of UFP might be through oxidative stress mechanism. They also designed one protocol which enable them to elucidate the potential adjuvant effect of UFP (<180 nm) on the secondary immune response (*82*). This "real-life" exposure to ambient PM could efficiently promote the secondary immune response and exacerbate the existing allergic airway inflammation/disease in already-sensitized animals. Similar to previous study, oxidative stress may still play a role in the enhanced secondary immune response. This report suggests that exposure to PM may be more deleterious to the already infected human population. Samuelsen et al. (*79*) compared the enhanced allergic sensitization of similar sized PM from wood smoke and road traffic with reference diesel exhaust particles (DEP) against allergen OVA, and concluded that PM from wood smoke showed comparable enhanced allergic sensitization with road traffic particles, but much less than DEP. Considering the exposure to ambient PM in childhood might increase children's asthma risk, more recently, Herbert et al. (*84*) developed one procedure for the first time to demonstrate that exposure the weanling BALB/c mice to ambient particulates and OVA sensitization in early life can induce the asthmatic inflammation development if keeping repeatable low-level inhalational exposure to an allergen. These findings are likely to be relevant to the induction of childhood asthma and help to reduce the risk of getting asthma inflammation. To test the hypothesis of diabetes patients having high risk of developing atherosclerosis upon chronic exposure to PM (*86*), Jacobs and coworkers recently have showed the positive allergic effect of atherosclerosis on non-smoking adult by testing the concentration of oxidized low-density lipoprotein (an atherosclerosis marker). The exposure to PM has an effect on both oxidative modification of lipoprotein process and the increased blood leukocytes and platelets process which contribute to the initiation and progress of atherosclerosis. Overall, the adverse health effects of PM are likely to depend on several factors, including its size and composition, the level and duration of exposure, and age and sensitivity of the exposed individuals, and so on.

Although there are a great number of reports related to the adverse adjuvant effect of ambient particulate matter, the immunological mechanism of adjuvant effect is still not understood completely (*8*) and there are still debates over different proposed mechanisms. So to elucidate the mechanisms of adjuvant effect still remains to be one of critical tasks for toxicological and immunological scientists and more carefully designed studies are needed for a better understanding the mechanisms. One of possible approaches might be design and preparation of engineered nanoparticles to mimic different properties of ambient PM. And the adjuvant effect of those engineered nanoparticles with well controlled properties might help to understand their allergic mechanism. These studies would serve and suggest how to prevent the adverse adjuvant effect of PM and thus reduce the health risk of exposure to PM for human beings.

Conclusion

Nanomaterials show significantly promising applications in the development of new vaccines. The adjuvants, nanoparticles can enhance the immune response greatly and reduce the vaccine/antigen dosage with great vaccine efficiency. This review discussed the application of nanoparticles as nano adjuvants in vaccination development, including a broad range of polymeric and inorganic nanoparticles. Given the promising results to date, nanoparticles can be used to discover and improve the safety and efficiency of new vaccines.

Due to safety concerns of nanomaterials and ambient particulate matter, recent progresses of as-prepared nanomaterials and particulate matter from air pollution as adjuvants to promote the adverse allergic sensitization of human diseases have also been discussed. This would promote people to take precautions about the ambient particulate matter in air and thus improve the public health.

References

1. Jebb, M.; Sudeep, P. K.; Pramod, P.; Thomas, K. G.; Kamat, P. V. *J. Phys. Chem. B* **2007**, *111*, 6839–6844.
2. Fukui, T.; Murata, K.; Ohara, S.; Abe, H.; Naito, M.; Nogi, K. *J. Power Sources* **2004**, *125*, 17–21.
3. Cuenot, S.; Frétigny, C.; Demoustier-Champagne, S.; Nysten, B. *Phys. Rev. B* **2004**, *69*, 165410.
4. Brigger, I.; Dubernet, C.; Couvreur, P. *Adv. Drug Delivery Rev.* **2002**, *54*, 631–651.
5. Nel, A.; Xia, T.; Madler, L.; Li, N. *Science* **2006**, *311*, 622–627.
6. Thomassen, L. C. J.; Aerts, A.; Rabolli, V.; Lison, D.; Gonzalez, L.; Kirsch-Volders, M.; Napierska, D.; Hoet, P. H.; Kirschhock, C. E. A.; Martens, J. A. *Langmuir* **2010**, *26*, 328–335.
7. Mahmoudi, M.; Hofmann, H.; Rothen-Rutishauser, B.; Petri-Fink, A. *Chem. Rev.* **2012**, *112*, 2323–2338.
8. Li, N.; Xia, T.; Nel, A. E. *Free Radical Biol. Med.* **2008**, *44*, 1689–1699.
9. Oyewumi, M. O.; Kumar, A.; Cui, Z. *Expert Rev. Vaccines* **2010**, *9*, 1095–1107.
10. Rice-Ficht, A. C.; Arenas-Gamboa, A. M.; Kahl-McDonagh, M. M.; Ficht, T. A. *Curr. Opin. Microbiol.* **2009**, *13*, 106–112.

11. Sasaki, S.; Okuda, K. DNA Vaccines: Methods and Protocols. In *Methods in Molecular Medicine*; Lowrie, D. B., Whalen, R. G., Eds.; Humana Press: Totowa, NJ, 2000; Vol. 21, pp 241–249.
12. Pettrovsky, N.; Aguilar, J. C. *Immunol. Cell Biol.* **2004**, *82*, 488–496.
13. De Souza Reboucas, J.; Esparza, I.; Ferrer, M.; Sanz, M. L.; Irache, J. M.; Gamazo, C. *J. Biomed. Biotechnol.* **2012**, *2012*, 474605.
14. Slutter, B.; Soema, P. C.; Ding, Z.; Verheul, R.; Hennink, W.; Jiskoot, W. *J. Controlled Release* **2010**, *14*, 207–214.
15. De Temmerman, M. L.; Rejman, J.; Demeester, J.; Irvine, D. J.; Gander, B.; De Smedt, S. C. *Drug Discovery Today* **2011**, *16*, 488–496.
16. Kumari, A.; Yadav, S. K.; Yadav, S. C. *Colloids Surf., B* **2010**, *75*, 1–18.
17. Csaba, N.; Garcia-Fuentes, M.; Alonso, M. J. *Adv. Drug Delivery Rev.* **2009**, *61*, 140–157.
18. Mandan, T.; Munshi, N.; De, T. K.; Maitra, A.; Usha Sarma, P.; Aggarwal, S. S. *Int. J. Pharm.* **1997**, *159*, 135–147.
19. Soppimath, K. S.; Aminabhavi, T. M.; Kulkarni, A. R.; Rudzinski, W. E. *J. Controlled Release* **2001**, *70*, 1–20.
20. Spiers, I. D.; Eyles, J. E.; Baillie, L. W. J.; Williamson, E. D.; Alpar, H. O. *J. Pharm. Pharmacol.* **2000**, *52*, 1195–1201.
21. Combadiere, B.; Mahe, B. *Comp. Immunol. Microb.* **2008**, *31*, 293–315.
22. Bershteyn, A.; Hanson, M. C.; Crespo, M. P.; Moon, J. J.; Li, A. V.; Suh, H.; Irvine, D. J. *J. Controlled Release* **2012**, *157*, 354–365.
23. Moon, J. J.; Suh, H.; Polhemus, M. E.; Ockenhouse, C. F.; Yadava, A.; Irvine, D. J. *PLoS One* **2012**, *7*, e31472.
24. Gutierro, I.; Hernández, R. M.; Igartua, M.; Gascán, A. R.; Pedraz, J. L. *Vaccine* **2002**, *21*, 67–77.
25. Kanchan, V.; Panda, A. K. *Biomaterials* **2007**, *28*, 5344–5357.
26. Jung, T.; Kamm, W.; Breitenbach, A.; Hungerer, K-D.; Hundt, E.; Kissel, T. *Pharm. Res.* **2001**, *18*, 352–360.
27. Vila, A.; Sanchez, A.; Evora, C.; Soriano, I.; Jato, J.; Alonso, M. J. *Aerosol. Med.* **2004**, *17*, 174–185.
28. Caputo, A.; Brocca-Cofano, E.; Castaldello, A.; Voltan, R.; Gavioli, R.; Srivastava, I. K.; Barnett, S. W.; Cafaro, A.; Ensoli, B. *Vaccine* **2008**, *26*, 1214–1227.
29. Ren, H.; Wulff, W. D. *J. Am. Chem. Soc.* **2011**, *133*, 5656–5659.
30. Jayakumar, R.; Nwe, N.; Tokura, S.; Tamura, H. *Int. J. Biol. Macromol.* **2007**, *40*, 175–181.
31. Gupta, N. K.; Tomar, P.; Sharma, V.; Dixit, V. K. *Vaccine* **2011**, *29*, 9026–9037.
32. Zhao, K.; Shi, X.; Zhao, Y.; Wei, H.; Sun, Q.; Huang, T.; Zhang, X.; wang, Y. *Vaccine* **2011**, *29*, 8549–8556.
33. Prego, C.; Paolicelli, P.; Diaz, B.; Vicente, S.; Sanchez, A.; Gonzalez-Fernandez, A.; Alonso, M. J. *Vaccine* **2010**, *28*, 2607–2614.
34. Vicente, S.; Diaz-Freitas, B.; Peleterio, M.; Sanchez, A.; Pascual, D. W.; Gonzalez-Fernandez, A.; Alonso, M. J. *PLoS One* **2013**, *8*, e62500.

35. Schroeder, U.; Graff, A.; Buchmeier, S.; Rigler, P.; Silvan, U.; Tropel, D.; Jockusch, B. M.; Aebi, U.; Burkhard, P.; Schoenenberger, C. A. *J. Mol. Biol.* **2009**, *386*, 1368–1381.
36. Ma, Y.; Zhuang, Y.; Xie, X.; Wang, C.; Wang, F.; Zhou, D.; Zeng, J.; Cai, L. *Nanoscale* **2011**, *3*, 2307–2314.
37. Bal, S. M.; Slütter, B.; van Riet, E.; Kruithof, A. C.; Ding, Z.; Kersten, G. F.; Jiskoot, W.; Bouwstra, J. A. *J. Controlled Release* **2010**, *142*, 374–383.
38. Florindo, H. F.; Pandit, S.; Lacerda, L.; Goncalves, L. M. D.; Alpar, H. O.; Almeida, A. J. *Biomaterials* **2009**, *30*, 879–891.
39. Gomez, S.; Gamazo, C.; Roman, B. S.; Ferrer, M.; Sanz, M. L.; Irache, J. M. *Vaccine* **2007**, *25*, 5263–5271.
40. Broos, S.; Lundberg, K.; Akagi, T.; Kadowaki, K.; Akashi, M.; Greiff, L.; Borrebaeck, C. A.; Lindstedt, M. *Vaccine* **2010**, *28*, 5075–5085.
41. Fifis, T.; Gamvrellis, A.; Crimeen-Irwin, B.; Pietersz, G. A.; Li, J.; Mottram, P. L.; McKenzie, I. F. C.; Plebanski, M. *J. Immunol.* **2004**, *173*, 3148–3154.
42. Song, R.; Harding, C. V. *J. Immunol.* **1996**, *156*, 4182–4190.
43. Vidard, L.; Kovacsovics-Bankowski, M.; Kraeft, S. K.; Chen, L. B.; Benacerraf, B.; Rock, K. L. *J. Immunol.* **1996**, *156*, 2809–2818.
44. Ren, H.; Wulff, W. D. *Org. Lett.* **2013**, *15*, 242–245.
45. Nandakumar, K. S. *J. R. Soc., Interface* **2011**, *8*, 1748–1759.
46. Shakya, A. K.; Kumar, A.; Klaczkowska, D.; Hultqvist, M.; Hagenow, K.; Holmdahl, R.; Nandakumar, K. S. *Am. J. Pathol* **2011**, *179*, 2490–2500.
47. Xiang, S. D.; Scholzen, A.; Minigo, G.; David, C.; Apostolopoulos, V.; Mottram, P. L.; Plebanski, M. *Methods* **2006**, *40*, 1–9.
48. Mao, H. Q.; Roy, K.; Troung-Le, V. L.; Janes, K. A.; Lin, K. Y.; Wang, Y.; August, J. T.; Leong, K. W. *J. Controlled Release* **2001**, *70*, 399–421.
49. Rao, C. N. R.; Ramakrishna Matte, H. S. S.; Voggu, R.; Govindaraj, A. *Dalton Trans.* **2012**, *41*, 5089–5120.
50. Sperling, R. A.; Gil, P. R.; Zhang, F.; Zanella, M.; Parak, W. J. *Chem. Soc. Rev.* **2008**, *37*, 1896–1908.
51. Eck, W.; Craig, G.; Sigdel, A.; Gerd, R.; Old, L. J.; Tang, L.; Brennan, M. F.; Allen, P. J.; Mason, M. D. *ACS Nano* **2008**, *2*, 2263–2272.
52. Liu, J.; Yu, M.; Zhou, C.; Yang, S.; Ning, X.; Zheng, J. *J. Am. Chem. Soc.* **2013**, *13*, 4978–4981.
53. Brown, S. D.; Nativo, P.; Smith, J-A.; Stirling, D.; Edwards, P. R.; Venuopal, B.; Flint, D. J.; Plumb, J. A.; Graham, D.; Wheate, N. J. *J. Am. Chem. Soc.* **2010**, *132*, 4678–4684.
54. Frey, A.; Mantis, N.; Kozlowski, P. A.; Quayle, A. J.; Bajardi, A.; Perdomo, J. J.; Robey, F. A.; Neutra, M. R. *Vaccine* **1999**, *17*, 3007–3019.
55. Maquieira, A.; Brun, E. M.; Garces-García, M.; Puchades, R. *Anal. Chem.* **2012**, *84*, 9340–9348.
56. Geva, M.; Frolow, F.; Eisenstein, M.; Addadi, L. *J. Am. Chem. Soc.* **2003**, *125*, 696–704.
57. Frey, A.; Neutra, M. R.; Robey, F. A. *Bioconjugate Chem.* **1997**, *8*, 424–433.
58. Chen, Y-S.; Hung, Y-C.; Lin, W-H; Huang, G. S. *Nanotechnology* **2010**, *21*, 195101.

59. Niikura, K.; Matsunaga, T.; Suzuki, T.; Kobayashi, S.; Yamaguchi, H.; Orba, Y.; Kawaguchi, A.; Hasegawa, H.; Kajino, K.; Ninomiya, T.; Ijiro, K.; Sawa, H. *ACS Nano* **2013**, *7*, 3926–3938.
60. Parween, S.; Gupta, P. K.; Chauhan, V. S. *Vaccine* **2011**, *29*, 2451–2460.
61. Cruz, L. J.; Rueda, F.; Cordobilla, B.; Simón, L.; Hosta, L.; Albericio, F.; Domingo, J. C. *Mol. Pharm.* **2011**, *8*, 104–116.
62. Wang, T.; Jiang, H.; Zhao, Q.; Wang, S.; Zou, M.; Cheng, G. *Int. J. Pharm.* **2012**, *436*, 351–358.
63. Mahony, D.; Cavallaro, A. S.; Stahr, F.; Mahony, T. J.; Qiao, S. Z.; Mitter, N. *Small*, *9*, 3138–3146.
64. Carvalho, L. V.; de C. Ruiz, R.; Scaramuzzi, K.; Marengo, E. B.; Matos, J. R.; Tambourgi, D. V.; Sant'Anna, O. A. *Vaccine* **2010**, *28*, 7829–7836.
65. Mercuri, L. P.; Carvalho, L. V.; Lima, F. A.; Quayle, C.; Fantini, M. C. A.; Tanaka, G. S.; Cabrera, W. H.; Furtado, M. F. D.; Tambourgi, D. V.; Matos, J. R.; Jaroniec, M.; Sant'Anna, O. A. *Small* **2006**, *2*, 254–256.
66. Villa, C. H.; Dao, T.; Ahearn, L.; Fehrenbacher, N.; Casey, E.; Rey, D. A.; Korontsvit, T.; Zakhaleva, V.; Batt, C. A.; Philips, M. R.; Scheinberg, D. A. *ACS Nano* **2011**, *5*, 5300–5311.
67. Pusic, K.; Xu, H.; Stridiron, A.; Aguilar, Z.; Wang, A.; Hui, G. *Vaccine* **2011**, *29*, 8898–8908.
68. Pusic, K.; Aguilar, Z.; McLoughlin, J.; Kobuch, S.; Xu, H.; Tsang, M.; Wang, A.; Hui, G. *FASEB J.* **2013**, *27*, 1153–1166.
69. Ren, H.; Wulff, W. D. *Org. Lett.* **2010**, *12*, 4908–4011.
70. Madl, A. K.; Pinkerton, K. E. *Crit. Rev. Toxicol.* **2009**, *39*, 629–658.
71. Han, B.; Guo, J.; Abrahaley, T.; Qin, L.; wang, L.; Zheng, Y.; Li, B.; Liu, D.; Yao, H.; Yang, J.; Li, C.; Xi, Z.; Yang, X. *PLoS One* **2011**, *6*, e17236.
72. Brandenberger, C.; Rowley, N. L.; Jackson-Humbles, D. N.; Zhang, Q.; Lewandowski, R. P.; Wagner, J. G.; Chen, W.; Kaminski, N. E.; Baker, G. L.; Worden, R. M.; Harkema, J. R. *Part. Fibre Toxicol.* **2013**, *10*, 26.
73. Inoue, K.; Takano, H.; Yanagisawa, R.; Hirano, S.; Sakurai, M.; Shimada, A.; Yoshikawa, T. *Environ. Health Perspect.* **2006**, *114*, 1325–1330.
74. Inoue, K.; Yanagisawa, R.; Koike, E.; Nishikawa, M.; Takano, H. *Free Radical Biol. Med.* **2010**, *48*, 924–934.
75. Su, C.-L.; Chen, T.-T.; Chang, C.-C.; Chuang, K.-J.; Wu, C.-K.; Liu, W.-T.; Ho, K. F.; Lee, K.-Y.; Ho, S.-C.; Tseng, H.-E.; Chuang, H.-C.; Cheng, T.-J. *Int. J. Nanomed.* **2013**, *8*, 2783–2799.
76. Inoue, K.; Takano, H.; Yanagisawa, R.; koike, E.; Shimada, A. *Toxicol. Appl. Pharmacol.* **2009**, *234*, 68–76.
77. Nawrot, T. S.; Nemmar, A.; Nemery, B. *Eur. Heart J.* **2006**, *27*, 2269–2271.
78. Bernstein, J. A.; Alexis, N.; Barnes, C.; Bernstein, I. L.; Bernstein, J. A.; Nel, A.; Peden, D.; Diaz-Sanchez, D.; Tarlo, S. M.; Williams, P. B. *J. Allergy Clin. Immunol.* **2004**, *114*, 1116–1123.
79. Samuelsen, M.; Nygaard, U. C.; Løvik, M. *Toxicology* **2008**, *246*, 124–131.
80. Alberg, T.; Cassee, F. R.; Groeng, E-C.; Dybing, E.; Løvik, M. *J. Toxicol. Environ. Health A* **2009**, *72*, 1–13.
81. Li, N.; Wang, M.; Bramble, L. A.; Schmitz, D. A.; Schauer, J. J.; Sioutas, C.; Harkema, J. R.; Nel, A. E. *Environ. Health Perspect.* **2009**, *117*, 1116–1123.

82. Li, N.; Harkema, J. R.; Lewandowski, R. P.; Wang, M.; Bramble, L. A.; Gookin, G. R.; Ning, Z.; Kleinman, M. T.; Sioutas, C.; Nel, A. E. *Am. J. Physiol. Lung Cell Mol. Physiol.* **2010**, *299*, L374–383.
83. Noah, T. L.; Zhou, H.; Zhang, H.; Horvath, K.; Robinette, C.; Kesic, M.; Meyer, M.; Diaz-Sanchez, D.; Jaspers, I. *Am. J. Respir. Crit. Care Med.* **2012**, *185*, 179–185.
84. Herbert, C.; Siegle, J. S.; Shadie, A. M.; Nikolaysen, S.; Garthwaite, L.; Hansbro, N. G.; Foster, P. S.; Kumar, R. K. *Dis. Models & Mech.* **2013**, *6*, 479–488.
85. Wagner, J. G.; Morishita, M.; Keeler, G. J.; Harkema, J. R. *Environ. Health* **2012**, *11*, 45.
86. Jacobs, L.; Emmerechts, J.; Hoylaerts, M. F.; Mathieu, C.; Hoet, P. H.; Nemery, B.; Tim, S.; Nawrot, T. S. *PLoS One* **2011**, *6*, e16200.

Chapter 2

Luminescent Gold Nanodots for Detection of Heavy Metal Ions, Proteins and Bacteria

Binesh Unnikrishnan[1] and Chih-Ching Huang*[,1,2]

[1]Institute of Bioscience and Biotechnology, National Taiwan Ocean University, 2, Pei-Ning Road, Keelung, 20224, Taiwan
[2]Center of Excellence for the Oceans, National Taiwan Ocean University, 2, Pei-Ning Road, Keelung, 20224, Taiwan
***E-mail: huanging@ntou.edu.tw**

Photoluminescent gold nanodots (Au NDs) are one among the most promising candidates for cell imaging, biological labeling, clinical and sensing applications. The great advantage of the gold based NDs is that they are more biocompatible compared to Cd or Pb or other heavy metal based photoluminescent NDs. Therefore, they attract great attention in the field of biological and clinical applications. In this review, we highlight the synthesis, optical properties and applications of some new types of luminescent Au NDs developed by our research group.

Introduction

Gold nanodots (Au NDs) are of great interest due to their fascinating properties such as biocompatibility, stability, photoluminescence, easy methods of preparation, small size, etc (*1*–*3*). Because the size of Au NDs is comparable to the de Broglie wavelength at the Fermi level, NDs exhibit optical and chemical properties which are dramatically different from those of nanoparticles (NPs) with diameters greater than 2 nm (*4*). The collective excitation of the electrons in the conduction band results in coherent oscillation of electrons and results in surface plasmon resonance (SPR) in the case of Au NPs with size >3 nm (*5*, *6*). Although no apparent surface plasmon resonance absorption, Au NDs (<2-nm) are very interesting in the field of optical biosensor applications because of their unique size- and surface-dependent photoluminescence (PL) properties. When the size of the Au NDs is less than 2 nm, their core show quantum confinement effects which produce discrete electronic energy levels and molecule-like electronic transitions (*7–9*). The emission energies of the electrons of the fluorescent Au NDs with the size less than 2 nm can be expressed by $E_{Fermi}/N^{1/3}$, as predicted by the jellium model (*10*). Moreover, the functional molecule or capping agent attached to the surface of the metal core of the NDs has significant influence on the quantum effect and thereby influence the PL properties (*11–13*). The emission wavelength of the AuNDs can thus be tuned by controlling the size and the capping agent.

Alkanethiol (RSH)–capped Au NDs are interesting because they display PL with large Stokes shifts and long lifetimes (*14*, *15*). Unike ultrasmall Au nanoclusters (Au NCs, $<Au_{100}$), the PL properties of these luminescent Au NDs are not only directly related to their sizes but also might arise from electronic transitions of surface Au(I) coupling to the ligand–metal charge transfer transition (*16*). Among the primary topics of this Chapter is a discussion of synthesis and optical properties of these new types of luminescent Au NDs.

Luminescent Au NDs have great practicality in several fields owing to their excellent physical and chemical properties. The biocompatibility is the major property that attracts the Au NDs over their high-quantum yield counterpart, the inorganic quantum dots. Therefore, Au NDs are an alternative to the toxic quantum dots such as CdSe, CdTe, and fluorescent dye molecules, in biological, clinical and biosensor applications. Thiolate stabilized monodisperse Au NDs have been found very stable and useful in many applications, such as sensing and cell imaging (*17–19*). The interaction of sulfur and gold atom is strong enough to immobilize thiol group containing molecules to the gold nanoparticles (Au NPs) to produce highly stable Au NDs. For that reason tremendous amount of work has been done in synthesis and characterization of Au NDs formed through the Au-S bond between the metal core and the shell (*17–19*). This Chapter reviews the synthesis and application of highly luminescent Au NDs protected with alkanethiol, which were recently published by our research group. Finally, the Chapter concludes with a brief summary of the present and future directions of research in luminescent Au NDs.

Preparation and Optical Properties of Alkanethiols–Protected Au NDs

Unlike large Au NPs (>3 nm), luminescent Au NDs capped with alkanethiols (RS−Au NDs) exhibit luminescence. The luminescence of Au NDs arises from both the Au(I)–thiol complexes and the Au cores (Figure 1). In luminescent Au NDs, Au ions provide available states and hybridize with organic ligands to form charge transfer bands (*20*). Luminescence is roughly attributed to metal–metal electronic transitions, coupled with strong ligand–metal/metal–metal unit charge transfer transitions (*21*, *22*). These Au(I) complexes generally exhibit large Stokes-shifted luminescence with lifetimes in the microsecond regime. Charge transfer between ligands and Au or Au–Au units is responsible for these large Stokes shifts and long-lifetime emissions. The emission center is usually assigned to metal–metal electronic transitions. In addition, at low temperature, the emission energy is inversely proportional to the Au–Au distance (*23*). Moreover, lowering the temperature can result in a decrease in the Au(I)–Au(I) bond length, corresponding to a blue shift of the emission energy (*24*).

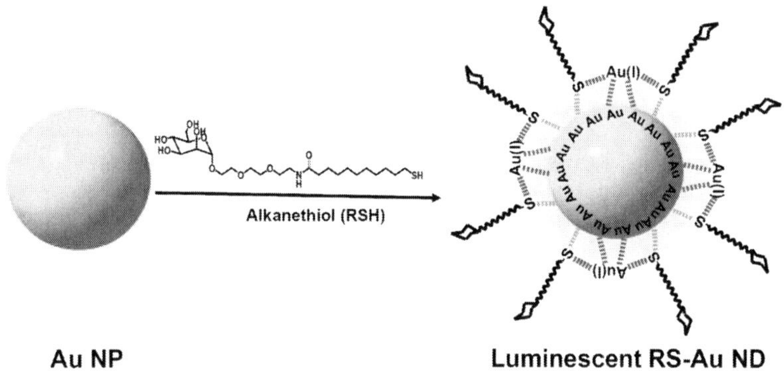

Figure 1. Schematic representation of the preparation of AuND/polynuclear Au(I)–thiol (core/shell) of luminescent RS-Au NDs from Au NPs and alkanethiols.

Luminescent 11-MUA-Au NDs

A general strategy for synthesizing luminescent small Au NDs using multidentate ligands featuring soft electron donor groups to coordinate with Au^+ ions has been reported (*25*). A representative procedure has been adopted from our pioneer work to demonstrate the synthesis of luminescent 11-mercaptoundecanoic acid (11-MUA) protected AuNDs (11-MUA−AuNDs) (*25*). In this method, first Au NPs is synthesized by the reduction of $HAuCl_4$ by tetrakis(hydroxymethyl)phosphonium chloride (THPC) (*26*) and then it is functionalized with 11-MUA at room temperature by simply mixing to form luminescent 11-MUA−Au NDs. The average particle size of the Au NPs synthesized by the reduction of $HAuCl_4$ by THPC has been found to be 2.9 nm. While the 11-MUA−Au NDs has an average particle diameter of 2.0 nm and are homogeneously dispersed. The decreased size of the Au NPs in the presence of 11-MUA is due to the very strongly covalent, distinctly directional Au-S bond and preference of dissociation into very smaller Au and gold thiolate clusters (on the basis of fragmentation energies), as opposed to breaking the Au-S bond and detaching the Au cluster from the thiolate radical (*27*, *28*).

Although the 11-MUA−Au NDs exhibits an absorption band centred at a wavelength of 375 nm, no apparent absorption band appears for the Au NPs (Figure 2A) (*29*). The absorption bands for 11-MUA−Au NDs originate from metal-center (Au $5d^{10}$ to 6sp interband transitions) and/or ligand-to-metal NP charge transfer (LMNCT; S→Au) mixed with ligand-to-metal–metal charge transfer (LMMCT; S→Au···Au) on the particle surface (*30*–*33*). The photoluminescent bands of the AuNPs and 11-MUA−Au NDs are centred at wavelengths of 618 and 520 nm, respectively (Figure 2B). Upon decreasing the size of the protected Au NDs, the emission wavelength undergoes a blue shift. The PL intensity of the 11-MUA−Au NDs is ca. 60,000-fold higher than that of the Au NPs. 11-MUA−Au NDs shows a QY of 3.1×10^{-2} through comparison with quinine (QY 53%). The PL lifetimes of 11-MUA-Au NDs (τ_1/τ_2) are 48/340 ns. Such a behavior of biexponential emission decay in the PL lifetime measurement was possibly attributed to the differential distribution of complicated luminescent pathways of polynuclear Au(I)–thiolate complexes, ligand-to-metal charge transfer (LMCT) or LMCT/MC triplets LMCT/metal centered (MC) triplets (*34–36*). The long radiative lifetimes are suggestive of contribution from the polynuclear Au(I)–thiolate complexes shell of a triplet parentage (*37*). It could be assigned to metal perturbed intraligand phosphorescence, a low-lying triplet state populated via intersystem crossing from the lowest singlet state (*38*). Thus, we suggested that the luminescence of Au NDs mainly originated from Au ND/polynuclear Au(I)–thiol (core/shell) complexes (Figure 1).

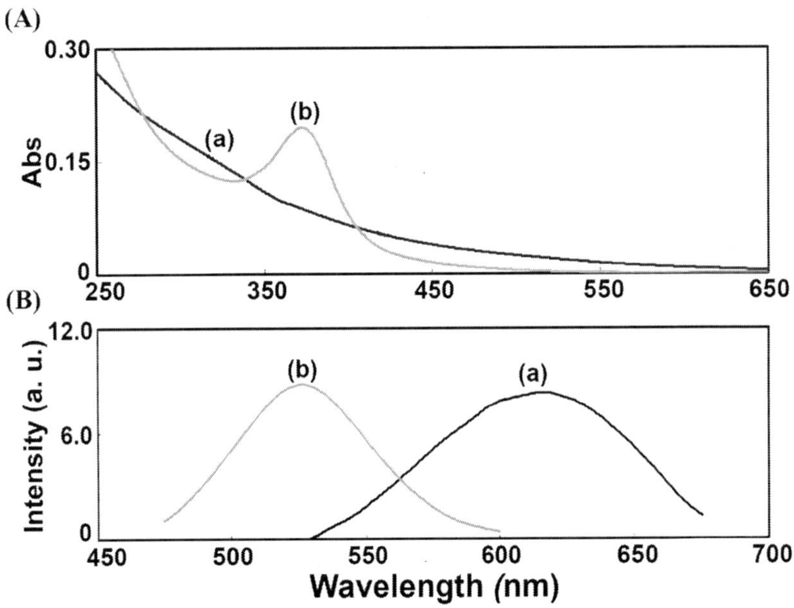

Figure 2. (A) UV-vis absorbance and (B) photoluminescencespectra of (a) Au NPs and (b) 11-MUA−Au NDs. The concentrations of the Au NPs and 11-MUA-Au NDs in (B) were 0.94 μM and 0.15 nM, respectively. Reprinted with permission from reference (29). Copyright 2008 American Chemical Society.

Luminescent Man-Au NDs

The water-soluble, biofunctional, luminescent mannose-capped Au NDs (Man–Au NDs) can be prepared through the reaction of 2.9-nm-diameter Au NPs with 11-mercapto-3,6,9-trioxecyl-α-d-mannopyranoside (Man-RSH) (*39*).

The Man-Au NDs prepared from Man-SH produce only a quantum yield of 8.6%. However, highly fluorescent Man–Au NDs (quantum yield 20.5%) could be obtained if it is prepared from 29,29′-dithio bis(3′,6′,9′,12′,15′,18′-hexaoxanonacosyl α-D-mannopyranoside)(Man–RSSR–Man) in presence of a reducing agent, $NaBH_4$ (*40*). It is well established that $NaBH_4$ can reduce the disulfide (R–SS–R) linkage into thiols (R–SH). As shown in Figure 3, the QY (20.5%) of Man–Au NDs prepared by the $NaBH_4$ (1.0 mM)-assisted method is over 5-fold higher than that of NDs prepared without $NaBH_4$ (4.8%). The higher QY of Man–Au NDs prepared using $NaBH_4$ (1.0 mM) is presumably due to their high surface coverage by Man–RS ligands. Figure 3 also shows that the PL intensity of Man−Au NDs decreases as the concentration of $NaBH_4$ (>1 mM), suggesting that the monolayer polynuclear Au(I)-thiol complexes on the ND surface are destroyed towing to the reduction of gold ions by higher concentration of $NaBH_4$.

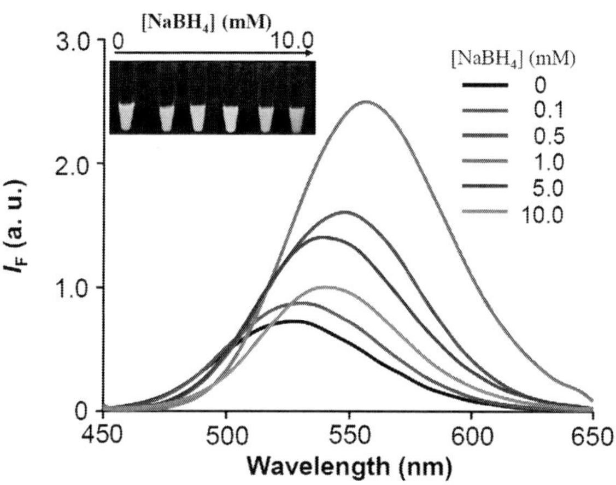

Figure 3. Luminescence spectra of 2.9 nm AuNP solutions after reaction with Man–RSSR–Man (2.5 mM) in 5 mM sodium tetraborate (pH 9.2) containing $NaBH_4$ at various concentrations (0–10 mM) for 48 h. Inset: photograph displaying the luminescence of the various Man–Au ND solutions after reaction for 48 h. Reprinted with permission from reference (40). Copyright 2011 Elsevier.

Effect of Ligand Length and Density of RSH

Huang *et al.* reported that the PL efficiency and wavelength of the Au NDs were various under capping with four different thiolates (N-2-mercaptopropionylglycine, 3-mercapto-1-propanesulfonic acid, mercaptosuccinic acid, and glutathione) (*41*). Guo *et al.* also reported similar observation that the PL intensity depends on the length of the capping ligands (*42*). Zhang *et al.* demonstrated that the electronic behavior of Au NDs with size ~2 nm can be tuned with the capping molecules such as dendrimers and thiols (*43*). In our study, we prepared a series of RS–Au NDs which luminesce at wavelengths within the range 501–613 nm, with quantum yields ranging from 0.0062 to 3.1% (*25*). The Au NPs were self-assembled with alkanethiol ligands of various lengths (C_2–C_{11}) by introducing alkanethiol stock solutions into the as-prepared Au NPs (2.9-nm). The PL wavelength and QY of these RS–Au NDs can be controlled by varying the size of the alkanethiol, including 2-mercaptoethanol (2-ME), 6-mercaptohexanol (6-MH), 11-mercaptoundecanol (11-MU), and 11-MUA. The as-prepared 11-MUA–Au NDs show highest QY (~3.1%) as a result of their high solubility and long chain 11-MUA molecules minimizing the luminescence quenching of Au NDs from collisions with quenchers in solution and at minimizing the number of internal nonradiative relaxation pathways (*44*). Not only the size of alkanethiols, we also demonstrated the preparation of Au and Au/Ag NDs whose luminescence could be controlled by varying the molar ratios of THPC to Au ions (for the Au NDs), and of Ag ions to Au ions (for the Au/Ag alloy NDs) (*45*).

To study the effect of concentration of thiol ligands on the PL of Au NDs, various Man–Au NDs were prepared under different Man-RSH/Au NPs molar ratios (*46*). Man–Au NDs with different sizes (2.9−1.5 nm) and PL wavelengths (618−525 nm) were obtained. The sizes of these Au NDs decreased upon increasing the concentration of Man-RSH from 1.0 mM to 10 mM. This phenomenon arose presumably from the formation of very strong Au–S bonds and dissociation into very small Au and Au−thiolate clusters in solution (*27, 28*). No Au NDs were present in the solution containing Man-RSH with a concentration higher than 20 mM, revealing that the Au NPs were all dissociated into Au–thiolate complexes under these conditions. As a result, the as-prepared solution did not absorb light at 360–380 nm and did not luminesce.

The 11-MUA–Au NDs were purified (denoted as *p*11-MUA–Au NDs) and were further used to study the impact of ligand density of (R-SH) on the PL of Au NDs (*47*). The PL intensity (excitation/emission: 375/525 nm) of the Au NDs decreased dramatically after purification through centrifugation as a result of a decrease in the 11-MUA ligand density on the Au NDs. However, the PL intensity of the purified Au NDs was recovered after introducing new mercaptoaliphatic acid ligands (Figure 4A). The ligand-recapped Au NDs are denoted herein as Au NDs@ligand; e.g., the 11-MUA recapped Au NDs are denoted as Au NDs@11-MUA). The PL intensity of the Au NDs@ligand species increased upon increasing the ligand density and chain length (Figure 4B). The PL intensity at 525 nm of the *p*11-MUA–Au NDs (10 nM) were investigated in the presence of various ligands, namely 2-mercaptoacetic acid [2-MAA, $HSCH_2CO_2H$], 3-mercaptopropanoic acid [3-MPA, $HS(CH_2)_2CO_2H$], 6-mercaptohexanoic acid [6-MHA, $HS(CH_2)_5CO_2H$], 11-MUA [$HS(CH_2)_{10}CO_2H$], 16-mercaptohexadecanoic acid [16-MHDA, $HS(CH_2)_{10}CO_2H$], and undecanoic acid [UDA, $CH_3(CH_2)_9COOH$]. Upon increasing the chain length of the mercaptoaliphatic acid from 2 to 16 carbon atoms, we observed an increase in PL, consistent with our previous finding of the PL intensity increasing upon increasing the chain length of the alkanethiol in the preparation of luminescent Au NDs (*25*). Mercaptoaliphatic acids with longer chain lengths can form highly ordered, layered supramolecular structures in the shells of Au NDs, with noticeable aurophilic Au(I)–thiol interactions, which in turn generate increased PL (*48*). Moreover, well-ordered layer structures in the shells of Au NDs can decrease the number of internal nonradiative relaxation pathways through the quencher in the solution (*49*). The longer alkanethiol chains of 16-MHDA increased the hydrophobicity, resulting in a decrease in the solubility and inducing larger degree of aggregation of the Au NDs. Although it was possible that UDA might also have inserted into the shells of the Au NDs through hydrophobic interactions between straight alkane chains, the lack of Au(I)–thiol interactions would limit the resulting PL enhancement.

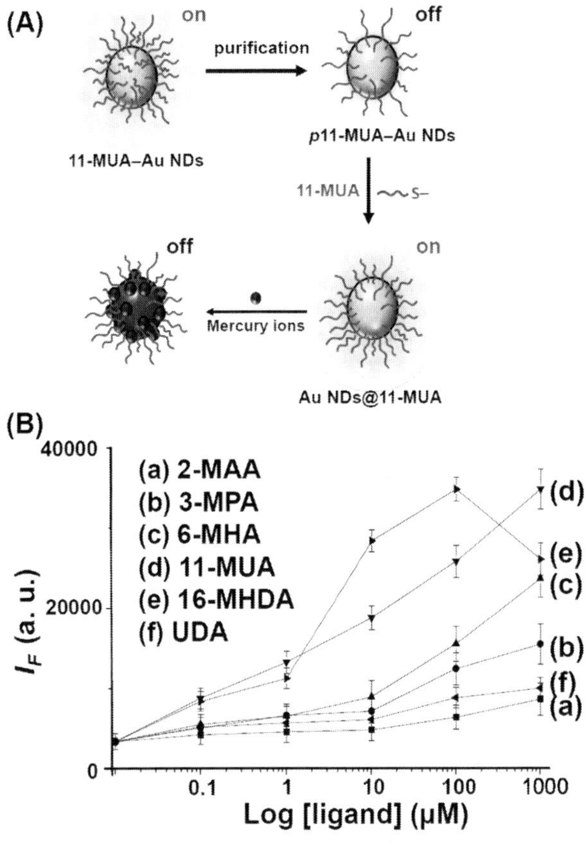

Figure 4. (A) Cartoon representation of the preparation of Au NDs@11-MUA forthe detection of Hg ions based on Hg-induced luminescence quenching. (B) PL intensities (525 nm) of p11-MUA-Au NDs (10 nM) in 5.0 mM sodiumphosphate (pH 7.0) in the presence of various capping ligands. Reprinted with permission from reference (47). Copyright 2013 Royal Society of Chemistry.

Applications of Luminescent Au NDs

Heavy Metal Ions

Water-soluble, luminescent, thiol-capped Au NCs and Au NDs have been widely employed to sense heavy metal ions, such as Hg^{2+}, Pb^{2+}, Cu^{2+}, and Ag^{+} (*50–55*). Mercury is well known for its toxicity to the brain, nervous system, endocrine system and kidneys, yet is widely distributed in the global environment (*56*). The 11-MUA–Au NDs are readily employed for the sensing of Hg^{2+} ions, based on a mechanism involving PL quenching through aggregation, induced by Hg^{2+}–carboxylate bonding and strong Au–Hg aurophilic interactions (*25*). Nevertheless, interference from Pb^{2+} and Cd^{2+} ions and thiol molecules was strong in the absence of a masking agent. Recently, using the Au NDs@11-MUA as a probe have been demonstrated that allowed the detection of total concentration of mercury ions {inorganic mercury [Hg^{2+}] plus organic mercury (methylmercury [$MeHg^{+}$], ethylmercury [$EtHg^{+}$] and phenylmercury [$PhHg^{+}$])} inaqueous solution (*47*). Figure 5 reveals that only the four Hg species in the solution of Au NDs@11-MUA resulted in apparent PL quenching; that is, the other remaining ions have insignificant effects under the same experimental conditions. We noted that purified 11-MUA-Au NDs probes allowed selective detection of Hg^{2+}, but only in the presence of the masking agent PDCA to avoid interference from Cd^{2+}, Ag^{+} or Pb^{2+} (*25*). In addition, AuNDs@11-MUA [core: Au NDs; shell: Au(I)-11-MUAcomplexes] enables the detection of the total concentration of Hg species through favorable Au–Hg aurophilic interactions since Au NDs@11-MUA nanosensor exhibits almost identical quenching ratios in the presence of each of the four Hg species (Figure 5). The selectivity of Au NDs@11-MUA under optimized 11-MUA concentration (10 μM) for the total mercury over other metal ions is remarkably high (1000-fold), with a limit of detection (LOD, S/N = 3) for mercury ions of 2.0 nM. The practicality of this approach have been demonstrated for the determination of mercury ions in complicated biological urine and plasma samples, as well as in a fish sample (DORM II). This simple, sensitive and selective approach appears to have practical potential for use in the clinical screening of mercury ions in biofluids. More recently, poly(N-isopropylacrylamide) microgels (PNIPAM MGs) incorporated with photoluminescent Au NDs have been developed for the detection of Hg^{2+} (*57*). Like Au NDs, Au NDs–PNIPAM MGs exhibit an absorption band at 375 nm that is assigned for ligands to metal charge transfer mixed with metal centered (ds/dp) states and PL at 520 nm originated from Au ND/polynuclear gold (I)–thiolate (core/shell) complexes when excited at 375 nm. The Au NDs–PNIPAM MGs are stable against salt (up to 500 mM NaCl). Based on Hg^{2+} induced PL quenching due to the formation of Au–Hg amalgam and Au NDs–PNIPAM MGs aggregates, the Au NDs–PNIPAM MGs probe allows selective detection of Hg^{2+} over a concentration range from 1 to 20 nM. This simple, sensitive, and selective approach has been validated for the determination of Hg in a representative fish sample (DORM II).

Proteins

Using two differently sized Au nanomaterials, acting separately as donor [aptamer-modified Au NPs (Apt–Au NPs)] and acceptor [platelet-derived growth factor-modified Au NDs (PDGF-Au NDs)], homogeneous PL quenching assays have been developed for the analysis of platelet-derived growth factors (PDGFs) and their receptor (PDGFR) (*29*). This molecular light switching system based on PDGF–Au NDs and Apt–Au NPs allowed the analysis of PDGFs as well as the PDGFR in separate homogeneous solutions as a result of competitive interactions of the PDGF–Au NDs and PDGFs with the Apt–Au NPs, and also the Apt–Au NPs and the PDGF receptor with the PDGF–Au NDs, respectively. The limits of detection (LODs) for PDGF AA and PDGF receptor were 80 pM and 0.25 nM, respectively.

Biofunctional Man–Au NDs are capable of sensing concanavalin A (Con A) through multivalent interactions (*39*). Con A is a member of the lectin family; it binds selectively to α-mannopyranosyl and α-glucopyranosyl residues (*58*). Although most lectin–carbohydrate interactions have low binding constants, usually over 10^3–10^7 M^{-1} range (*59*), the extremely high local concentrations of carbohydrates found on the surfaces of carbohydrate-captured nanoparticles make them particularly useful for probing glycoproteins (*60–64*). Man-Au ND/Con A aggregates are formed through specific interactions between the mannose units and the multivalent Con A. The recorded PL intensity of the supernatant after centrifuge is inversely proportional to the concentration of Con A. This turn-off assay by Man–Au NDs allows detection of Con A as low as 0.7 nM with remarkable selectivity (>100-fold) over other proteins (*39*). A similar sensing strategy has been applied for the detection of human immunoglobulin G (hIgG) (*65*). The specific interactions that occur between protein A and hIgG allowed selective detection of hIgG using protein A-conjugated Au NDs (PA–Au NDs), with an LOD ~ 10 nM and a remarkable selectivity (>50-fold) over other tested proteins. Recently, Man–Au NDs have been employed for the detection of the thyroid-cancer-marker thyroglobulin (Tg) (*66*). In the presence of Tg, the association between the Man–Au NDs and Con A is weakened as a result of the competition between Tg and Man–Au NDs for interactions with Con A, leading to an increase in luminescence (Figure 6A). The LOD of Con A/Man-Au NDs for Tg is 48 pM. This highly sensitive detection of Tg by the Con A/G-Man–Au NDs is caused mainly by the low background signal and dramatic increase in the intensity of the Tg-induced turn-on luminescence signal of the Con A/G-Man–Au NDs. For selective detection of Tg in serum samples, antibody-conjugated Au NPs (Ab–AuNPs, 13.3 nm) are also effective selectors for enriching Tg from complicated serum samples. This approach allows the removal of most of the background luminescence and nonspecific glycoproteins, leading to the detection of Tg at concentrations as low as 12 pM (Figure 6B). Furthermore, a time-resolved luminescence assay using Man–Au NDs with a lifetime of >100 ns has been demonstrated for the detection of Tg (*46*). The LOD of this assay for Tg is 90 pM in the presence of BSA (50 μM).

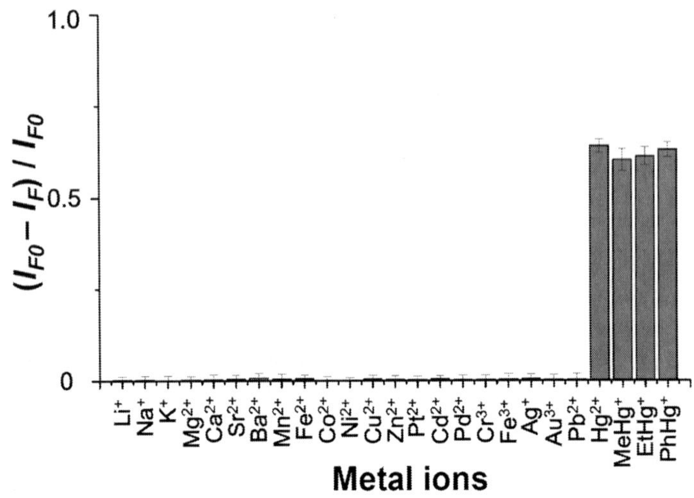

Figure 5. Selectivities of the Au NDs@11-MUA$_{10\ \mu M}$ (10 nM) sensor for different metal ions (100 μM) and four Hg species (1.0 μM) in 5.0 mM sodium phosphate (pH 7.0). Reprinted with permission from reference (47). Copyright 2013 Royal Society of Chemistry.

Recently, the photoluminescent 11-MUA–Au NDs have been used to detect hemoglobin (Hb) through PL quenching based on redox reactions between the 11-MUA–Au NDs and the Fe(II) ions of hemin units (Figure 7A) (*67*). This 11-MUA–Au ND assay is also sensitive and selective for other heme-containing proteins, including cytochrome c, and myoglobin. The hemin-induced PL quenching of 11-MUA–Au NDs strongly supports the quenching mechanism. As shown in Figure 7B, the hemin (1.0 μM) caused a significant decrease in the PL of the 11-MUA–Au NDs at 520 nm (ca. 55% of its original value) and reached saturation within 30 s after hemin was added to the Au NDs solution (100 nM). The binding energy (BE) for the Au $4f_{7/2}$ electrons in the 11-MUA–Au NDs in the absence of hemin was 84.7 eV, within the range from 84.0 eV for bulk Au to 85.0 eV for a polynuclear Au(I)–11-MUA complex. In contrast, the BE for the Au $4f_{7/2}$ electrons in the 11-MUA–Au NDs in the presence of 1.0 μM hemin was 84.3 eV, suggesting a decrease in the oxidation state (reduction) of the Au^+ ions on the surfaces of the NDs. The Stern–Volmer quenching constants (K_{sv}) for hemin, cytochrome c, Hb, and myoglobin were 5.6×10^7, 1.7×10^7, 1.6×10^7, and 6.2×10^6 M^{-1}, respectively, in good agreement with the order of their sizes and reduction potentials. The LOD of this 11-MUA–Au NDs probe for Hb in biological buffer was 0.5 nM. In addition, this 11-MUA–Au ND probe provided quantitative results that correlated with those obtained using a commercial HiCN assay for detection of Hb in five human plasma samples, revealing its great potential for the diagnosis of diseases associated with changes in the levels of Hb.

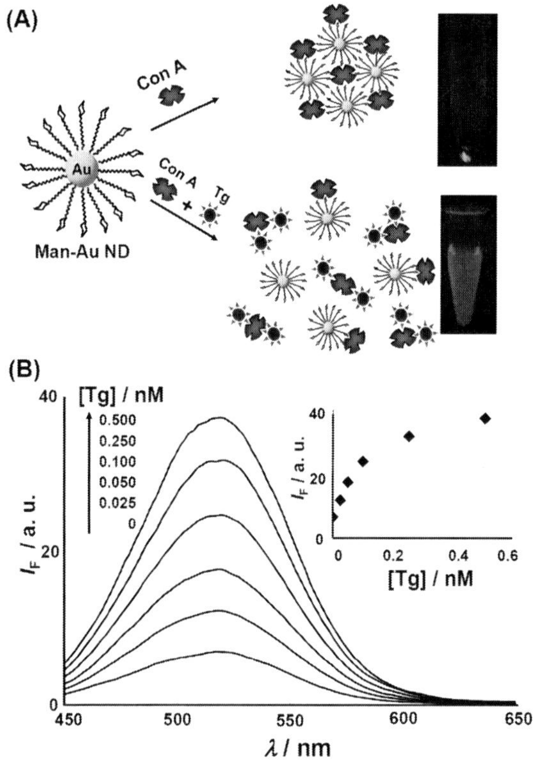

Figure 6. (A)Schematic representation of a Tg nanosensor that is operated based on Tg-mediated modulation of the interaction between Con A and Man−Au NDs. (B) Validation of the use of Ab−Au NPs selector and Con A/Man−Au ND probe for the luminescent detection of Tg (0–0.5 nM) in a serum sample spiked with standard Tg solutions. Reprinted with permission from reference (66). Copyright 2010 John Wiley & Sons.

Bacteria

Detection of bacteria is a crucial issue for effective clinical diagnosis and treatment, pharmaceutical production, and for ensuring the safety of the food supply and environment (*68–70*). Although cell plating and culture-based analyses give reliable results for bacteria screening, they are very time consuming (>24 h) and labor intensive (*71*). Therefore, detection of bacteria in food, environmental, and clinical samples without the requirement of time-consuming culture processes would allow rapid screening and diagnosis. Recently, a simple, one-pot, aqueous self-assembly procedure for the preparation of Man-Au NDs have been developed for *Escherichia coli* (*E. coli*)-labeling and as an antibacterial agent (*40*). Based on the selective binding of Man–Au NDs to the type 1 fimbriae in the pili of *E. coli*, the *E coli*-Man–Au ND complex are readily formed after incubation of *E. coli* with Man–Au NDs. After mannose displaces the Man-Au NDs from the *E. coli* surface, the solution is centrifuged. The PL intensity of displaced Man-Au NDs in the supernatant solution depends on the concentration of the *E coli*. Furthermore, by adding Man-RSSR-Man to this supernatant the ligand density of the Man–Au NDs can be increased significantly and thus amplify their PL signal. The increased ligand density reduces the quenching by other species in the solution and thus shows enhanced PL. Monitoring the luminescence of Man–Au NDs allows the detection of *E. coli* at levels as low as 150 CFU/mL. (CFU)/mL. In addition, Man–Au NDs were also found to be efficient antibacterial agents, selectively inhibiting the growth of *E. coli* through Man–Au ND-induced agglutination (*40*). The agglutination was due to strong binding and cross-linking of *E. coli* by the ultra high mannose density Au NDs to multivalent interactions between the surface confined mannose units and FimH lectin domains on type 1 pili. Figure 8 displays the growth curve of *E. coli* (1.0 × 10^8 CFU/mL) in sterile Luria-Bertani (LB) media that had been pretreated with Man–Au NDs. A very low growth rate of *E. coli* was observed in the presence of Man–Au NDs (>250 nM). The inset to Figure 8 reveals the very high turbidity of *E. coli* grown in the LB medium for 10 h in the absence of Man–Au NDs; in contrast, the culture turbidity was very low in their presence. This high degree of growth inhibition for *E. coli* resulted from the multivalent interactions provided by the ultra high density of mannose ligands on Au NDs. Compared with other Glyco-NPs that often have sizes larger than 10 nm, both the ligand density and local concentration of Man–Au NDs Au NDs (1.8 nm) are ultra high.

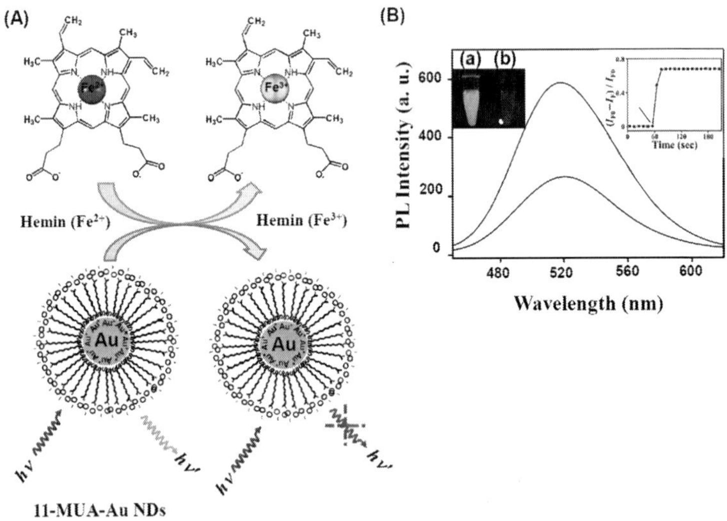

Figure 7. (A) Schematic representation of the photoluminescence quenching mechanism of 11- MUA−Au NDs in the presence of hemin. (B) PL spectra of 100 nM 11-MUA−AuNDs in 5 mM HEPES buffer (pH7.4) in the (a) absence and (b) presence of hemin (1.0μM). Reprinted with permission from reference (67). Copyright 2013 Elsevier.

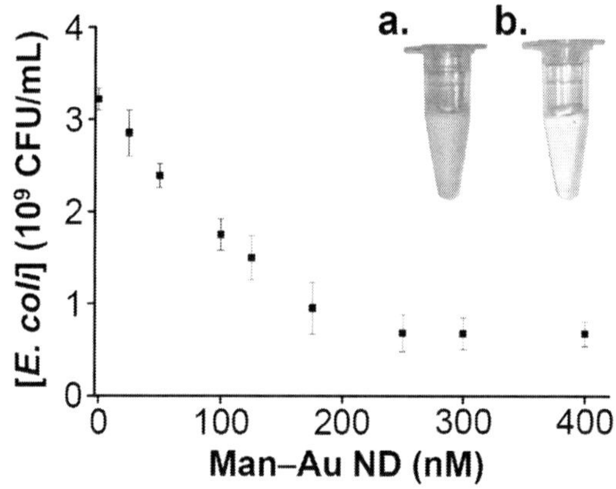

Figure 8. Effect of Man−Au NDs (0−400 nM) on the growth of E. coli (1.0 × 10^8 CFU/mL). Inset: photographs of E. coli (1.0 × 10^8 CFU/mL) grown for 10 h in the LB medium in the (a) absence and (b) presence of Man−Au ND (250 nM). Reprinted with permission from reference (40). Copyright 2011 Elsevier.

Conclusion and Perspective

In this review, we have summarized the synthesis and properties of some new type of Au NDs and it reveals the fascinating properties and promising future of these particles in their biosensing applications in environmental and biological samples. Owing to their biocompatibility, the search for new types of Au NDs is growing steadily. Unlike the inorganic luminescent NDs in which the shell is made up of toxic inorganic materials, the luminescent Au NDs is protected and stabilized by molecules called capping ligands. Different kinds of capping ligands such as alkanethiols, proteins, DNA, thiol derived carbohydrates, aptamers, etc. are used in the synthesis of Au NDs.

The molecular length of capping ligand, charge of the ligand and the ligand density on the Au NP surface etc. greatly influence the stability, size, and PL properties of the NDs. Therefore, adequate understanding of the interaction of the ligand with the Au NPs and their effect on the properties of Au NDs are essential for their practical applications. To synthesize highly stable and fluorescent Au NDs is a challenge. Therefore, efforts are in progress to develop easy methods of preparation and understand the mechanism of luminescence properties of Au NDs. The size must be controlled to below 2 nm to get highly luminescent Au NDs. Though the PL wavelength is tunable through modification of the chain length of the ligand molecule, the number of atoms in the core of the Au NDs may not be uniform. Hence, further research has to be made to understand and control the Au NDs formation to synthesize Au NDs with uniform size and optical properties.

Also, this review concludes that the luminescence of Au NDs mainly originate from Au ND/polynuclear Au(I)–thiol (core/shell) complexes. Therefore, further research will lead to discover new types of Au NDs with high luminescence properties and large Stokes shift. More studies with biocompatible ligands for Au NDs will enable the invention of biocompatible Au NDs for in vivo applications.

A complete knowledge and control of the charge transfer between the Au ND core and the surface ligand is lacking. For this reason, more research has to be done in this filed to understand the ligand-to-metal NP charge transfer (LMNCT; S→Au) and ligand-to-metal–metal charge transfer. Control on this property of Au NDs will open ways to develop Au NDs with interesting properties

Acknowledgments

This study was supported by the National Science Council of Taiwan under Contracts 101-2628-M-019-001-MY3.

References

1. Lin, C.-A. J.; Yang, T.-Y.; Lee, C.-H.; Huang, S. H.; Sperling, R. A.; Zanella, M.; Li, J. K.; Shen, J.-L.; Wang, H.-H.; Yeh, H.-I.; Parak, W. J.; Chang, W. H. *ACS Nano* **2009**, *3*, 395–401.
2. Shang, L.; Dong, S.; Nienhaus, G. U. *Nano Today* **2011**, *6*, 401–418.
3. Chevrier, D. M.; Chatt, A.; Zhang, P. *J. Nanophotonics* **2012**, *6*, 064504.
4. Zheng, J.; Nicovich, P. R.; Dickson, R. M. *Annu. Rev. Phys. Chem.* **2007**, *58*, 409–431.
5. Link, S.; Ei-Sayed, M. A. *Annu. Rev. Phys. Chem.* **2003**, *54*, 331–366.
6. Yang, Y.; Matsubara, S.; Nogami, M.; Shi, J.; Huang, W. *Nanotechnology* **2006**, *17*, 2821–2827.
7. Qian, H.; Zhu, M.; Wu, Z.; Jin, R. *Acc. Chem. Res.* **2012**, *45*, 1470–1479.
8. Liu, X.; Li, C.; Xu, J.; Lv, J.; Zhu, M.; Guo, Y.; Cui, S.; Liu, H.; Wang, S.; Li, Y. *J. Phys. Chem. C* **2008**, *112*, 10778–10783.
9. Chen, S.; Ingram, R. S.; Hostetler, M. J.; Pietron, J. J.; Murray, R. W.; Schaaff, T. G.; Khoury, J. T.; Alvarez, M. M.; Whetten, R. L. *Science* **1998**, *280*, 2098–2101.
10. Zheng, J.; Zhang, C. W.; Dickson, R. M. *Phys. Rev. Lett.* **2004**, *93*, 077402.
11. Yuan, Z.; Peng, M.; He, Y.; Yeung, E. S. *Chem. Commun.* **2011**, *47*, 11981–11983.
12. Wu, Z.; Jin, R. *Nano Lett.* **2010**, *10*, 2568–2573.
13. Aldeek, F.; Muhammed, M. A. H.; Palui, G.; Zhan, N.; Mattoussi, H. *ACS Nano* **2013**, *7*, 2509–2521.
14. Hostetler, M. J.; Wingate, J. E.; Zhong, C.-J.; Harris, J. E.; Vachet, R. W.; Clark, M. R.; Londono, J. D.; Green, S. J.; Stokes, J. J.; Wignall, G. D.; Glish, G. L.; Porter, M. D.; Evans, N. D.; Murray, R. W. *Langmuir* **1998**, *14*, 17–30.
15. Zhang, J.; Fu, Y.; Conroy, C. V.; Tang, Z.; Li, G.; Zhao, R. Y.; Wang, G. *J. Phys. Chem. C* **2012**, *116*, 26561–26569.
16. Negishi, Y.; Nobusada, K.; Tsukuda, T. *J. Am. Chem. Soc.* **2005**, *127*, 5261–5270.
17. Chevrier, D. M.; MacDonald, M. A.; Chatt, A.; Zhang, P.; Wu, Z.; Jin, R. *J. Phys. Chem. C* **2012**, *116*, 26947–26947.

18. Lin, C.-A. J.; Lee, C.-H.; Hsieh, J.-T.; Wang, H.-H.; Li, J. K.; Shen, J.-L.; Chan, W.-H.; Yeh, H.-I; Chang, W. H. *J. Med. Biol. Eng.* **2009**, *6*, 276–283.
19. Templeton, A. C.; Wuelfing, W. P.; Murray, R. W. *Acc. Chem. Res.* **2000**, *33*, 27–36.
20. Zhou, C.; Sun, C.; Yu, M.; Qin, Y.; Wang, J; Kim, M.; Zheng, J. *J. Phys. Chem. C* **2010**, *114*, 7727–7732.
21. Vogler, A.; Kunkely, H. *Coord. Chem. Rev.* **2001**, *219–221*, 489–507.
22. Koshevoy, I. O.; Smirnova, E. S.; Haukka, M.; Laguna, A.; Chueca, J. C.; Pakkanen, T. A.; Tunik, S. P.; Ospino, I.; Crespo, O. *Dalton Trans.* **2011**, *40*, 7412–7422.
23. Coker, N. L.; Krause Bauer, J. A.; Elder, R. C. *J. Am. Chem. Soc.* **2004**, *126*, 12–13.
24. Forward, J. M.; Bohmann, D.; Fackler, J. P.; Staples, R. J. *Inorg. Chem.* **1995**, *34*, 6330–6336.
25. Huang, C.-C.; Yang, Z.; Lee, K.-H.; Chang, H.-T. *Angew. Chem., Int. Ed.* **2007**, *46*, 6824–6828.
26. Duff, D. G.; Baiker, A. *Langmuir* **1993**, *9*, 2301–2309.
27. Krüger, D.; Fuchs, H.; Rousseau, R.; Marx, D.; Parrinello, M. *J. Chem. Phys.* **2001**, *115*, 4776–4786.
28. Konôpka, M.; Rousseau, R.; Štich, I.; Marx, D. *J. Am. Chem. Soc.* **2004**, *126*, 12103–12111.
29. Huang, C.-C.; Chiang, C.-K.; Lin, Z.-H.; Lee, K.-H.; Chang, H.-T. *Anal. Chem.* **2008**, *80*, 1497–1504.
30. Zhao, J.; Wu, W.; Sun, J.; Guo, S. *Chem. Soc. Rev.* **2013**, *42*, 5323–5351.
31. Li, C.-K.; Lu, X.-X.; Wong, K. M.-C.; Chan, C.-L.; Zhu, N.; Yam, V. W.-W. *Inorg. Chem.* **2004**, *43*, 7421–7430.
32. Castiñeiras, A.; Pedrido, R. *Dalton Trans.* **2012**, *41*, 1363–1372.
33. Yam, V. W.-W.; Cheng, E. C.-C. *Chem. Soc. Rev.* **2008**, *37*, 1806–1813.
34. Au, V. K.-M.; Zhu, N.; Yam, V. W.-W. *Inorg. Chem.* **2013**, *52*, 558–567.
35. Pyykkö, P. *Inorg. Chim. Acta* **2005**, *358*, 4113–4130.
36. Pyykkö, P. *Chem. Soc. Rev.* **2008**, *37*, 1967–1997.
37. Stellwagen, D.; Weber, A.; Bovenkamp, G. L.; Jin, R.; Bitter, J. H.; Kumar, C. S. S. R. *RSC Adv.* **2012**, *2*, 2276–2283.
38. Iwamura, M.; Nozaki, K.; Takeuchi, S.; Tahara, T. *J. Am. Chem. Soc.* **2013**, *135*, 538–541.
39. Huang, C.-C.; Chen, C.-T.; Shiang, Y.-C.; Lin, Z.-H.; Chang, H.-T. *Anal. Chem.* **2009**, *81*, 875–882.
40. Tseng, Y.-T.; Chang, H.-T.; Chen, C.-T.; Chen, C.-H.; Huang, C.-C. *Biosens. Bioelectron.* **2011**, *27*, 95–100.
41. Huang, T.; Murray, R. W. *J. Phys. Chem. B* **2001**, *105*, 12498–12502.
42. Guo, Y.; Wang, Z.; Shao, H.; Jiang, X. *Analyst* **2012**, *137*, 301–304.
43. Zhang, P.; Sham, T. K. *Appl. Phys. Lett.* **2002**, *81*, 736–738.
44. Arvapally, R. K.; Sinha, P.; Hettiarachchi, S. R.; Coker, N. L.; Bedel, C. E.; Patterson, H. H.; Elder, R. C.; Wilson, A. K.; Omary, M. A. *J. Phys. Chem. C* **2007**, *111*, 10689–10699.
45. Huang, C.-C.; Liao, H.-Y.; Shiang, Y.-C.; Lin, Z.-H.; Yang, Z.; Chang, H.-T. *J. Mater. Chem.* **2009**, *19*, 755–759.

46. Huang, C.-C.; Lin, T.-Y.; Yu, W.-S.; Chen, C.-T.; Chang, H.-T. *J. Biomed. Nanotechnol.* **2009**, *5*, 579–585.
47. Chang, H.-Y.; Chang, H.-T.; Hung, Y.-L.; Hsiung, T.-M.; Lin, Y.-W.; Huang, C.-C. *RSC Adv.* **2013**, *3*, 4588–4597.
48. Cha, S.-H.; Kim, J.-U.; Kim, K.-H.; Lee, J.-C. *Chem. Mater.* **2007**, *19*, 6297–6303.
49. Luo, Z.; Yuan, X.; Yu, Y.; Zhang, Q.; Leong, D. T.; Lee, J. Y.; Xie, J. *J. Am. Chem. Soc.* **2012**, *134*, 16662–16670.
50. Hu, D.; Sheng, Z.; Gong, P.; Zhang, P.; Cai, L. *Analyst* **2010**, *135*, 1411–1416.
51. Muhammed, M. A. H.; Verma, P. K.; Pal, S. K.; Retnakumari, A.; Koyakutty, M.; Nair, S.; Pradeep, T. *Chem. Eur. J.* **2010**, *16*, 10103–10112.
52. Chai, F.; Wang, T.; Li, L.; Liu, H.; Zhang, L.; Su, Z.; Wang, C. *Nanoscale Res. Lett.* **2010**, *5*, 1856–1860.
53. Lin, Y.-H.; Tseng, W.-L. *Anal. Chem.* **2010**, *82*, 9194–9200.
54. Lin, Y.-W.; Huang, C.-C.; Chang, H.-T. *Analyst* **2011**, *136*, 863–871.
55. Chen, P.-C.; Yeh, T.-Y.; Ou, C.-M.; Shih, C.-C.; Chang, H.-T. *Nanoscale* **2013**, *5*, 4691–4695.
56. Holmes, P.; James, K. A. F.; Levy, L. S. *Sci. Total Environ.* **2009**, *408*, 171–182.
57. Chen, L.-Y.; Ou, C.-M.; Chen, W.-Y.; Huang, C.-C.; Chang, H.-T. *ACS Appl. Mater. Interfaces* **2013**, *5*, 4383–4388.
58. Cairo, C. W.; Gestwicki, J. E.; Kanai, M.; Kiessling, L. L. *J. Am. Chem. Soc.* **2002**, *124*, 1615–1619.
59. Mann, D. A.; Kanai, M.; Maly, D. J.; Kiessling, L. L. *J. Am. Chem. Soc.* **1998**, *120*, 10575–10581.
60. Hartmann, M.; Betz, P.; Sun, Y.; Gorb, S. N.; Lindhorst, T. K.; Krueger, A. *Chem. Eur. J.* **2012**, *18*, 6485–6492.
61. Gao, J.; Liu, D.; Wang, Z. *Anal. Chem.* **2008**, *80*, 8822–8827.
62. Shinde, S. B.; Fernandes, C. B.; Patravale, V. B. *J. Controlled Release* **2012**, *159*, 164–180.
63. Sánchez-Pomales, G.; Morris, T. A.; Falabella, J. B.; Tarlov, M. J.; Zangmeister, R. A. *Biotechnol. Bioeng.* **2012**, *109*, 2240–2249.
64. Dube, D. H.; Champasa, K.; Wang, B. *Chem. Commun.* **2011**, *47*, 87–101.
65. Shiang, Y.-C.; Lin, C.-A.; Huang, C.-C.; Chang, H.-T. *Analyst* **2011**, *136*, 1177–1182.
66. Huang, C.-C.; Hung, Y.-L.; Shiang, Y.-C.; Lin, T.-Y.; Lin, Y.-S.; Chen, C.-T.; Chang, H.-T. *Asian J. Chem.* **2010**, *5*, 334–341.
67. Chen, L.-Y.; Huang, C.-C.; Chen, W.-Y.; Lin, H.-J.; Chang, H.-T. *Biosens. Bioelectron.* **2013**, *43*, 38–44.
68. Miriagou, V.; Cornaglia, G.; Edelstein, M.; Galani, I.; Giske, C. G.; Gniadkowski, M.; Malamou-Lada, E.; Martinez-Martinez, L.; Navarro, F.; Nordmann, P.; Peixe, L.; Pournaras, S.; Rossolini, G. M.; Tsakris, A.; Vatopoulos, A.; Cantón, R. *Clin. Microbiol. Infect.* **2010**, *16*, 112–122.
69. Terry, L. A.; White, S. F.; Tigwell, L. J. *J. Agric. Food Chem.* **2005**, *53*, 1309–1316.
70. Hageskal, G.; Lima, N.; Skaar, I. *Mycol. Res.* **2009**, *113*, 165–172.
71. Pan, L.; Gu, J.-D. *J. Polym. Environ.* **2007**, *15*, 57–65.

Chapter 3

Gold Nanomaterials Based Absorption and Fluorescence Detection of Mercury, Lead, and Copper

Po-Cheng Chen, Prathik Roy, Li-Yi Chen, Ya-Na Chen, and Huan-Tsung Chang[*]

Department of Chemistry, National Taiwan University, 1, Section 4, Roosevelt Road, Taipei 106, Taiwan
[*]E-mail: changht@ntu.edu.tw

Gold nanomaterials (Au NMs) such as Au nanoparticles (Au NPs) and fluorescent Au nanodots (Au NDs) have been widely used for monitoring the levels of potentially toxic metal (PTM) ions (e.g., Hg^{2+}, Pb^{2+}, Cu^{2+}) in aquatic ecosystems. We focus on the absorption and fluorescence assays for Hg^{2+}, Pb^{2+}, and Cu^{2+} ions using functionalized Au NPs and Au NDs. The preparation of functionalized Au NPs and Au NDs are discussed. The optical sensing systems is divided into four parts including Au NP-based colorimetric systems, catalytic Au NPs-based systems, Au NPs-based quenching systems, and Au NDs-based systems. Factors controlling the selectivity and sensitivity of the nanosensors toward the analytes are discussed, including the size of Au NMs, length and sequence of DNA, nature of ligands, and sensing conditions. To show the merits of Au NMs, their practicality for the detection of Hg^{2+}, Pb^{2+}, and Cu^{2+} ions are highlighted.

Introduction

Heavy metal pollutants are a great environmental issue, mainly because they can cause severe effects even at very low concentrations (*1–3*). Mercury, lead, and copper are three most particularly interesting potential toxic metals (PTMs) since their common use in many ways; for example copper used in electricity, lead used in plumbing, and mercury used in battery. These metal species have strong affinity toward important biologically active compounds such as thiol compounds like glutathione (GSH) and proteins, leading to changes in their biological functions. It has been known that mercury tends to accumulate in human beings, resulting in drastically negative effects on the nervous system and organs such as liver and bones. Serious mercury pollution even caused the death of thousands of Japanese people in 1956. Owing to their high toxicity, most countries have set high standards to control the pollution of these three metals in the ecosystems. For example, the maximum levels of copper, lead, and mercury in drinking water recommended by the United States Environment Protection Agency (EPA) are 1300/20312, 15/72, and 2/10 ppb/nM, respectively.

Many techniques such as electrochemistry (*4*, *5*), inductively coupled plasma mass spectroscopy (ICP–MS) (*6*, *7*), and atomic absorption/emission spectroscopy (AAS/AES) (*8*, *9*) have been widely employed for the determination of the levels of the three metal ions. They are sensitive, but require high-cost instruments, large amounts of samples and reagents, and/or experienced operators. In addition, they can not provide detailed information about various metal species unless they are in conjunction with separation systems. In order to detect trace amounts of analytes from complicated samples, a tedious sample pretreatment process to remove matrices and concentrate the analytes of interest is usually conducted. Thus, inexpensive, rapid, convenient assays for the determination of the levels of metal species are still highly demanded.

In the past two decades, we have witnessed great progress in the development of nanomaterials (NMs) based sensing systems for the detection of metal ions, with advantages of sensitivity, selectivity, rapidness, and/or potential in-field analysis (*10–15*). NMs have sizes ranging from 1 – 100 nm and optical properties different from their corresponding bulk materials (*16–18*). The optical properties of NMs depend on their size, shape, composition, surface, and their surrounding environment (*19–22*). Gold nanoparticles (Au NPs) have attracted considerable attention, mainly because of their stability, biocompatibility, facile preparation, ease in conjugation, and strong surface plasmon resonance (SPR) absorption (extinction) in the visible region. Relative to organic dyes, Au NPs have great molar absorption coefficients (10^8–10^{10} M^{-1} cm^{-1}), allowing development of very sensitive sensing systems (*23*, *24*). Through strong Au–S or Au–N bonding or through physical adsorption, Au NPs are ready for conjugation with various recognition elements such as amphiphilic polymers, silanols, sugars, nucleic acids, and proteins, providing greater stability and/or selectivity (*25–27*). Analytes induced aggregation of Au NPs, leading to changes in the color (absorption), have been commonly applied to the detection of small molecules such as metal ions and anions, and biopolymers such as peptides, proteins and DNA (*23*, *28–33*). The color changes of Au NPs (10–50 nm) from ruby red to blue/purple are easily observed by the naked eye. Aggregation either occurs through interparticles crosslinking or non-crosslinking. Noncrosslinking aggregation is based on the de-stabilization of Au NPs through changes in their zeta potential. For example, salt induced aggregation of Au NPs. On the other hand, crosslinking aggregation takes place through covalent bonding between the target analytes and ligands modified on Au NPs (*34–36*). For instance, thiol conjugated Au NPs have been used for the sensitive and selective detection of Hg^{2+} ions (*37*, *38*). Owing to greater selectivity provided, crosslinking aggregation sensing systems are more common. De-aggregation of Au NPs is sometimes applied towards the detection of analytes such as Pb^{2+} ions (*39*, *40*). Au nanorods (NRs) with strong longitudinal SPR are also attractive materials for detection of heavy metal ions such as Cu^{2+}, Pb^{2+}, Hg^{2+} ions through aggregation induced change in their longitudinal SPR (*41–45*).

Fluorescent Au NMs (typically below 3 nm) represent another type of interesting sensing materials for the detection of metal ions. Fluorescent Au NMs are often called Au nanodots (NDs) or clusters (NCs). The size of Au NDs is relatively larger than Au NCs. Top-down and bottom-up approaches are usually applied to the preparation of Au NDs and Au NCs, respectively. The fluorescence property of Au NCs is mainly detrermined by their size. Unlike larger sizes of Au NPs (> 5 nm), Au NDs consisting of several to hundreds of gold atoms lack an apparent SPR band but possess molecular-like fluorescence properties due to confinement effect (*46–48*). The fluorescence properties of Au NDs are strongly dependent on their size, surface ligands (complexes), and oxidation states (*49*, *50*). For simplicity, Au NDs are use to represent Au NDs and Au NCs in this chapter. Various templates such as poly(amidoamine) dendrimers (PAMAM), polyethylenimine, bovine serum albumin (BSA), and small thiol compounds such as GSH have been used to prepare Au NDs, with emission colors ranging from blue to the near-infrared (near-IR) region (*51–55*). Relative to quantum dots (such as CdSe), Au NDs possess advantages of biocompatibility, large Stokes shift (>100 nm), and long fluorescence lifetimes (>20 ns), but with lower quantum yields (Φ) (*50*, *56–58*).

Great advances in the Au NPs and NDs based sensing systems for the detection of DNA, proteins, and small molecules like glucose, as well as for the targeting of cancer cells have been made in recent years, it is thus impossible to cover all of them. To highlight some of those systems, many excellent review papers have been published in the last three years (*13*, *29*, *30*, *57–61*). In this chapter, we focus on the detection of Hg^{2+}, Pb^{2+}, and Cu^{2+} ions using Au NPs and NDs through absorption and fluorescence detection modes. Preparation and optical properties of Au NPs and Au NDs are highlighted to show how the sensing materials are designed and work.

Synthesis and Optical Properties of Au NPs and NDs

Synthesis

Spherical and hydrophilic Au NPs are usually prepared from the reduction of Au^{3+} ions ($HAuCl_4$) with reducing agents such as citrate, which was introduced by Turkevitch in 1951 (*62*). Citrate ions not only act as a reducing agent but also a capping agent that stabilize the as-prepared Au NPs. By varying the citrate/Au^{3+} molar ratios from 5.0 to 0.5, different sizes of Au NPs (diameters 16−147 nm) can be prepared (*63*). For example, 32- and 56- nm Au NPs are easily prepared by adding 1% trisodium citrate solutions (0.5 mL and 0.3 mL, respectively) rapidly to 0.01% $HAuCl_4$ solutions (50 mL) under reflux, in which the mixtures react for 8 min. To further stabilize and/or functionalize Au NPs, the citrate ions on their surfaces are readily replaced by other desired ligands such as organic molecules, hydrophilic polymers, oligonucleotides, peptides, and proteins. In addition to citric acid, various reducing agents such as catechin, carbohydrates, gallic acid, and amino acids have been used to prepare differently sized and shaped Au NPs (*64–69*).

Most hydrophobic (and some hydrophilic) Au NPs having diameters of less than 2.0 nm are commonly reduced from Au^{3+} ions by sodium borohydride ($NaBH_4$) in toluene containing tetraoctylammonium bromide or dodecanethiol (*70*, *71*). In order to prepare small sizes of Au NPs, large thiol/gold mole ratios, fast addition rate of the reducing agent, and low reaction temperature (< 10 °C) are required. One-step aqueous preparation of highly monodisperse Au NPs with diameters below 2 nm can be prepared using thioether- or thiol-functionalized polymers such as alkyl thioether end-functionalized poly(methacrylic acid) (PMAA) (*72*). By varying the molar ratio of thioether end-functionalized PMAA and Au^{3+}, the size of the Au NPs can be finely tuned with core diameters between 1.5 and 4 nm (*72*, *73*).

There are two general approaches—top-down and bottom-up— for the preparation of Au NDs. Top-down approaches involve the preparation of small size (<3 nm) of Au NPs from Au^{3+} using tetrakis(hydroxymethyl)phosphonium chloride as a capping and reducing agent. The Au NPs are then etched by thiol compounds such as 11-mercaptoundecanoic acid (11-MUA) (*57*) and dihydrolipoic acid (DHLA) under alkaline conditions (*61*, *74*). By selecting different sizes and functional groups of thiol compounds, the sizes and emission colors of Au NDs can be tuned easily (*57*, *75*). Bottom-up approaches are applied to prepare Au NDs from Au^{3+} in the presence of templates such as BSA and lysozyme (Lys) (*53*, *76*). The proteins also act as reducing agents; tyrosine and histidine residues in the proteins possess strong reduction ability under alkaline conditions (pH > 10.0) (*77*, *78*). By varying the nature of proteins, reaction temperature (energy), pH values, and protein/Au^{3+} molar ratio, different sizes and thus emission colors of Au NDs can be prepared (*79–82*). For instance, BSA−Au NCs with different sizes of core are fabricated, with evidence provided by matrix assisted laser desorption ionization mass spectrometry (MALDI-MS) (*83*). In addition to the use of proteins, small compounds such as p-mercaptobenzoic acid, tiopronin and phenylethylthiol have been used to prepare Au NDs (NCs) (*84–86*). Because of their relatively poor solubility and stability, as well as lower value of Φ, their applications for the detection of metal ions are less common.

Optical Properties

Spherical Au NPs display a distinct SPR absorption band in the visible region (around 520 nm), which is due to the collective oscillation of conduction band electrons at their surfaces in resonance with the incident electromagnetic radiation. The SPR absorption band of Au NPs is dependent on not only their size and shape, but also the dielectric constant and the refractive index of the surrounding medium. The SPR band can be easily tuned from 517 to 575 nm by changing the diameters of spherical Au NPs from 9 to 99 nm (*87*). In addition to spherical Au NPs, Au NRs having an additional strong SPR band (longitudinal) in the NIR region (634−743 nm), depending on their aspect ratios (length/width), are found to be useful for sensing of metal ions and photothermal therapy of cancer cells (*88*). By taking the advantage of the size dependence SPR properties, many sensing optical systems using Au NPs and NRs have been developed (*29*, *35*, *38*, *89*, *90*).

Because of having very high extinction coefficients (e.g. 2×10^8 cm^{-1} mol^{-1} L at 518 nm for 13 nm Au NPs), Au NPs relative to conventional molecules are more effective quenchers. Au NPs causing significant quenching of organic fluorophores and fluorescent NMs through surface energy transfer (SET) have been demonstrated (*24*, *91*). Although the molar extinction coefficient increases upon increasing the size of Au NPs, 13 nm Au NPs are found to be more common, mainly because of their greater stability. Because the SET efficiency is highly dependent on the distance between the signal elements and the surfaces of Au NPs, analytes that induce changes in their distance can be detected through fluorescence quenching mechanisms.

Au NDs have sizes comparable to Fermi wavelength of an electron (0.5 nm for gold) and discrete energy levels, showing molecular-like optical properties (*50*). The emission energy of Au NDs can be well-fitted by the simple scaling relation of $E_{Fermi}/N^{1/3}$ (jellium approximation), where E_{Fermi} is the Fermi energy of bulk Au and N is the number of gold atoms, respectively (*49*). By conducting electron spray ionization (ESI) mass spectrometry, many Au NDs with magic numbers of Au atoms such as Au_5, Au_8, Au_{13}, Au_{23} and Au_{31} have been identified (*51*). The fluorescence lifetime of polymers (proteins) capped Au NDs is in the order of nanoseconds, which is evidence of the intraband (sp–sp) band transition. Because their fluorescence properties are strongly dependent on their sizes (the number of Au atoms), control of the polymer (protein)/Au^{3+} molar ratio is important. It is also important to use high concentrations (e.g. 0.5 mM) of the polymers (proteins) as templates to provide stable and fluorescent Au NDs. The Φ values of Au NDs are usually less than 10%; for example they are 6% for BSA−Au NDs at 640/470 nm (emission/excitation wavelengths), 5.6% for Lys−Au NDs at 657/360 nm, 3.5% for pepsin−Au NDs at 670/360 nm, and 6% for lactoferrin−Au NDs at 650/445 nm respectively (*53*, *76*, *81*, *92*). On comparison to polymers (proteins) capped Au NDs, alkanethiolate Au NDs prepared from etching of small sizes of Au NPs by thiol compounds have less size-dependent fluorescence but greater ligand-dependent properties (*93*). The manifesting long life time (hundred ns~μs) and large stoke shifts (> 170 nm) of alkane−thiolate Au NDs are attributed to electronic transitions of Au(I) coupling to the ligand–metal charge transfer transition (*94–96*). The interband transition (from d band electrons to the sp conduction band) and the relaxed radiative recombination across the energy gap between the highest and lowest occupied molecular orbitals (HOMO–LUMO gap) within the sp conduction band are responsible for fluorescence (*97–99*). As a result, the emission colors can be easily tuned by the selection of various thiol compounds. For example, the emission colors of 2-mercaptoethanol, 6-mercaptohexanol, and 11-MUA capped Au NDs are reddish, yellow, and green, with Φ values of 0.006%, 1.5%, and 3.1%, respectively. In addition, these Au NDs (e.g. 11-MUA Au NDs) possess strong absorption located at around 375 nm as a result of metal-centered and/or ligand−metal charge transfer (LMCT; S→Au) transitions (*57*, *100*).

The fluorescence properties of Au NDs are dependent on the nature (size, species, and oxidation state) of cores and shells (complexes). Thus analytes inducing changes in the size, core, and shell can be detected through either fluorescence "turn-on" or "turn-off" mechanisms. While the latter mechanism is more common, it is usually less sensitive. The fluorescence properties of Au NDs are quite sensitive to the changes in the environmental conditions such as ionic strength. Thus buffer providing a constant pH value and ionic strength is important when applying Au NDs for analytical purposes. Although biocompatible Au NDs are conjugated with recognition elements easily, their low Φ values and poor photo-stability limit their use in cell imaging.

Detection of Pb^{2+}, Hg^{2+}, and Cu^{2+} Using Au NPs and NDs

Au NP-Based Colorimetric Systems

The sensitivity and selectivity of Au NP-based sensors are highly dependent on the size and concentration of the Au NPs, the nature and density of the recognition elements, the pH, ionic strength, and composition of the buffer solution, and/or temperature. We provide several interesting Au NP-based colorimetric sensing systems for the detection of Pb^{2+}, Hg^{2+}, and Cu^{2+} to show these important factors and to highlight the features of Au NPs in the advanced sensing technologies.

Small Ligands Stabilized Au NPs

Detection of the PTM ions using small ligands stabilized Au NPs can be through their specific interactions with Au atoms/ions and/or with the small ligand on their surfaces (*37*, *38*, *101*, *102*). The signal change (SPR absorption wavelength and absorbance) can be either due to the changes in the surface properties (zeta potential, refractive index, and dielectric constant) or the particle size (aggregation).

The change in the SPR absorption due to the formation of Au−Hg alloys through strong interaction between Au and Hg (d^{10}−d^{10} interaction), Au NPs have been used to detect Hg^{2+}, with an LOD of 3.3 fM (*90*). Although 11-MUA−Au NPs are sensitive for the detection of Pb^{2+}/Hg^{2+} ions through the aggregation of Au NPs due to the coordination between carboxylic acids of 11-MUA with Pb^{2+}/Hg^{2+} ions, the selectivity is not great (*38*). In order to improve the selectivity for the detection of Hg^{2+}, 3-mercaptopropionic acid (3-MPA) functionalized Au NPs (MPA−Au NPs) in the presence of 2,6-pyridinedicarboxylic acid (PDCA) was demonstrated, with minimum interference from Mn^{2+}, Pb^{2+}, and Cd^{2+} ions (Figure 1) (*37*). The study also revealed that the composition of the buffer solution plays an important role in determining the selectivity and sensitivity of the MPA−Au NP sensor because it controls the surface density of the Au NPs and their complexation with metal ions. Under optimal conditions of buffer composition (Tris−borate) and the MPA ligand density on the Au NPs, as well as excellent cooperation of PDCA, the MPA−Au NPs provide an LOD of 100 nM for Hg^{2+} ions.

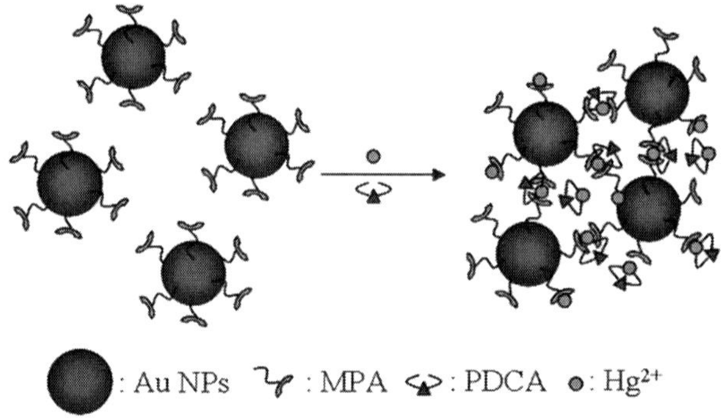

Figure 1. Schematic representation of a homogenous assay using MPA−Au NPs in the presence of PDCA for the highly selective and sensitive detection of Hg^{2+}. Reprinted with permission from reference (37). Copyright 2007 Royal Society of Chemistry.

By taking advantage of stable complexes of $Hg(SCN)_n$ relative to that of $Au(SCN)_n$, hydrophilic (11-mercapto-undecyl)-trimethyl-ammonium (MTA) capped Au NPs (QA−Au NPs) provide selectivity toward the detection of Hg^{2+} ions in acidic solution at ambient temperature without the addition of any other masking agents (*103*). Destabilizing a well-dispersed aqueous solution of Au NPs through removal of thiolates on their surfaces by Hg^{2+} results in a naked eye-based assay for Hg^{2+}, with an LOD 1.0 μM. With the assistance of solar light irradiation for 30 s, the displacement reaction of QA−Au NPs and Hg^{2+} is accelerated, decreasing the LOD down to 30 nM (6 ppb). Similarly, *N*-alkylaminopyrazole ligands functionalized Au NPs are selective and sensitive for the detection of Hg^{2+} ions, in which the ligand plays a triple functionality—inducing Au NPs formation, stabilizing the formed Au NPs, and binding with Hg^{2+} (*104*).

By using 2-mercaptoethanol (2-ME) to stabilize Au NPs, a sensing system for the detection of Pb^{2+} ions has been proposed (*105*). Owing to the formation of stable $Au(2\text{-}ME)^{2-}$ complexes (log β $[Au(2\text{-}ME)^{2-}]$ = ~30) on the surfaces, the Au NPs are suspended well in aqueous solution. Through accelerated leaching of Au^+ ions on the surfaces of Au NPs in the presence of Pb^{2+} by thiosulfate ($S_2O_3^{2-}$) ions, 2-ME stabilized Au NPs are sensitive for the detection of Pb^{2+} ions as a result of decreases in the SPR absorption at 520 nm. Relative to Au NPs, Pb–Au alloys formed on the surfaces of Au NPs are more easily etched by $S_2O_3^{2-}$ and 2-ME. This probe provides a linear range over 2.5 nM to 10 μM and an LOD of 0.5 nM for the detection of Pb^{2+} ions.

Citrate-capped Au NPs are selective towards Pb^{2+} ions at pH 11.2 (*106*). At such a high pH, most interference metal ions precipitates, while $Pb(OH)^{3-}$ are still capable of coordination with carboxylic groups of citrates, leading to aggregation of Au NPs (*106*). Gallic acid stabilized Au NPs are selective for Pb^{2+} ions through the multivalent coordination behavior between phenolic hydroxyl groups of gallic acid and Pb^{2+} ions (*65*, *107*). In addition, GSH and crown ethers are also found to stabilize Au NPs and provide chelating capability for Pb^{2+} ions (*108*, *109*). Crown ether functionalized Au NPs relative to the GSH functionalized ones provide better selectivity for Pb^{2+} ions through its coordination with Pb^{2+} ions, but poor sensitivity. In order to further improve the sensitivity, improvement of the stability of crown ether functionalized Au NPs is an important issue.

By taking advantage of Cu^{2+} accelerating the leaching rate of silver coated Au NPs by thiosulfate, a sensitive and selective sensing system has been developed for the detection of Cu^{2+} ions (*110*). By measuring the decrease in the SPR absorption at 400 nm as a result of decrease in the size of silver coated Au NPs, the probe allows detection of Cu^{2+} ions down to 1.0 nM, with a wide linear detection range (5–800 nM). Dopamine dithiocarbamate (DDTC) functionalized Au NPs allow detection of Cu^{2+} ions at pH 9.0, based on the fact that Cu^{2+} ions have multivalent coordination with the phenolic hydroxyl groups of DDTC. Through the specific interaction, the solution color changes from purple to blue, which can be detected by the naked eye. By measuring the increase in the SPR absorption at 665 nm, this probe allows the detection of Cu^{2+} ions at 14.9 μM (*111*).

DNA-Stabilized Au NPs

DNA (Oligonucleotide) functionalized Au NPs (DNA–Au NPs) have been prepared for the detection of various biopolymers and small analytes, mainly through the analytes induced changes in the conformation of the DNA and zeta potential of the DNA–Au NPs, leading to change in their stability (*112–116*). The interaction between DNA and metal ions is highly dependent on the sequence and size of DNA used to prepared DNA–Au NPs. DNA molecules can be bound to the surfaces of Au NPs through electrostatic attraction or Au–S bonding when DNA and thiol derivative DNA are used, respectively. Relative to bare Au NPs, DNA–Au NPs are much more stable in the high-ionic-strength media, mainly because of greater negative charges provided by DNA on the surfaces of Au NPs. For example, DNA–Au NPs are stable in solution containing NaCl at the concentration up to 500 mM. Such high stability makes DNA–Au NPs useful for the detection of analytes in complicated systems.

Through thymine–mercury–thymine (T–Hg^{2+}–T) coordination, T_{33} conjugated Au NPs are sensitive and selective for the detection of Hg^{2+} ions, with an LOD of 250 nM (*117*). Due to the formation of T–Hg–T coordinates, the DNA changes its random-coil structure to hairpin-like structure, leading to loss of some DNA molecules and decreases in the zeta potential. As a result, they become unstable (aggregates) in the solution containing 50 mM NaCl (Figure 2). By controlling the negative charge density of the DNA strands and thereby varying their structures-absorbed onto Au NP surfaces, the stability of the T_n/Au NPs (non-covalent bonding) increases in the order T_{80}/Au NPs > T_{33}/Au NPs > T_7/Au NPs (*115*). The solution of Au NPs is more sensitive toward Hg^{2+} in the presence of T_{33} than in the presence of T_{80} or T_7, mainly because of the greater capability of T_{80} to stabilize the Au NPs and because of the difficulty encountered by T_7 in forming folded structures, respectively. This approach provides an LOD of 250 nM for Hg^{2+} ions as obseved by the naked eye.

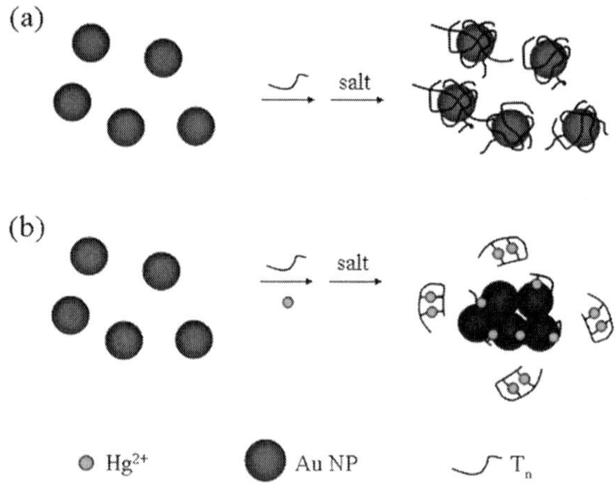

Figure 2. Schematic representation of an Hg^{2+} nanosensor based on high affinity between polythymine (T_n) and 13-nm-diameter Au NPs. Reprinted with permission from reference (117). Copyright 2008 Royal Society of Chemistry.

Interestingly, the melting point (T_m) of the DNA−Au NP aggregates is extraordinarily sharp, occurring over a temperature range much narrower than that of the original DNA. By taking advantage of the unique characteristic of DNA−Au NPs, a selective and sensitive (100 nM) colorimetric detection method has been developed for Hg^{2+} (*118*). The Au NPs maintain aggregation at higher temperature ($T < T_m$) in the presence of Hg^{2+} ions. However, Au NP aggregates are re-dispersed in the presence of other metal ions. Alternatively, detection of Hg^{2+} ions through destabilizing DNA duplex (Au NP aggregated networks) at ambient temperature has been demonstrated, with an LOD of 1 μM by the naked eye (*119*).

Thrombin-binding aptamer (TBA) containing six T units and nine guanosine (G) units with a random coil structure is changed into a hairpin-like structure and a G−quadruplex structure upon the addition of Hg^{2+} and Pb^{2+} ions (*120, 121*). By taking advantage of T−Hg−T coordination and Pb^{2+} induced formation of stable G−quadruplex, unmodified Au NPs as a probe and 15-mer TBA aptamer (5′ -GGTTGGTGTGGTTGG-3′) as a sensing element were employed for developing a colorimetric assay for the detection of Hg^{2+} and Pb^{2+} ions (*122*). Upon the addition of Hg^{2+} or Pb^{2+}, the conformation changes of TBA aptamer facilitate the salt-induced Au NP aggregation. By using the characteristic wavelength (800 nm) of Hg^{2+} induced aggregated Au NPs, Pb^{2+} interference in the determination of Hg^{2+} is minimized. This approach provides an LOD of 200 nM for Hg^{2+} ions (*122*).

DNAzymes and Au NPs have been used to develope several sensitive sensing systems for the detection of Pb^{2+} ions (*39*, *123*, *124*). DNAzymes (also called deoxyribozymes, catalytic DNA, or DNA enzymes) require specific cofactors (metal ions such as Pb^{2+} ions) to catalyze many important biological reactions (*125*, *126*). These sensing systems take advantage of specific interaction of DNAzymes with Pb^{2+} and its capability for cleavage of DNA (substrate), leading to changes in the color of the Au NP solution. For example, DNAzyme catalyzes the hydrolysis of substrate DNA in the presence of Pb^{2+} ions (Figure 3) (*127*, *128*). After the release of ssDNA products from substrate DNA, these ssDNA products have access to the surfaces of Au NPs to protect Au NPs from aggregation under high ionic strength environment. This approach provides an LOD of 3 nM for Pb^{2+} ions in the presence of 100 mM NaCl. In order to further accelerate the reaction rate (estimated k_{obs} ~50 min^{-1} at pH 7.0) and improve the sensitivity, DNAzymes and Au NPs for the detection of Pb^{2+} ions have been employed on lateral flow devices containing conjugation pad, control zone, and test zone (*123*). This easy-to-use dipstick test allows visualization of Pb^{2+} ions at a concentration down to *ca.* 0.5 μM and by the naked eye.

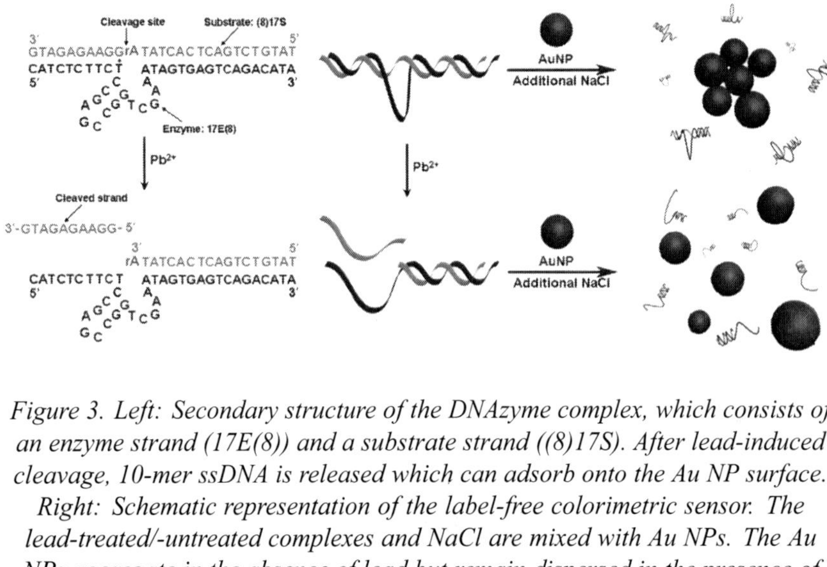

Figure 3. Left: Secondary structure of the DNAzyme complex, which consists of an enzyme strand (17E(8)) and a substrate strand ((8)17S). After lead-induced cleavage, 10-mer ssDNA is released which can adsorb onto the Au NP surface. Right: Schematic representation of the label-free colorimetric sensor. The lead-treated/-untreated complexes and NaCl are mixed with Au NPs. The Au NPs aggregate in the absence of lead but remain dispersed in the presence of lead. Reprinted with permission from reference (127). Copyright 2008 John Wiley & Sons.

By taking advantage of click chemistry (Cu^{2+} ions as catalyst), DNA−Au NPs are selective for the detection of Cu^{2+} ions (*129*). In the presence of Cu^{2+} ions, a click cyclization reaction between the alkyne moiety and the azide group occurs, forming three-stranded DNA−Au NP aggregates and a color change from red to purple. This approach provides an LOD of 20 µM for Cu^{2+} ions and exceptional selectivity over other metal ions.

Catalytic Au NPs-Based Systems

Through Pb^{2+} ions and Hg^{2+} ions induced the formation of Pb−Au /Hg−Au alloys on the surfaces of Au NPs that possess catalytic activity for hydrogen peroxide (peroxidase mimicking), sensitive and selective assays using bare Au NPs and nonfluorescent Amplex UltraRed (AUR) for the two metal ions have been developed (*130–132*). After the reaction with hydroxyl radicals generated from hydrogen oxide, AUR forms highly fluorescent product at 584 nm when excited at 540 nm. By controlling pH values, the formation of Hg−Au and Pb−Au alloys on their surfaces can be controlled (Figure 4), leading to pH-dependent catalytic activity of the Au NPs in the H_2O_2-mediated oxidation of AUR (*131*). It is worth pointing out that the selectivity towards Hg^{2+} ions and Pb^{2+} ions can be easily achieved in pH 7.0 and 9.0, respectively. Linear responses to the concentrations of Hg^{2+} and Pb^{2+} ions are over the ranges 5.0 nM–1.0 μM and 5.0 nM–5.0 μM, respectively. The LOD of Hg^{2+} and Pb^{2+} are both 4 nM. For the improvement of sensitivity, oligonucleotide (T30695) which is easily foldable into stable intramolecular G−quadruplex structures in the presence of Pb^{2+} ions are anchored on the surface of Au NPs (*133*). Owing to this specific interaction, more Pb^{2+} ions are found on the surfaces of Au NPs, leading to improved catalytic activity towards H_2O_2. This system provides better selectivity and sensitivity (LOD: 50 pM) towards Pb^{2+} ions.

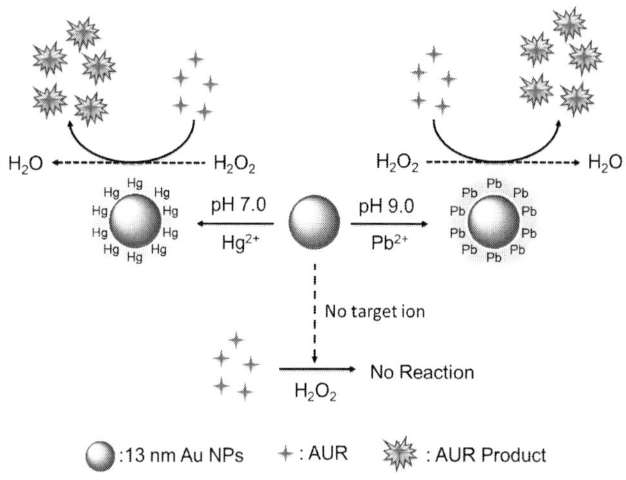

Figure 4. Schematic representation of the sensing mechanism of the H_2O_2−AUR−Au NP probe for the detection of Hg^{2+} and Pb^{2+} ions. Reprinted with permission from reference (131). Copyright 2012 Elsevier

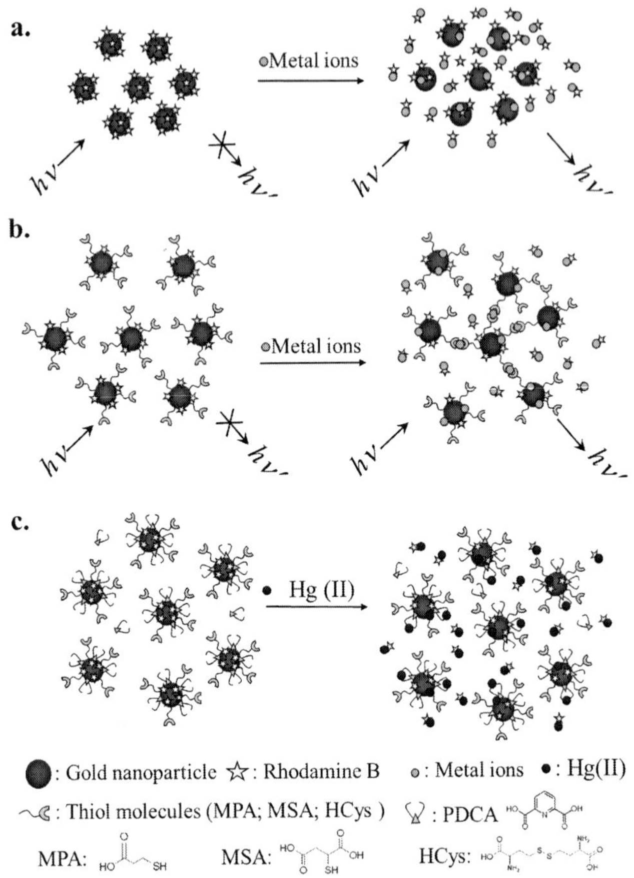

Figure 5. Schematic representation of a homogeneous assay employing MPA−Au NPs and PDCA for the highly selective and sensitive detection of Hg^{2+} ions. Reprinted with permission from reference (137). Copyright 2006 American Chemical Society.

Au NPs-Based Quenching Systems

Having a very high molar absorptivity, Au NPs are efficient quenchers. Au NPs has a Stern−Volmer quenching (K_{SV}) constant that is greater by several orders of magnitude than that of typical small molecule dye-quencher pairs (*24*, *134*). Based on their high quenching efficiency, various sensitive fluorescence sensing systems using Au NPs have been developed for the detection of metal ions (*135*, *136*). Both energy-transfer and electron-transfer processes are considered to be major deactivation pathways for the excited fluorophores on Au NPs (*91*). Fluorescence "turn on" sensor relative to "turn-off" one has been considered as more sensitive, mainly because of ease in measuring low-concentration contrast relative to a "dark" background. A sensitive "turn on" fluorescent sensor for

Hg^{2+} ions in aqueous solution has been demonstrated (Figure 5), based on Hg^{2+} ions displacing rhodamine B (RB) units on the surfaces of Au NPs (*137*). The fluorescence of RB molecules in bulk solution is high, which significantly decreases when they are adsorbed on the surface of Au NPs. RB molecules are released from the Au NP surface in the presence of Hg^{2+} ions, thereby restoring the fluorescence of RB with 400-fold enhancement. In order to improve the selectivity toward Hg^{2+} against other metal ions, PDCA is essential when using MPA–Au NPs. The selectivity of the PDCA/RB–Au NP–MPA system for Hg^{2+} ions is remarkably high (at least 50-fold); the LOD for Hg^{2+} ions in pond water sample is 10 nM.

By taking advantage of a specific interaction between T-rich ssDNA and Hg^{2+} ions, a highly selective and sensitive approach by using DNA-functionalized Au NPs and OliGreen (a dye that binds specifically to DNA) has been developed for the detection of Hg^{2+} ions (*115*). The conformations of these DNA derivatives change from linear to hairpin structures when Hg^{2+} ions interact with thymidine units of the DNA molecules, causing the release of some of the DNA molecules from the surface of the Au NPs, primarily as a result of steric effect. Once DNA molecules are released from the Au NP surfaces to the bulk solution, OliGreen subsequently interacts specifically with these free DNA molecules, resulting in an increase in the fluorescence at 525 nm upon excitation at 480 nm. The fluorescence of DNA–Oligreen complexes are more intense (~1000-fold) than that of free Oligreen, providing an LOD of 25 nM for Hg^{2+} ions. Although the DNA molecules on the surfaces of the Au NPs also interact with OliGreen, a negligible degree of fluorescence occurs because of quenching caused by the Au NPs. Alternatively, a sensing system using dye labeled DNA and bare Au NPs has been developed for the detection of Hg^{2+} ions (*138*). In the absence of Hg^{2+} ions, dye labeled DNA tends to adsorb on Au NPs surface, which leads to fluorescence quenching. In the presence of Hg^{2+} ions, through strong T–Hg^{2+}–T coordination, dsDNA formed has lower affinity towards Au NPs and is released to the bulk solution, leading to lower degree of fluorescence quenching (stronger fluorescence). We point out that ssDNA relative to dsDNA has greater interaction with Au NPs.

Au NDs-Based Systems

Based on the aggregation-induced luminescence quenching of 11-MUA–Au NDs by Hg^{2+} ions in aqueous solution, a sensitive and selective approach for the detection of Hg^{2+} ions has been developed (*46*). Through strong and specific Au–Hg aurophilic interactions, a significant decrease in fluorescence occurs as shown in Figrue 6 (*139*). This sensing system provides an LOD of 5.0 nM for Hg^{2+} ions, with selectivity at least 10-fold against Pb^{2+} and Cd^{2+} in the presence of PDCA. The LOD is lower than the maximum level (10 nM) of mercury permitted in drinking water by EPA. Aggregation-induced luminescence quenching of GSH–Au NDs was applied to the detection of Cu^{2+} ions with an LOD of 3.6 nM (*140*). By simply etching nonfluorescent GSH–Au NPs with 11-MUA, functionalized 11-MUA/GSH–Au NDs have been prepared and used for the detection of Pb^{2+} ions, with high selectivity and sensitivity (LOD 2 nM) (*141*). BSA−Au NDs prepared via core etching of mercaptosuccinic acid protected Au NPs by BSA are applied to sensing of Cu^{2+} ions (LOD 15 nM), based on the aggregation of BSA−Au NDs induced by the complexation of BSA with the Cu^{2+} ions (*142*).

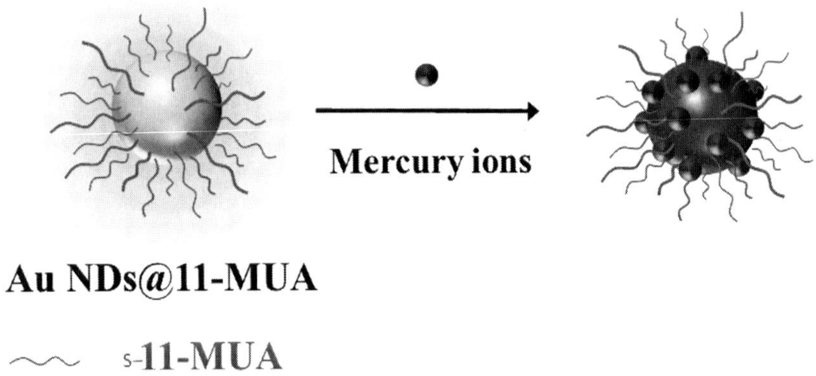

Figure 6. Schematic representation of the sensing mechanism of Au NDs@11-MUA for the detection of Hg^{2+} ions. Adapted with permission from reference (139). Copyright 2013 Royal Society of Chemistry.

BSA−Au NDs that exhibit intense red emission ($\lambda_{em\ max}$ = 640 nm) with Φ ~ 6% are useful for the detection of Hg^{2+} ions, based on fluorescence quenching as a result of the formation of Au−Hg amalgam through the high-affinity metallophilic Hg^{2+}−Au^{+} interaction (d^{10}−d^{10} interaction) (*143*). This approach provides extremely high sensitivity (LOD 0.5 nM) and high selectivity toward Hg^{2+} ions over other environmentally relevant metal ions in aqueous media. Other than the surface state of Au^{+}, the conformation of BSA plays an important role in determining the selectivity and sensitivity of Hg^{2+} ions (*83*). Lys−Au NDs have been found sensitive for detecting Hg^{2+} ions and CH_3Hg^{+} based on fluorescence quenching, with their LODs of 3 pM and 4 nM, respectively (*144*). When using BSA and Lys as templates and reducing agents for the preparation of protein−Au NDs, the nature of proteins affect the optical properties and the sensitivity for the detection of Hg^{2+} ions (*76*). The BSA−Au NDs (size 0.8 nm; λ_{em} 640 nm) relative to the Lys−Au NCs (size 1 nm; λ_{em} 657 nm) provide greater sensitivity for Hg^{2+} ions (0.5 vs. 10 nM). By taking advantages that the fluorescent Au NDs are easily internalized by cells and their near IR emission is easily distinguished from the autofluorescence of cells, BSA−Au NDs and DHLA−Au NDs have been used separately to monitor the concentrations of Hg^{2+} ions inside cells (*145*, *146*).

Table I. Comparison of Representative Au NPs and Au NDs Assays for the Detection of PTM Ions

Target	*Method*	*Probe*	*LOD (nM)*	*Samples*	*Reference*
Hg^{2+}	Colorimetric	MPA–Au NPs	100	N/A	(*37*)
		MPA–homocysteine–Au NPs	10	N/A	(*102*)
		Apt/Au NPs	0.6	Tap and lake water	(*147*)
		DNA–Au NPs	0.1	Tap and river water	(*113*)
	Fluorometric	PDCA/RB–Au NP–MPA	10	River water and battery	(*137*)
		Oligreen–DNA–Au NPs	25	Pond water	(*115*)
		DNA–Au NPs	40	Tap and river water	(*138*)
		11-MUA–Au NDs	5.0	Pond water	(*57*)
		BSA–Au NCs	0.5	N/A	(*143*)
		Lys–Au NCs	10	N/A	(*76*)
		Lys VI–Au NCs	0.03	Sea water	(*144*)
		Quantum Dot/DNA/Au NPs	2	River water	(*136*)

Continued on next page.

Table I. (Continued). Comparison of Representative Au NPs and Au NDs Assays for the Detection of PTM Ions

Target	*Method*	*Probe*	*LOD (nM)*	*Samples*	*Reference*
Pb^{2+}	Colorimetric	Gallic acid–Au NPs	10	N/A	(*65*)
		2-ME/$S_2O_3^{2-}$–Au NPs	0.5	Soil and pond water	(*105*)
		Peptide–Au NPs	242	N/A	(*32*)
		Au NPs, DNAzyme, and substrate	3.0	Pond water	(*127*)
	Fluorometric	GSH–Au NDs	2.0	Lake water	(*141*)
		DNA–Au NPs	0.05	blood samples	(*130*)
		Au NPs	4.0	Pond water and blood samples	(*131*)
Cu^{2+}	Colorimetric	Azide/alkyne functionalized –Au NPs	50000	N/A	(*148*)
		DNA–Au NPs	20000	N/A	(*129*)
	Fluorometric	GSH–Au NDs	3.6	N/A	(*140*)
		BSA–Au NCs	15	N/A	(*142*)

Conclusion and Perspective

This chapter describes several functional Au NPs and Au NDs based sensors that allow the sensitive and selective detection of Hg^{2+}, Pb^{2+}, and Cu^{2+} ions through analytes induced changes in colors, absorption and fluorescence. Table I summarizes several Au NPs and NDs based optical approaches for the sensitive and selective detections of Hg^{2+}, Pb^{2+}, and Cu^{2+} ions, emphasizing their LODs and practicality. These methods require use of minimum amounts of samples and reagents. The main strategy of developing Au NM based assays is through governing the surface properties of Au NMs and their interparticle interactions. The surface ligands play vital roles in determining the selectivity of these assays. Advantages of Au NPs and Au NDs are their biocompatibility, ease in preparation and conjugation with recognition elements, and stability against oxidation. Although Au NPs are usually less sensitive, they can be used easily to develop sensing system based on color changes observed by the naked eye. For colored samples, Au NDs are more suitable, mainly because of strong absorption background which leads to serious interference and consequently reduces the sensitivity. Relative to Au NPs, Au NDs are suitable for the detection of the metal ions at low concentrtations but they are also quite sensitive to the sample matrix. For cell images, Au NDs over Au NPs are advantageous, but their photostability must be further enhanced.

Although these sensing systems have shown successful from certain viewpoints, each of them does has some limitation (disadvantages). Some sensors are not quite stable in critical conditions (extremely high and low pH values, and high-salt media). Thus, developing stable Au NPs and Au NDs sensors are still needed for the analysis of complicated environmental samples. Great amounts of proteins (eg. Human serum albumin) and small thiol molecules (e.g. GSH) in blood samples result in the aggregation of Au NPs or fluorescence quenching of Au NDs, leading to loss in sensitivity, selectivity, and reproducibility. Large amounts of proteins and expensive DNA are required for the preparation of protein–Au NDs and DNA–Au NPs, respectively, increasing the cost of the analysis. It is extremely difficult to purify the small sizes of Au NDs. In addition, poor photostability and weak fluorescence of Au NDs are problematic for developing sensitive cell imaging systems.

Relative to ICP–MS, these optical sensing systems for the detection of metal ions provide disadvantages of narrower linearity and lower throughputs (multiple element analysis). In order to enhance the sensitivity and throughput, strategies for the preparation of Au NPs and Au NDs based bi- and tri-metallic nanocomposites are needed. In addition, fluorescence in conjunction with other sensitive techniques such as fluorescence anisotropy (*149*), surface enhancement Raman scattering (SERS) (*150*, *151*) and MALDI–MS (*152*) are worthy to be tested. Nonspecific adsorption of cations, anions, and proteins on the surfaces of Au NPs and Au NPs can be minimized by introducing hydrophilic polymers such as poly(ethylene glycol). We have recently shown that Au NDs prepared in hydrogels are stable in high-salt media and useful for the detection of Hg^{2+} ions in complicated environmental systems (*153*). Thus, we can foresee that functional Au NDs– and Au NPs–hydrogel sensing systems will become popular for the analysis of complicated samples. In order to further improve the sensitivity, we have recently purified Au NDs through a simple centrifugation/wash process when using Al_2O_3 NPs as a support for the Au NDs (*154*). The purified AuNDs−Al_2O_3 NPs relative to 11-MUA−Au NDs provides greater sensitivity for Ag^+ ions. With the advancement in nanotechnology, versatile NMs will be prepared for the detection of these PTM ions with gret promising. For instance, the development of Ag NCs with higher QY than Au NDs reveals high selectivity and sensitivity towards Cu^{2+} ions (*155, 156*). The specificity of DNA−Au NPs for the metal ions can be improved if more suitable aptamers are screened and synthesized. In order to prevent use of great amounts of proteins for the preparation of Au NDs, small organic molecules such as GSH that have been used to prepared fluorescent Au NCs can be tested. However, problems of poor solubility and weak fluorescence have to be overcome. Preparation of Au NDs in silica and polymers may be worthy to be tried. With great advances in the nanotechnologies, it is our strong belief that low-cost, high-sensitive, and specific NPs based sensors will be developed for the detection of metal ions.

Acknowledgments

This study was supported by the National Science Council of Taiwan under Contracts NSC 101-2113-M-002-002-MY3.

References

1. Bryan, G. W.; Langston, W. J. *Environ. Pollut.* **1992**, *76*, 89–131.
2. Echeverria, D.; Aposhian, H. V.; Woods, J. S.; Heyer, N. J.; Aposhian, M. M.; Bittner, A. C.; Mahurin, R. K.; Cianciola, M. *FASEB J.* **1998**, *12*, 971–980.
3. Jarup, L. *Br. Med. Bull.* **2003**, *68*, 167–182.
4. Güell, R.; Aragay, G.; Fontàs, C.; Anticó, E.; Merkoçi, A. *Anal. Chim. Acta* **2008**, *627*, 219–224.
5. Palchetti, I.; Laschi, S.; Mascini, M. *Anal. Chim. Acta* **2005**, *530*, 61–67.
6. Caroli, S.; Forte, G.; Iamiceli, A. L.; Galoppi, B. *Talanta* **1999**, *50*, 327–336.
7. Goullé, J.-P.; Mahieu, L.; Castermant, J.; Neveu, N.; Bonneau, L.; Lainé, G.; Bouige, D.; Lacroix, C. *Forensic Sci. Int.* **2005**, *153*, 39–44.
8. Gasparik, J.; Vladarova, D.; Capcarova, M.; Smehyl, P.; Slamecka, J.; Garaj, P.; Stawarz, R.; Massanyi, P. *J. Environ. Sci. Health., Part A* **2010**, *45*, 818–823.
9. Pohl, P. *TrAC, Trends Anal. Chem.* **2009**, *28*, 117–128.
10. Cho, E. S.; Kim, J.; Tejerina, B.; Hermans, T. M.; Jiang, H.; Nakanishi, H.; Yu, M.; Patashinski, A. Z.; Glotzer, S. C.; Stellacci, F. *Nat. Mater.* **2012**, *11*, 978–985.
11. Dong, Y.; Wang, R.; Li, G.; Chen, C.; Chi, Y.; Chen, G. *Anal. Chem.* **2012**, *84*, 6220–6224.
12. Roy, P.; Lin, Z.-H.; Liang, C.-T.; Chang, H.-T. *J. Hazard. Mater.* **2012**, *243*, 286–291.

13. Zhang, T.; Cheng, Z.; Wang, Y.; Li, Z.; Wang, C.; Li, Y.; Fang, Y. *Nano Lett.* **2010**, *10*, 4738–4741.
14. Wu, C.-S.; Khaing Oo, M. K.; Fan, X. *ACS Nano* **2010**, *4*, 5897–5904.
15. Xu, H.; Zeng, L.; Xing, S.; Shi, G.; Chen, J.; Xian, Y.; Jin, L. *Electrochem. Commun.* **2008**, *10*, 1893–1896.
16. Ray, P. C. *Chem. Rev.* **2010**, *110*, 5332–5365.
17. Roduner, E. *Chem. Soc. Rev.* **2006**, *35*, 583–592.
18. Talapin, D. V.; Lee, J.-S.; Kovalenko, M. V.; Shevchenko, E. V. *Chem. Rev.* **2009**, *110*, 389–458.
19. Alivisatos, A. P. *Science* **1996**, *271*, 933–937.
20. Baker, S. N.; Baker, G. A. *Angew. Chem., Int. Ed.* **2010**, *49*, 6726–6744.
21. Djurišić, A. B.; Leung, Y. H. *Small* **2006**, *2*, 944–961.
22. Loh, K. P.; Bao, Q.; Eda, G.; Chhowalla, M. *Nat. Chem.* **2010**, *2*, 1015–1024.
23. De, M.; Ghosh, P. S.; Rotello, V. M. *Adv. Mater.* **2008**, *20*, 4225–4241.
24. Fan, C.; Wang, S.; Hong, J. W.; Bazan, G. C.; Plaxco, K. W.; Heeger, A. J. *Proc. Natl. Acad. Sci. U.S.A.* **2003**, *100*, 6297–6301.
25. Wilson, R. *Chem. Soc. Rev.* **2008**, *37*, 2028–2045.
26. Sapsford, K. E.; Algar, W. R.; Berti, L.; Gemmill, K. B.; Casey, B. J.; Oh, E.; Stewart, M. H.; Medintz, I. L. *Chem. Rev.* **2013**, *113*, 1904–2074.
27. Rosi, N. L.; Mirkin, C. A. *Chem. Rev.* **2005**, *105*, 1547–1562.
28. Agasti, S. S.; Rana, S.; Park, M.-H.; Kim, C. K.; You, C.-C.; Rotello, V. M. *Adv. Drug Delivery Rev.* **2010**, *62*, 316–328.
29. Liu, D.; Wang, Z.; Jiang, X. *Nanoscale* **2011**, *3*, 1421–1433.
30. Saha, K.; Agasti, S. S.; Kim, C.; Li, X.; Rotello, V. M. *Chem. Rev.* **2012**, *112*, 2739–2779.
31. Si, S.; Kotal, A.; Mandal, T. K. *J. Phys. Chem. C* **2006**, *111*, 1248–1255.
32. Slocik, J. M.; Zabinski, J. S.; Phillips, D. M.; Naik, R. R. *Small* **2008**, *4*, 548–551.
33. Wang, Z.; Ma, L. *Coord. Chem. Rev.* **2009**, *253*, 1607–1618.
34. Daniel, M.-C.; Astruc, D. *Chem. Rev.* **2003**, *104*, 293–346.
35. Ghosh, S. K.; Pal, T. *Chem. Rev.* **2007**, *107*, 4797–4862.
36. Zhao, W.; Brook, M. A.; Li, Y. *ChemBioChem* **2008**, *9*, 2363–2371.
37. Huang, C.-C.; Chang, H.-T. *Chem. Commun.* **2007**, 1215–1217.
38. Kim, Y.; Johnson, R. C.; Hupp, J. T. *Nano Lett.* **2001**, *1*, 165–167.
39. Liu, J.; Lu, Y. *J. Am. Chem. Soc.* **2003**, *125*, 6642–6643.
40. Liu, J.; Lu, Y. *J. Am. Chem. Soc.* **2004**, *126*, 12298–12305.
41. Wang, Y.; Li, Y. F.; Wang, J.; Sang, Y.; Huang, C. Z. *Chem. Commun.* **2010**, *46*, 1332–1334.
42. Wang, S.; Chen, Z.; Chen, L.; Liu, R.; Chen, L. *Analyst* **2013**, *138*, 2080–2084.
43. Placido, T.; Aragay, G.; Pons, J.; Comparelli, R.; Curri, M. L.; Merkoçi, A. *ACS Appl. Mater. Interfaces* **2013**, *5*, 1084–1092.
44. Chen, G.; Jin, Y.; Wang, W.; Zhao, Y. *Gold Bull.* **2012**, *45*, 137–143.
45. Durgadas, C. V.; Lakshmi, V. N.; Sharma, C. P.; Sreenivasan, K. *Sens. Actuators, B* **2011**, *156*, 791–797.
46. Wilcoxon, J.; Martin, J.; Parsapour, F.; Wiedenman, B.; Kelley, D. *J. Chem. Phys.* **1998**, *108*, 9137.

47. Chen, S. W.; Ingram, R. S.; Hostetler, M. J.; Pietron, J. J.; Murray, R. W.; Schaaff, T. G.; Khoury, J. T.; Alvarez, M. M.; Whetten, R. L. *Science* **1998**, *280*, 2098–2101.
48. Jin, R. *Nanoscale* **2010**, *2*, 343–362.
49. Zheng, J.; Nicovich, P. R.; Dickson, R. M. *Annu. Rev. Phys. Chem.* **2007**, *58*, 409–431.
50. Zheng, J.; Zhou, C.; Yu, M.; Liu, J. *Nanoscale* **2012**, *4*, 4073–4083.
51. Zheng, J.; Zhang, C. W.; Dickson, R. M. *Phys. Rev. Lett.* **2004**, *93*, 077402-1–077402-4.
52. Negishi, Y.; Nobusada, K.; Tsukuda, T. *J. Am. Chem. Soc.* **2005**, *127*, 5261–5270.
53. Xie, J. P.; Zheng, Y. G.; Ying, J. Y. *J. Am. Chem. Soc.* **2009**, *131*, 888–889.
54. Zheng, J.; Petty, J. T.; Dickson, R. M. *J. Am. Chem. Soc.* **2003**, *125*, 7780–7781.
55. Duan, H.; Nie, S. *J. Am. Chem. Soc.* **2007**, *129*, 2412–2413.
56. Lin, C.-A. J.; Lee, C.-H.; Hsieh, J.-T.; Wang, H.-H.; Li, J. K.; Shen, J.-L.; Chan, W.-H.; Yeh, H.-I.; Chang, W. H. *J. Med. Biol. Eng.* **2009**, *29*, 276–283.
57. Huang, C. C.; Yang, Z.; Lee, K. H.; Chang, H. T. *Angew. Chem., Int. Ed.* **2007**, *46*, 6824–6828.
58. Huang, C.-C.; Chen, C.-T.; Shiang, Y.-C.; Lin, Z.-H.; Chang, H.-T. *Anal. Chem.* **2009**, *81*, 875–882.
59. Shang, L.; Dong, S. J.; Nienhaus, G. U. *Nano Today* **2011**, *6*, 401–418.
60. Shiang, Y.-C.; Huang, C.-C.; Chen, W.-Y.; Chen, P.-C.; Chang, H.-T. *J. Mater. Chem.* **2012**, *22*, 12972–12982.
61. Shang, L.; Azadfar, N.; Stockmar, F.; Send, W.; Trouillet, V.; Bruns, M.; Gerthsen, D.; Nienhaus, G. U. *Small* **2011**, *7*, 2614–2620.
62. Turkevich, J.; Stevenson, P. C.; Hillier, J. *Discuss. Faraday Soc.* **1951**, *11*, 55–75.
63. Kimling, J.; Maier, M.; Okenve, B.; Kotaidis, V.; Ballot, H.; Plech, A. *J. Phys. Chem. B* **2006**, *110*, 15700–15707.
64. Huang, H.; Yang, X. *Carbohydr. Res.* **2004**, *339*, 2627–2631.
65. Huang, K.-W.; Yu, C.-J.; Tseng, W.-L. *Biosens. Bioelectron.* **2010**, *25*, 984–989.
66. Nune, S. K.; Chanda, N.; Shukla, R.; Katti, K.; Kulkarni, R. R.; Thilakavathy, S.; Mekapothula, S.; Kannan, R.; Katti, K. V. *J. Mater. Chem.* **2009**, *19*, 2912–2920.
67. Selvakannan, P.; Mandal, S.; Phadtare, S.; Pasricha, R.; Sastry, M. *Langmuir* **2003**, *19*, 3545–3549.
68. Sudeep, P. K.; Joseph, S. T. S.; Thomas, K. G. *J. Am. Chem. Soc.* **2005**, *127*, 6516–6517.
69. Wangoo, N.; Bhasin, K. K.; Mehta, S. K.; Suri, C. R. *J. Colloid Interface Sci.* **2008**, *323*, 247–254.
70. Brust, M.; Fink, J.; Bethell, D.; Schiffrin, D. J.; Kiely, C. *J. Chem. Soc., Chem. Commun.* **1995**, 1655–1656.
71. Brust, M.; Walker, M.; Bethell, D.; Schiffrin, D. J.; Whyman, R. *J. Chem. Soc., Chem. Commun.* **1994**, 801–802.

72. Wang, Z.; Tan, B.; Hussain, I.; Schaeffer, N.; Wyatt, M. F.; Brust, M.; Cooper, A. I. *Langmuir* **2007**, *23*, 885–895.
73. Hussain, I.; Graham, S.; Wang, Z.; Tan, B.; Sherrington, D. C.; Rannard, S. P.; Cooper, A. I.; Brust, M. *J. Am. Chem. Soc.* **2005**, *127*, 16398–16399.
74. Lin, C.-A. J.; Yang, T.-Y.; Lee, C.-H.; Huang, S. H.; Sperling, R. A.; Zanella, M.; Li, J. K.; Shen, J.-L.; Wang, H.-H.; Yeh, H.-I.; Parak, W. J.; Chang, W. H. *ACS Nano* **2009**, *3*, 395–401.
75. Huang, C.-C.; Hung, Y.-L.; Shiang, Y.-C.; Lin, T.-Y.; Lin, Y.-S.; Chen, C.-T.; Chang, H.-T. *Chem. Asian J.* **2010**, *5*, 334–341.
76. Wei, H.; Wang, Z. D.; Yang, L. M.; Tian, S. L.; Hou, C. J.; Lu, Y. *Analyst* **2010**, *135*, 1406–1410.
77. Xie, J.; Lee, J. Y.; Wang, D. I. C.; Ting, Y. P. *ACS Nano* **2007**, *1*, 429–439.
78. Yang, X.; Shi, M.; Zhou, R.; Chen, X.; Chen, H. *Nanoscale* **2011**, *3*, 2596–2601.
79. Liu, H.; Zhang, X.; Wu, X.; Jiang, L.; Burda, C.; Zhu, J.-J. *Chem. Commun.* **2011**, *47*, 4237–4239.
80. Yan, L.; Cai, Y.; Zheng, B.; Yuan, H.; Guo, Y.; Xiao, D.; Choi, M. M. F. *J. Mater. Chem.* **2012**, *22*, 1000–1005.
81. Kawasaki, H.; Hamaguchi, K.; Osaka, I.; Arakawa, R. *Adv. Funct. Mater.* **2011**, *21*, 3508–3515.
82. Yue, Y.; Liu, T.-Y.; Li, H.-W.; Liu, Z.; Wu, Y. *Nanoscale* **2012**, *4*, 2251–2254.
83. Chen, P.-C.; Chiang, C.-K.; Chang, H.-T. *J. Nanopart. Res.* **2013**, *15*, 1336-1–1336-10.
84. Donkers, R. L.; Lee, D.; Murray, R. W. *Langmuir* **2004**, *20*, 1945–1952.
85. Jadzinsky, P. D.; Calero, G.; Ackerson, C. J.; Bushnell, D. A.; Kornberg, R. D. *Science* **2007**, *318*, 430–433.
86. Wang, G. L.; Huang, T.; Murray, R. W.; Menard, L.; Nuzzo, R. G. *J. Am. Chem. Soc.* **2005**, *127*, 812–813.
87. Link, S.; El-Sayed, M. A. *J. Phys. Chem. B* **1999**, *103*, 4212–4217.
88. Huang, X.; El-Sayed, I. H.; Qian, W.; El-Sayed, M. A. *J. Am. Chem. Soc.* **2006**, *128*, 2115–2120.
89. Marinakos, S. M.; Chen, S.; Chilkoti, A. *Anal. Chem.* **2007**, *79*, 5278–5283.
90. Rex, M.; Hernandez, F. E.; Campiglia, A. D. *Anal. Chem.* **2006**, *78*, 445–451.
91. Kamat, P. V.; Barazzouk, S.; Hotchandani, S. *Angew. Chem., Int. Ed.* **2002**, *41*, 2764–2767.
92. Xavier, P. L.; Chaudhari, K.; Verma, P. K.; Pal, S. K.; Pradeep, T. *Nanoscale* **2010**, *2*, 2769–2776.
93. Negishi, Y.; Takasugi, Y.; Sato, S.; Yao, H.; Kimura, K.; Tsukuda, T. *J. Am. Chem. Soc.* **2004**, *126*, 6518–6519.
94. Pyykkö, P. *Angew. Chem., Int. Ed.* **2004**, *43*, 4412–4456.
95. Pyykkö, P. *Inorg. Chim. Acta* **2005**, *358*, 4113–4130.
96. Yam, V. W.-W.; Cheng, E. C.-C. *Chem. Soc. Rev.* **2008**, *37*, 1806–1813.
97. Templeton, A. C.; Chen, S.; Gross, S. M.; Murray, R. W. *Langmuir* **1998**, *15*, 66–76.
98. Negishi, Y.; Tsukuda, T. *Chem. Phys. Lett.* **2004**, *383*, 161–165.

99. Menard, L. D.; Gao, S.-P.; Xu, H.; Twesten, R. D.; Harper, A. S.; Song, Y.; Wang, G.; Douglas, A. D.; Yang, J. C.; Frenkel, A. I.; Nuzzo, R. G.; Murray, R. W. *J. Phys. Chem. B* **2006**, *110*, 12874–12883.
100. Forward, J. M.; Bohmann, D.; Fackler, J. P.; Staples, R. J. *Inorg. Chem.* **1995**, *34*, 6330–6336.
101. Lin, Y.-W.; Huang, C.-C.; Chang, H.-T. *Analyst* **2011**, *136*, 863–871.
102. Tan, Z.-q.; Liu, J.-f.; Liu, R.; Yin, Y.-g.; Jiang, G.-b. *Chem. Commun.* **2009**, 7030–7032.
103. Liu, D.; Qu, W.; Chen, W.; Zhang, W.; Wang, Z.; Jiang, X. *Anal. Chem.* **2010**, *82*, 9606–9610.
104. Aragay, G.; Pons, J.; Ros, J.; Merkoçi, A. *Langmuir* **2010**, *26*, 10165–10170.
105. Chen, Y.-Y.; Chang, H.-T.; Shiang, Y.-C.; Hung, Y.-L.; Chiang, C.-K.; Huang, C.-C. *Anal. Chem.* **2009**, *81*, 9433–9439.
106. Guan, J.; Jiang, L.; Zhao, L.; Li, J.; Yang, W. *Colloids Surf., A* **2008**, *325*, 194–197.
107. Yoosaf, K.; Ipe, B. I.; Suresh, C. H.; Thomas, K. G. *J. Phys. Chem. C* **2007**, *111*, 12839–12847.
108. Alizadeh, A.; Khodaei, M.; Karami, C.; Workentin, M.; Shamsipur, M.; Sadeghi, M. *Nanotechnology* **2010**, *21*, 315503–315511.
109. Chai, F.; Wang, C.; Wang, T.; Li, L.; Su, Z. *ACS Appl. Mater. Interfaces* **2010**, *2*, 1466–1470.
110. Lou, T.; Chen, L.; Chen, Z.; Wang, Y.; Chen, L.; Li, J. *ACS Appl. Mater. Interfaces* **2011**, *3*, 4215–4220.
111. Mehta, V. N.; Kumar, M. A.; Kailasa, S. K. *Ind. Eng. Chem. Res.* **2013**, *52*, 4414–4420.
112. Elghanian, R.; Storhoff, J. J.; Mucic, R. C.; Letsinger, R. L.; Mirkin, C. A. *Science* **1997**, *277*, 1078–1081.
113. He, Y.; Zhang, X.; Zeng, K.; Zhang, S.; Baloda, M.; Gurung, A. S.; Liu, G. *Biosens. Bioelectron.* **2011**, *26*, 4464–4470.
114. Huang, C.-C.; Huang, Y.-F.; Cao, Z.; Tan, W.; Chang, H.-T. *Anal. Chem.* **2005**, *77*, 5735–5741.
115. Liu, C.-W.; Huang, C.-C.; Chang, H.-T. *Langmuir* **2008**, *24*, 8346–8350.
116. Wang, Z.; Lu, Y. *J. Mater. Chem.* **2009**, *19*, 1788–1798.
117. Liu, C.-W.; Hsieh, Y.-T.; Huang, C.-C.; Lin, Z.-H.; Chang, H.-T. *Chem. Commun.* **2008**, 2242–2244.
118. Lee, J.-S.; Han, M. S.; Mirkin, C. A. *Angew. Chem., Int. Ed.* **2007**, *46*, 4093–4096.
119. Xue, X.; Wang, F.; Liu, X. *J. Am. Chem. Soc.* **2008**, *130*, 3244–3245.
120. Liu, C.-W.; Huang, C.-C.; Chang, H.-T. *Anal. Chem.* **2009**, *81*, 2383–2387.
121. Smirnov, I.; Shafer, R. H. *J. Mol. Biol.* **2000**, *296*, 1–5.
122. Wang, Y.; Yang, F.; Yang, X. *Biosens. Bioelectron.* **2010**, *25*, 1994–1998.
123. Mazumdar, D.; Liu, J.; Lu, G.; Zhou, J.; Lu, Y. *Chem. Commun.* **2010**, *46*, 1416–1418.
124. Pelossof, G.; Tel-Vered, R.; Willner, I. *Anal. Chem.* **2012**, *84*, 3703–3709.
125. Pan, T.; Uhlenbeck, O. C. *Nature* **1992**, *358*, 560–563.
126. Cuenoud, B.; Szostak, J. W. *Nature* **1995**, *375*, 611–614.
127. Wang, Z.; Lee, J. H.; Lu, Y. *Adv. Mater.* **2008**, *20*, 3263–3267.

128. Wei, H.; Li, B.; Li, J.; Dong, S.; Wang, E. *Nanotechnology* **2008**, *19*, 095501-1–099501-5.
129. Xu, X.; Daniel, W. L.; Wei, W.; Mirkin, C. A. *Small* **2010**, *6*, 623–626.
130. Li, C.-L.; Huang, C.-C.; Chen, W.-H.; Chiang, C.-K.; Chang, H.-T. *Analyst* **2012**, *137*, 5222–5228.
131. Wang, C.-I.; Huang, C.-C.; Lin, Y.-W.; Chen, W.-T.; Chang, H.-T. *Anal. Chim. Acta* **2012**, *745*, 124–130.
132. Wu, Y.-S.; Huang, F.-F.; Lin, Y.-W. *ACS Appl. Mater. Interfaces* **2013**, *5*, 1503–1509.
133. Li, T.; Dong, S.; Wang, E. *J. Am. Chem. Soc.* **2010**, *132*, 13156–13157.
134. Dulkeith, E.; Morteani, A. C.; Niedereichholz, T.; Klar, T. A.; Feldmann, J.; Levi, S. A.; van Veggel, F. C. J. M.; Reinhoudt, D. N.; Möller, M.; Gittins, D. I. *Phys. Rev. Lett.* **2002**, *89*, 203002-1–203002-4.
135. Darbha, G. K.; Ray, A.; Ray, P. C. *ACS Nano* **2007**, *1*, 208–214.
136. Li, M.; Wang, Q.; Shi, X.; Hornak, L. A.; Wu, N. *Anal. Chem.* **2011**, *83*, 7061–7065.
137. Huang, C.-C.; Chang, H.-T. *Anal. Chem.* **2006**, *78*, 8332–8338.
138. Wang, H.; Wang, Y.; Jin, J.; Yang, R. *Anal. Chem.* **2008**, *80*, 9021–9028.
139. Chang, H.-Y.; Chang, H.-T.; Hung, Y.-L.; Hsiung, T.-M.; Lin, Y.-W.; Huang, C.-C. *RSC Adv.* **2013**, *3*, 4588–4597.
140. Chen, W.; Tu, X.; Guo, X. *Chem. Commun.* **2009**, 1736–1738.
141. Yuan, Z.; Peng, M.; He, Y.; Yeung, E. S. *Chem. Commun.* **2011**, *47*, 11981–11983.
142. Habeeb Muhammed, M. A.; Verma, P. K.; Pal, S. K.; Retnakumari, A.; Koyakutty, M.; Nair, S.; Pradeep, T. *Chem. Eur. J.* **2010**, *16*, 10103–10112.
143. Xie, J. P.; Zheng, Y. G.; Ying, J. Y. *Chem. Commun.* **2010**, *46*, 961–963.
144. Lin, Y.-H.; Tseng, W.-L. *Anal. Chem.* **2010**, *82*, 9194–9200.
145. Pu, K.-Y.; Luo, Z.; Li, K.; Xie, J.; Liu, B. *J. Phys. Chem. C* **2011**, *115*, 13069–13075.
146. Shang, L.; Yang, L.; Stockmar, F.; Popescu, R.; Trouillet, V.; Bruns, M.; Gerthsen, D.; Nienhaus, G. U. *Nanoscale* **2012**, *4*, 4155–4160.
147. Li, L.; Li, B.; Qi, Y.; Jin, Y. *Anal. Bioanal. Chem.* **2009**, *393*, 2051–2057.
148. Zhou, Y.; Wang, S.; Zhang, K.; Jiang, X. *Angew. Chem., Int. Ed.* **2008**, *47*, 7454–7456.
149. Yin, B.-C.; Zuo, P.; Huo, H.; Zhong, X.; Ye, B.-C. *Anal. Biochem.* **2010**, *401*, 47–52.
150. Wang, G.; Lim, C.; Chen, L.; Chon, H.; Choo, J.; Hong, J.; demello, A. J. *Anal. Bioanal. Chem.* **2009**, *394*, 1827–1832.
151. Fan, Y.; Long, Y. F.; Li, Y. F. *Anal. Chim. Acta* **2009**, *653*, 207–211.
152. Lin, Y.-W.; Chen, W.-T.; Chang, H.-T. *Rapid Commun. Mass Spectrom.* **2010**, *24*, 933–938.
153. Chen, L.-Y.; Ou, C.-M.; Chen, W.-Y.; Huang, C.-C.; Chang, H.-T. *ACS Appl. Mater. Interfaces* **2013**, *5*, 4383–4388.
154. Chen, P.-C.; Yeh, T.-Y.; Ou, C.-M.; Shih, C.-C.; Chang, H.-T. *Nanoscale* **2013**, *5*, 4691–4695.
155. Lan, G.-Y.; Huang, C.-C.; Chang, H.-T. *Chem. Commun.* **2010**, *46*, 1257–1259.

156. Su, Y.-T.; Lan, G.-Y.; Chen, W.-Y.; Chang, H.-T. *Anal. Chem.* **2010**, *82*, 8566–8572.

Chapter 4

Identification of the Amount of Binding Sites and Dissociation Constants of a Ligand-Receptor Complex Using AlGaN/GaN High Electron Mobility Transistors

Chih-Cheng Huang,[1] Geng-Yen Lee,[2] Jen-Inn Chyi,[2] Hui-Teng Cheng,[3] Chen-Pin Hsu,[1] Yu-Fen Huang,[4] and Yu-Lin Wang[1,*]

[1]Institute of Nanoengineering and Microsystems, National Tsing Hua University, Hsinchu, 300, Taiwan, R.O.C.
[2]Department of Electrical engineering, National Central University, Jhongli City, Taoyuan County 32001, Taiwan, R.O.C.
[3]Department of Nephrology, National Taiwan University Hospital, Hsinchu branch, Hsinchu 300, Taiwan, R.O.C.
[4]Department of Biomedical Engineering and Environmental Science, National Tsing Hua University, Hsinchu 300, Taiwan, R.O.C.
[*]E-mail: ylwang@mx.nthu.edu.tw

Ligand-receptor binding site model and the operational model for the depletion mode field-effect-transistors (FETs) are incorporated together for the transistor-based sensors to elucidate the binding affinity between ligands and receptors. AlGaN/GaN high electron mobility transistors (HEMTs) were immobilized with antibodies to detect a short peptide consisting of 20 amino acids. The drain current change of the transistor caused by ligand-receptor binding was regarded as signals and fitted into the binding-site models for analysis of binding affinity. The dissociation constants of the ligand-receptor pairs and the number of binding sites on receptors were revealed form the analysis. The results are very consistent to data reported by other methods from literature. The incorporation of the HEMTs and the binding-site models is demonstrated to be useful for studying the binding affinity between ligands and receptors.

Keywords: dissociation constants; GaN; HEMTs; binding-site; transistors; biosensors; ligand; receptor

Introduction

Ligand-receptor binding plays a critical role in biological systems. Elucidation of binding affinity between ligands and receptors is important in drug development (*1*), ligand selection (*2*), antibody (*3*) and nuclei acid developments. Revealing the dissociation constants and the number of binding-sites of a ligand-receptor complex can allow us to quantitatively display the binding affinity. Enzyme-linked immunosorbent assay (ELISA) (*4*), ultra-violet/visible light (UV/VIS) spectrum (*5*), surface plasmon resonance (SPR) (*6*), isothermal titration calorimetry (ITC) (*7*), surface acoustic wave devices (SAW) (*8*, *9*) and quartz crystal microbalance (QCM) (*10*) have been used to study the binding affinity between ligands and receptors. However, these techniques also have some limitations or drawbacks. For example, labeling, is frequently used in ELISA and UV/VIS spectrum which may alter the activity of biomolecules. A large sample amount is usually required in ITC. The commercialized SPR sensor system has been widely used for ligand-receptor binding affinity study. However, the price for the SPR equipment is still high. On the other hand, biosensors cannot only detect biomolecules, but also can be used to study the binding affinity of the ligand-receptor complex (*8–13*). Among different biosensors, field-effect-transistor (FET)-based sensors are particularly attractive due to the advantages such as low-cost, high sensitivity, label-free and real-time detection. Si-nanowire, carbon nanotubes (CNTs) and graphene were fabricated as FETs and have been used to study the protein-protein interaction by using Langmuir equation for a one-binding-site model (*11–13*) to extract the dissociation. However, a one-binding-site model may not be enough for any receptor, which may have more than one binding site. Besides, the dissociation constants for different binding sites may also be different and have to be extracted, respectively. To the best of our knowledge, FET sensors have not yet been used for identifying the number of binding sites on a receptor and the dissociation constants of the receptor-ligand complex, simultaneously. Although FETs have been used to extract dissociation constants, the incorporation of the FET operation model with binding-site model have not yet been reported. The incorporation is crucial and will be proposed in our study. Among various FETs, AlGaN/GaN HEMTs have been demonstrated for gas, chemical, and biomolecular detection (*14–18*). GaN is biocompatible, chemically and thermally stable. Reliable detections of biological agents in real-time have been proved by using AlGaN/GaN HEMT-based biosensors (*16*, *19–22*). The miniaturized sensors only require a small amount of samples and have low manufacturing costs, due to the matured microfabrication technique.

In this study, the binding-site model, incorporated with the depletion mode FET operation model, is discussed and established theoretically. The relationship between the surface coverage ratio and the current change from the HEMT is theoretically created. The surface coverage ratio, which is the ratio of the ligand-bound receptors to all receptors on the sensor surface, has been introduced into the binding-site model and combined with the FET operation model to reveal the number of binding sites on a receptor and the dissociation constants at different binding sites for that system. The HEMTs and binding-site models are demonstrated by elucidating the number of binding sites and dissociation constants for an antibody-antigen pair. We demonstrated that our sensors cannot only detect the analytes of interest, but also can be used to study the binding affinity between ligands and receptors, simultaneously. This technology is able to facilitate the understanding of the fundamental inter-biomolecular reaction and one of the promising application is for drug development for diseases.

Formation of the Theory

Theoretical Ligand-Receptor Binding-Site Models

The chemical reaction between the surface-immobilized receptors and the free ligands in bulk solution can be expressed as the following equations.

$$[R]+[L]\leftrightarrow[RL] \tag{1}$$

$$K=K_A=\frac{1}{K_D}=\frac{[RL]}{[R][L]} \tag{2}$$

Where K is the equilibrium constant for formula (1), K_A is the association constant, and K_D is the dissociation constant. $[R]$ is the concentration of the unbound receptors immobilized on the sensor surface. $[L]$ is the ligand concentration in the bulk solution. $[RL]$ is the concentration of the ligand-receptor complexes on the sensor surface.

Because the total amount of receptors on surface is limited, the maximum concentration of receptors on surface equals to $[R]_{max}$. Therefore, the total receptors including ligand-bound or unbound ones equal to the maximum receptor concentration.

$$[R]_{\max}=[R]+[RL] \tag{3}$$

Here we can define a parameter, the surface coverage ratio α, expressed as,

$$\alpha\equiv\frac{[RL]}{[R]_{\max}} \tag{4}$$

By combining equation (2), (3), and (4), the surface coverage ratio can be shown as,

$$\alpha\equiv\frac{[RL]}{[R]_{\max}}=\frac{1}{1+\dfrac{K_D}{[L]}} \tag{5}$$

Rearranging equation (5), a linear equation can be obtained as,

$$\frac{[L]}{[RL]} = \frac{[L]}{[R]_{\max}} + \frac{K_D}{[R]_{\max}} \quad (6)$$

Equation (6) is the well known Langmuir isotherm equation.

Now, if we assume when ligands bind receptors, the formation of the ligand-receptor complexes bring signals from the transducer, which is the transistor in our case, then the signal P under a certain ligand concentration $[L]$ is proportional to the amount of the complexes $[RL]$ on the sensor surface. Therefore, when the receptors are saturated with the ligands, the maximum signal $P_{\max}$ is proportional to the maximum concentration of receptor $[R]_{\max}$ on surface. Accordingly, the Langmuir equation can also be expressed as,

$$\frac{[L]}{P} = \frac{[L]}{P_{\max}} + \frac{K_D}{P_{\max}} \quad (7)$$

Equation (7) is a linear equation of ligand concentration $[L]$ in bulk solution with the slope $1/P_{\max}$ and the intercept in y-axis $K_D/P_{\max}$. When signals are measured at different ligand concentration, the dissociation constant can be extracted from the line fitting of equation (7). Figure 1 (left) shows an example of the line fitting and extraction of dissociation constant from the Langmuir equation.

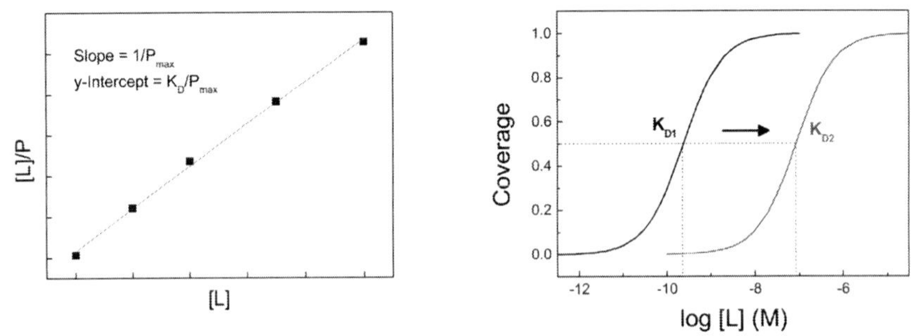

Figure 1. (left) The line fitting of the Langmuir equation to experimental data. (right) Ideal surface coverage curve for one-binding-site model.

The Langmuir isotherm equation is largely used to obtain dissociation constant between ligand and receptor (*11–13*). For many biological receptors, they have more than one binding site for a certain ligand. However, the Langmuir equation (eqn. 6 or eqn. 7), is obvious for one-binding-site receptor only. For some receptors with the unknown number of binding sites, it will be dangerous to just simply extract the dissociation constant from the Langmuir equation, because it leads to just one dissociation constant and the receptor is regarded to have one binding site only. To avoid this problem, a multiple-binding-site model is necessary. The multiple-binding-site model adopts the idea of the surface coverage ratio that we have mentioned in equation (4) and equation (5). From equation (5), the concentration of ligand-receptor complex is given as,

$$[RL] = \frac{1}{1+\frac{K_D}{[L]}} \times [R]_{\max} = \alpha \times [R]_{\max} \quad (8)$$

Since we know $[RL] \propto P$ and $[R]_{\max} \propto P_{\max}$, the equation (8) can be expressed as,

$$P = \frac{1}{1+\frac{K_D}{[L]}} \times P_{\max} = \alpha \times P_{\max} \quad (9)$$

We then can measure signals at different ligand concentrations from very dilute, to medium, and to very high ligand concentration until the signal saturates. The signal saturates because almost all receptors on surface are bound with the ligand and nearly no free receptors are available for binding.

The ideal surface coverage ratio as a function of the ligand concentration is shown as Figure 1 (right). This is the typical curve for the one-binding-site model. Notice that the surface coverage ratio is only dependent on the ratio of dissociation constant to the ligand concentration. Changing the dissociation constant does not alter the shape of the curve, but shifts it laterally on the x-axis, as shown in Figure 1 (right). When the surface coverage ratio equals to 0.5, the ligand concentration equals to the dissociation constant. This does not only work in surface-immobilized sensors, but is also applicable to any system with limited amount of receptors. In fact, it is a well-known definition in traditionally analytical chemistry.

For a one-binding-site system, by fitting experimental data points into this ideal surface coverage ratio model, the dissociation constant can be easily extracted. From equation (9), it is obvious that the experimental coverage ratio just equals to $P/P_{\max}$. When K_D equals to 10 folds or 10% of $[L]$, the surface coverage ratio goes to around 9.09% and 90.91%, respectively. Beyond 90.91% or below 9.09% of the surface coverage ratio, increasing or decreasing the concentration of ligand does not change the surface coverage ratio much. Therefore, we conclude that for a one-binding-site model, the most significant change of the surface coverage ratio is within the range of the ligand concentration between one order higher and one order lower than the value of the dissociation constant. Between 9.09% and 90.91% of the surface coverage ratios, if the ligand concentration ranges across two orders of magnitude, the receptor has one binding site. Within this range of coverage ratio, if the experimental data points cross more than two orders, such as three or four orders, then the receptor may have two dissociation constants, or more.

Here a two-binding-site model is proposed as the following. In the two-binding-site model, the chemical equations are described as below,

$$[R]_{site1} + [L] \leftrightarrow [RL]_{site1} \qquad K_{D1} \quad (10)$$

$$[R]_{site2} + [L] \leftrightarrow [RL]_{site2} \qquad K_{D2} \quad (11)$$

Where the K_{D1} and K_{D2} are the dissociation constants for the two binding sites on a receptor.

For a system with more than one dissociation constants, the signals from experiments can be regarded as the sum of the signal contributed from each individual binding site, which has its own dissociation constant. Therefore, the overall signals for a two-binding-site ligand-receptor complex can be expressed as the linear combination of the signal from each single binding site. The signal from each binding site is proportional to the coverage ratio of each single binding site and to the maximum signal of each binding site when the binding sites are saturated by ligands as well. The signal from a single binding site is simply regarded as a one-binding-site system. In other words, the two-binding-site model is mathematically composed of two one-binding-site models, as shown as followings,

$$P = P_1 + P_2 \tag{12}$$

$$P = \frac{P_{\max 1}[L]}{K_{D1} + [L]} + \frac{P_{\max 2}[L]}{K_{D2} + [L]} = \frac{P_{\max 1}}{\frac{K_{D1}}{[L]} + 1} + \frac{P_{\max 2}}{\frac{K_{D2}}{[L]} + 1} \tag{13}$$

$$P = \alpha_1 P_{\max 1} + \alpha_2 P_{\max 2} \tag{14}$$

Where α_1 and α_2 are the surface coverage ratio at binding site 1 and site 2 of the receptor. $P_{\max1}$ and $P_{\max2}$ are the maximum (saturated) signal contributed by site 1 and site 2, respectively. If we assume that ligands bound at the two different binding sites of a receptor equally affect either the channel conductance or current of the transistor, it is quite reasonable to allow $P_{\max1} = P_{\max2} = P_{\max}/2$. If the two binding sites on a receptor are expected to give different maximum (saturated) signals, then the $P_{\max1}$ and $P_{\max2}$ can be extracted individually. For even more complex multiple-binding-site systems, such as three- or four-binding-site receptors, the models can be built up as the similar procedures as shown in the two-binding site model.

Incorporation of FET Operational Model into the Binding-Site Model

For our depletion mode transistor-based sensors, which are operated in the linear region of FETs under a small drain bias (0.5V), the signal is defined as the drain current change, that is $P = \Delta I$. Therefore, $P_{\max} = \Delta I_{\max}$, $P_{\max1} = \Delta I_{\max1}$, and $P_{\max2} = \Delta I_{\max2}$ are resulted. In the previous section, we have assumed $[RL] \propto P$ and $[R]_{\max} \propto P_{\max}$. The surface coverage by ligand-receptor binding has linear relationship with the signal, which is the current change caused by the ligand-receptor binding. Therefore, it is required that $[RL] \propto \Delta I$ has to be demonstrated in order to satisfy the assumption used in the binding-site model. In the linear region of an FET operation, the current-voltage characteristics is shown as below.

$$V_D < V_G - V_T \tag{15}$$

$$I_D \cong \left(\frac{W\mu_n C}{L}\right)\left(V_G - V_T - \frac{1}{2}V_D\right)V_D \tag{16}$$

$$g_m \cong (\frac{\partial I_D}{\partial V_G})_{V_D} = (\frac{W\mu_n C}{L})V_D \tag{17}$$

Where V_G, V_T, and V_D are gate voltage, threshold voltage, and drain voltage, respectively. W, L, μ_n, C, and g_m are gate width, gate length, electron mobility, gate dielectric capacitance per unit area, and transconductance, respectively. We can assume that the ligand-receptor binding causes the gate voltage change as ΔV_G. From the equation (17), at a constant drain voltage, the drain current change is clearly proportional to the gate voltage change, that is

$$\Delta I_D \propto \Delta V_G. \tag{18}$$

Now how can we link the concentration of the ligand-receptor complex to the drain current change? Here we assume that each ligand binds on a receptor causing surface polarity change, leading to charge generation on the gate dielectric surface. Therefore, the total charge generated on the dielectric surface is proportional to the concentration of the ligan-receptor complex, which can be expressed as following equations,

$$\Delta Q = cq[RL] - cq[R]_{\max} \tag{19}$$

$$\Delta Q = C\Delta V_G \tag{20}$$

Where c is a proportional constant and q is unit charge. The charge generated on the dielectric surface will further induce gate voltage change. Noticed that in the equation (19), the charge change ΔQ is the charge generated on the dielectric surface at a certain ligand-receptor complex concentration compared to that of sensor without any ligand bound on the surface. Because the second term in the equation (19) is a constant, we can simply regard it as a standard point. Therefore, in the equation (20), the ΔV_G actually represent the gate voltage change for a certain ligand-receptor complex concentration on the sensor surface. From the equation (18), (19) and (20), we can derive the linear relationship between the drain current change and the concentration of the ligand-receptor complex, that is,

$$\Delta I_D \propto [RL]. \tag{21}$$

Now with the demonstration of the equation (21), the FET operational model has been successfully incorporated into the binding-site model, which will be used for our sensors.

Experimental

Fabrication of AlGaN/GaN HEMTs

The HEMT structure consisted of a 3 μm-thick undoped GaN buffer, 150 Å-thick undoped $Al_{0.25}Ga_{0.75}N$ and 10 Å-thick undoped GaN cap layer. The AlGaN layer was designed to induce high spontaneous polarization, which generated a two-dimensional electron gas beneath the AlGaN layer. The epilayers were grown by metal-organic chemical vapor deposition (MOCVD) on sapphire substrates.

Mesa isolation was performed using an Inductively Coupled Plasma (ICP) etching system with Cl_2/BCl_3 gases under ICP power of 300 W at 2 MHz and a process pressure of 10 mTorr. Ohmic contacts (60 × 60 μm2) separated with gaps of 30 μm consisted of e-beam deposited Ti/Al/Ni/Au and was annealed at 850 °C, 45 sec under flowing N_2. 100Å-thick gold was deposited on the gate region. Photoresist of 1.8 μm (Shipley S1818) was used to encapsulate the source/drain regions, with only the gate region open to allow the liquid solutions to cross the surface.

Immobilization of Antibody on HEMTs

1-ethyl-3-(3-N,Ndimethylaminopropyl)-carbodiimide hydrochloride (EDC) was purchased from Sigma-Aldrich. N-hydroxysulfosuccinimide (sulfo-NHS) was purchased from Thermo Fisher Scientific Pierce Biotechnology. 20mM of EDC and 50mM of sulfo-NHS were prepared in 2-(morpholino)ethanesulfonic acid (MES) buffer solution consisting of 10mM of 6-Mercaptohexanoic acid at pH=5 and allowed 15 minutes to react thoroughly at room temperature. The mixture was then titrated to increase pH value to 7.4. The anti-ferritin heavy chain (FHC) antibody purchased from Santa Cruz Biotechnology Inc. (cat. #sc-14416) was then added into the mixture and allowed 2 hours to react at room temperature. Tris(2-carboxyethyl)phosphine hydrochloride (TCEP) purchased from Sigma-Aldrich (cat. #C4706) was then added into the antibody-containing mixture, and then allowed for 2 hours to react. 100mM of TCEP and 1mM of 6-Mercaptohexanoic acid were prepared in the final mixture. The HEMT devices were then submerged into this mixture and stored at 4°C for 12 hours. This resulted in binding of the thiolated-antibody to the gold surface on the gate area of the HEMTs. The HEMTs were then rinsed with Phosphate buffered saline (PBS), followed by being submerged in bovine serum albumin (BSA) solution for 3 hours for surface blocking. After incubation, the sensor was thoroughly rinsed off with PBS and dried by a nitrogen blower. The schematics of the sensor is depicted as in Figure 2.

Sensor Measurements

The characteristics of source-drain current-voltage of the sensor was measured at 20°C using an Agilent B1500 parameter analyzer with the gate region exposed. The source-drain bias was fixed at 0.5V for the detection of the antigen, a short peptide, consisting of 20 amino acids, which was purchased from Santa Cruz Biotechnology Inc. (cat. #sc-14416P). This short peptide can specifically bind to the FHC antibody. Different concentrations of the peptides were sequentially dropped on the sensor in a real-time detection. A background test was also conducted with different concentrations of the short peptide at a constant bias of 0.5V. The chip for the background test was processed with the similar surface modification procedure (6-Mercaptohexanoic acid immobilized and BSA blocking) on the HEMT but no antibodies were immobilized on the chip.

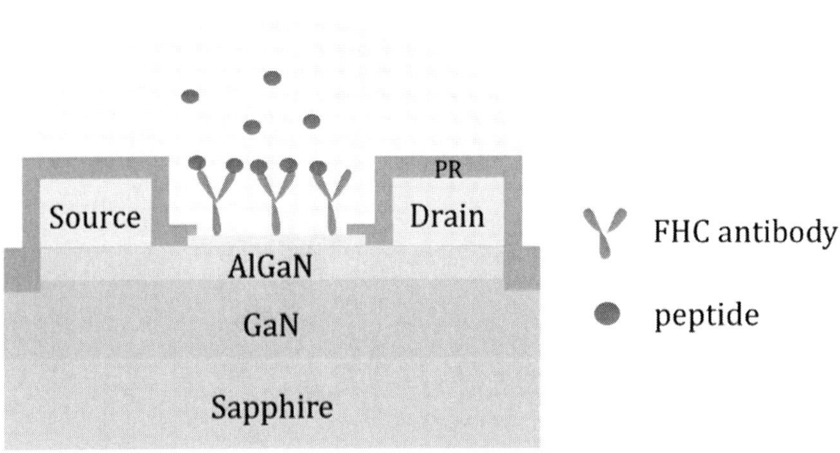

Figure 2. Schematic of the AlGaN/GaN HEMT sensor. The Au-coated gate area was functionalized with FHC antibodies. Small peptides are captured by the antibodies and sensed by the HEMT.

Results and Discussion

Binding Affinity between Peptide and the Antibody

Figure 3 (left) shows the real time detection of the peptide for the sensor immobilized with FHC antibody. PBS (0.1x) solution at pH=7.4 was initially dropped on the sensor. When an additional PBS solution was added on the sensor, an abrupt deep peak appeared and then quickly recovered to the baseline. This abrupt peak occurred due to the mechanical disturbance resulted from dropping the PBS droplet with a micropipette by hands. No net current change was observed for 100 seconds, after the current went back to the original baseline. There was no significant current change, when the target concentration of 0.5pM of the peptide was added onto the sensor. In sharp contrast, when the target concentration of 5pM of the peptide was dropped, a considerable current change was observed. The detection of the peptide ranges from 0.5pM to 5μM. The most significant current change was observed at 50pM. The drain current gradually saturated as the peptide concentration increased. Finally, no more current change was observed at higher concentration of the peptide, as shown in Figure 3 (left). The background test (control experiment) was conducted with different concentrations of the short peptide showing no significant current change, which demonstrated that there was no any non-specific binding on the chip (data not shown here). Therefore, the signals generated from the sensors were attributed to the specific antibody-antigen binding. Figure 3 (right) shows the current change versus the target concentration of the peptide. From Figure 3 (right), the limit of detection of this sensor was estimated about 10pM of the peptide in the buffer solution.

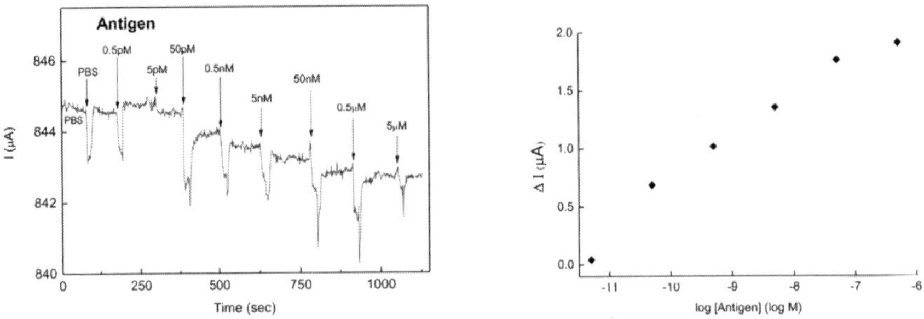

Figure 3. (left) Real-time detection of the peptide from 0.5pM to 5μM at a constant bias of 500 mV. (right) The current change versus the peptide target concentration. Reproduced with permission from reference (23). Copyright 2013 Elsevier.

To estimate the dissociation constant, the Langmuir isotherm equation was used to extract the dissociation constant. Here the current changes were introduced to the Langmuir equation for calculating the dissociation constant.

$$\frac{[Ag]}{\Delta I} = \frac{[Ag]}{\Delta I_{max}} + \frac{K_D}{\Delta I_{max}} \tag{22}$$

Where ΔI is the current change at the bulk concentration of antigen $[Ag]$, and ΔI_{max} is the saturated current change.

The dissociate constant can be extracted from the linear regression by using equation (3), as shown in Figure 4 (left). The y-axis and the x-axis are $[Ag]/\Delta I$ and $[Ag]$, respectively. The slope and the y-intercept are $1/\Delta I_{max}$ and $K_D/\Delta I_{max}$, respectively. The line-fitting shown in Figure 4 (left) gives a very good linear regression correlation coefficient (r^2=0.99992). The maximum current change extracted from the fitting was about 1.91μA, which is very close to the experimental observed one, 1.90μA. The extracted dissociation constant was then calculated as 1.37×10^{-9}M, which is also in the reasonable range for most IGg antibody-antigen complexes (*3, 8, 9*). Figure 4 (right) shows the curve of surface coverage ratio from equation (5) using the dissociation constant extracted from Figure 4 (left) and the $\Delta I/\Delta I_{max}$ obtained from the experimental results shown as dots versus the antigen concentration in log scale. It is clear that the curve of the surface coverage ratio does not match well with those experimental dots. In Figure 4 (right), it is obvious that between 9.09% and 90.91% of the surface coverage, the experimental data points cover almost 4 orders of the antigen concentration. Therefore, it is not possible to fit these experimental points well into the surface coverage ratio with the one-binding-site model. The wide range of the antigen concentration within that coverage ratio (9.09% ~ 90.91%) may be resulted from multiple binding sites of the receptor, as we have discussed previously. Compared with Figure 4 (left), Figure 4 (right) is obviously easier to decide whether one binding-site or multiple-binding-site model should be used to find out the number of binding sites and the dissociation constants. In this case, the total current change is assumed to be the sum of the current change resulted from antibody-antigen complex at site 1 and at site 2, respectively.

$$\Delta I = \Delta I_1 + \Delta I_2 \tag{23}$$

$$\Delta I = \frac{\Delta I_{\max 1}[Ag]}{K_{D1} + [Ag]} + \frac{\Delta I_{\max 2}[Ag]}{K_{D2} + [Ag]} = \frac{\Delta I_{\max 1}}{\frac{K_{D1}}{[Ag]} + 1} + \frac{\Delta I_{\max 2}}{\frac{K_{D2}}{[Ag]} + 1} \tag{24}$$

$$\Delta I = \alpha_1 \Delta I_{\max 1} + \alpha_2 \Delta I_{\max 2} \tag{25}$$

Where the K_{D1} and K_{D2} are the dissociation constants for the two binding sites on a receptor. α_1 and α_2 are the surface coverage ratio at site 1 and site 2 of the antibody.

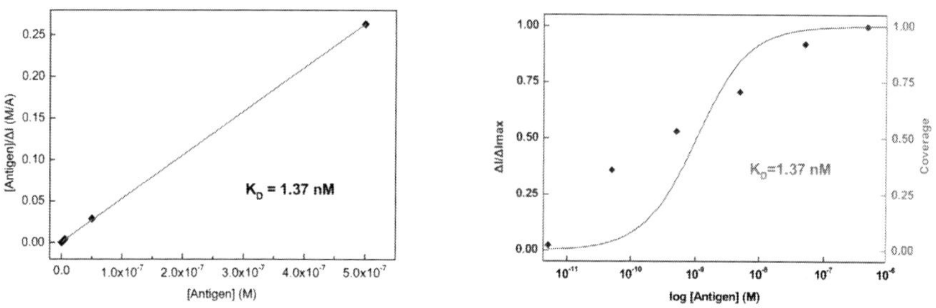

Figure 4. (left) [Ag]/ΔI versus [Ag] and the dissociation constant extracted from the linear regression using Langmuir adsorption isotherm; (right) Surface coverage ratio as a function of the antigen concentration calculated with the dissociation constant extracted from liner regression of Langmuir equation. The $\Delta I/\Delta I_{max}$ from the experiment are shown as dots versus the antigen concentration. Reproduced with permission from reference (23). Copyright 2013 Elsevier.

If we assume that antigens bind at the two different binding sites of an antibody equally affect the conductance of the transistor, it is fairly acceptable to allow $\Delta I_{max1} = \Delta I_{max2} = \Delta I_{max}/2$. Then the total surface coverage ratio α can be expressed as the average of the two individual ones ($\alpha = (\alpha_1 + \alpha_2)/2$). We then fit this two-binding-site model into the experimental data points to find out the two dissociation constants, K_{D1} and K_{D2}.

Figure 5 shows the average surface coverage ratio with error bars (standard deviation) from three measurements versus the antigen concentration. The model and the experimental data points perfectly fit together. The K_{D1} and K_{D2} are 4.404×10^{-11}M and 1.596×10^{-9}M, respectively. The correlation coefficient, r^2 is 0.9948. These two binding constants are in a reasonable range of regular antibody-antigen binding constants (*3*, *8*). The antigen, the peptide, consisting of only 20 amino acids is pretty small and the antibody is allowed to bind two antigens on its two binding sites. Because the two binding sites of an antibody are usually regarded as identical structures, the site 1 and the site 2 in this reaction should not be recognized as two different sites. Instead, they are more preferably regarded as the binding sequence with the antigens. The difference between these two dissociation constants is probably ascribed to the stereo hindrance resulted from the first antigen-antibody complex for the second one. It is well known that antibodies have two almost identical binding sites, as our results predict. Therefore, this technique is able to determine the number of the binding sites of a receptor for its analyte. If the high sensitivity region covers 5~6 orders of magnitude, we may further need to consider a three-binding-site model. If we look back at the detection limit of the sensor, which is between 5pM~50pM of the antigen, is just consistent with the lower-half high sensitivity region of (between one order lower and the order of the dissociation constant). This result shows that the AlGaN/GaN HEMTs are still very sensitive in low antigen concentration. However, the detection limit for the antigen of this sensor is not simply depending on the transistor, but also much on the affinity of the antibody-antigen complex, that is, its dissociation constant.

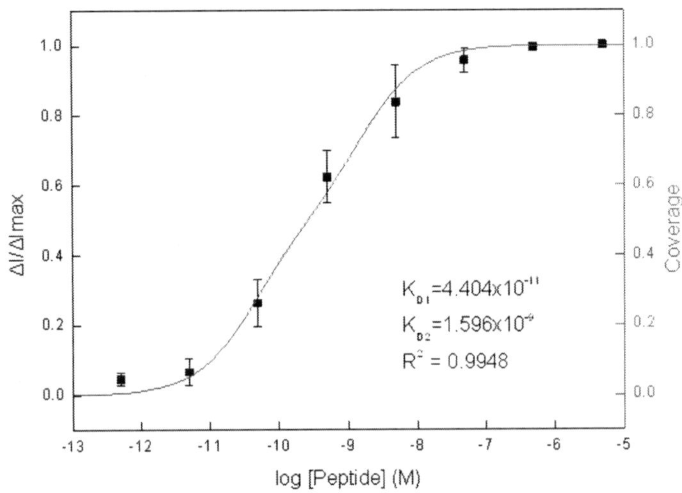

Figure 5. Surface coverage ratio as a function of the antigen concentration with equal maximum current changes for the two binding complexes on an antibody. Reproduced with permission from reference (23). Copyright 2013 Elsevier

Summary

The incorporation of ligand-receptor binding-site models and the transistor-based sensors made of AlGaN/GaN HEMTs is demonstrated for studying antibody-antigen and drug-HIV RT enzyme interactions. The analysis of the surface coverage ratio with the current change generated from the ligand-receptor binding, has led to the discovery of the binding affinity between ligands and receptors. A short peptide was tested with antibody-immobilized HEMTs. The amount of binding site and the dissociation constants were resolved for both cases and the results are consistent with reports from literatures. This platform is demonstrated to effectively reduce the cost and time for fundamental biological study and for drug developments. This technique is very promising and expected to be applied in many biological studies and biomedical applications.

Acknowledgments

This work was partially supported by National Science Council grant (NSC 99-2218-E-007 -019 -MY2 & NSC 101-2221-E-007-102-MY3) and by the research grant (100N2049E1) at National Tsing Hua University. We thank Dr. Jen-Inn Chyi and Dr. Hui-Teng Cheng for providing us the AlGaN/GaN epi wafers and the FHC antibody respectively, and for consulting as well.

References

1. Bertucci, C.; Cimitan, S. *J. Pharm. Biomed. Anal.* **2003**, *32* (4-5), 707–714.
2. Morrill, P. R.; Millington, R. B.; Lowe, C. R. *J. Chromatogr. B* **2003**, *793* (2), 229–251.
3. Kim, D. H.; Seo, S. M.; Paek, S. H.; Lim, G. S. *Anal. Biochem.* **2012**, *420* (1), 54–60.
4. Orosz, F.; Ovadi, J. *J. Immunol. Methods* **2002**, *270* (2), 155–162.
5. del Toro, M.; Gargallo, R.; Eritja, R.; Jaumot, J. *Anal. Biochem.* **2008**, *379* (1), 8–15.
6. Homola, J. *Chem. Rev.* **2008**, *108* (2), 462–493.
7. Chiad, K.; Stelzig, S. H.; Gropeanu, R.; Weil, T.; Klapper, M.; Mullen, K. *Macromolecules* **2009**, *42* (19), 7545–7552.
8. Mitsakakis, K.; Gizeli, E. *Biosens. Bioelectron.* **2011**, *26* (11), 4579–4584.
9. Dragusanu, M.; Petre, B. A.; Slamnoiu, S.; Vlad, C.; Tu, T. T.; Przybylski, M. *J. Am. Soc. Mass Spectrom.* **2010**, *21* (10), 1643–1648.
10. Liu, Y.; Tang, X. L.; Liu, F.; Li, K. *Anal. Chem.* **2005**, *77* (13), 4248–4256.
11. Lin, T. W.; Hsieh, P. J.; Lin, C. L.; Fang, Y. Y.; Yang, J. X.; Tsai, C. C.; Chiang, P. L.; Pan, C. Y.; Chen, Y. T. *Proc. Natl. Acad. Sci. U.S.A.* **2010**, *107* (3), 1047–1052.
12. Chen, Y. N.; Vedala, H.; Kotchey, G. P.; Audfray, A.; Cecioni, S.; Imberty, A.; Vidal, S.; Star, A. *ACS Nano* **2012**, *6* (1), 760–770.
13. Wang, W. U.; Chen, C.; Lin, K. H.; Fang, Y.; Lieber, C. M. *Proc. Natl. Acad. Sci. U.S.A.* **2005**, *102* (9), 3208–3212.
14. Chen, K. H.; Kang, B. S.; Wang, H. T.; Lele, T. P.; Ren, F.; Wang, Y. L.; Chang, C. Y.; Pearton, S. J.; Dennis, D. M.; Johnson, J. W.; Rajagopal, P.; Roberts, J. C.; Piner, E. L.; Linthicum, K. J. *Appl. Phys. Lett.* **2008**, *92* (19), 192103-1–192103-3.
15. Kang, B. S.; Wang, H. T.; Ren, F.; Pearton, S. J.; Morey, T. E.; Dennis, D. M.; Johnson, J. W.; Rajagopal, P.; Roberts, J. C.; Piner, E. L.; Linthicum, K. J. *Appl. Phys. Lett.* **2007**, *91* (25), 252103-1–252103-3.
16. Pearton, S. J.; Ren, F.; Wang, Y. L.; Chu, B. H.; Chen, K. H.; Chang, C. Y.; Lim, W.; Lin, J. S.; Norton, D. P. *Prog. Mater. Sci.* **2010**, *55* (1), 1–59.

17. Wang, Y. L.; Chu, B. H.; Chen, K. H.; Chang, C. Y.; Lele, T. P.; Tseng, Y.; Pearton, S. J.; Ramage, J.; Hooten, D.; Dabiran, A.; Chow, P. P.; Ren, F. *Appl. Phys. Lett.* **2008**, *93* (26), 262101-1–262101-3.
18. Yu, X.; Li, C.; Low, Z. N.; Lin, J.; Anderson, T. J.; Wang, H. T.; Ren, F.; Wang, Y. L.; Chang, C. Y.; Pearton, S. J.; Hsu, C. H.; Osinsky, A.; Dabiran, A.; Chow, P.; Balaban, C.; Painter, J. *Sens. Actuators, B* **2008**, *135* (1), 188–194.
19. Chen, K. H.; Wang, H. W.; Kang, B. S.; Chang, C. Y.; Wang, Y. L.; Lele, T. P.; Ren, F.; Pearton, S. J.; Dabiran, A.; Osinsky, A.; Chow, P. R. *Sens. Actuators, B* **2008**, *134* (2), 386–389.
20. Chu, B. H.; Kang, B. S.; Ren, F.; Chang, C. Y.; Wang, Y. L.; Pearton, S. J.; Glushakov, A. V.; Dennis, D. M.; Johnson, J. W.; Rajagopal, P.; Roberts, J. C.; Piner, E. L.; Linthicum, K. J. *Appl. Phys. Lett.* **2008**, *93* (4), 042114-1–042114-3.
21. Kang, B. S.; Ren, F.; Wang, L.; Lofton, C.; Tan, W. H. W.; Pearton, S. J.; Dabiran, A.; Osinsky, A.; Chow, P. P. *Appl. Phys. Lett.* **2005**, *87* (2), 023508-1–023508-3.
22. Wang, Y. L.; Chu, B. H.; Chen, K. H.; Chang, C. Y.; Lele, T. P.; Papadi, G.; Coleman, J. K.; Sheppard, B. J.; Dungen, C. F.; Pearton, S. J.; Johnson, J. W.; Rajagopal, P.; Roberts, J. C.; Piner, E. L.; Linthicum, K. J.; Ren, F. *Appl. Phys. Lett.* **2009**, *94* (24), 243901-1–243901-3.
23. Huang, C. C.; Lee, G. Y.; Chyi, J. I.; Cheng, H. T.; Hsu, C. P.; Hsu, Y. R.; Hsu, C. H.; Huang, Y. F.; Sun, Y. C.; Chen, C. C.; Li, S. S.; Yeh, J. A.; Yao, D. J.; Ren, F.; Wang, Y. L. *Biosens. Bioelectron.* **2013**, *41*, 717–722.

Interaction of Nanomaterials with Contaminants for Treatment

Chapter 5

Catalytic Impact of Activated Carbon on the Formation of Nitrosamines from Different Amine Precursors

Ching-Hua Huang,* Lokesh P. Padhye,† and Yung-Li Wang

School of Civil and Environmental Engineering, Georgia Institute of Technology, Atlanta, Georgia 30332, USA
†Current Address: Center for Environmental Science and Engineering, Indian Institute of Technology Bombay, Powai, Mumbai, 400076, India
***Phone: +1 404 8947694, Fax: +1 404 3857087, E-mail: ching-hua.huang@ce.gatech.edu**

N-nitrosamines are emerging carcinogenic contaminants and can be generated from transformation of amines. Activated carbon (AC) materials with micro- and nano-porous structures are commonly used for extraction of nitrosamines in water, and for removal of micropollutants in water and wastewater treatment. Recent research demonstrated that AC materials may catalyze transformation of amine precursors to yield nitrosamines under environmentally relevant conditions. AC catalyzes the formation of nitrosamine primarily via two different types of reactions: (I) catalyzing nitrosation of amine with nitrite or nitrate to yield nitrosamine, and (II) catalyzing oxidation of amine in the presence of molecular oxygen and nitrogen to generate nitrosamine. The AC-catalyzed nitrosamine formation is highly influenced by AC's properties, amine precursors' structures and reaction conditions. Considering the widespread usage of AC in environmental and industrial applications, a better understanding of the catalytic impact of AC on nitrosamine formation is critical to minimize undesirable nitrosamine formation where co-existence of AC and amine precursors is encountered.

Introduction

N-nitrosamines (Figure 1) are a group of emerging disinfection by-products (DBPs) that have received much attention in recent years due to their potential high toxicity (*1–4*). Nitrosamines are probable human carcinogens, with water concentrations as low as 0.2 ng/L associated with a 10^{-6} lifetime cancer risk (*5*) (Table 1). Although there are no federal regulations controlling nitrosamines in drinking water in the U.S. or Canada, the U.S. Environmental Protection Agency (USEPA) has added *N*-nitrosodimethylamine (NDMA), nitrosopyrrolidine (NPYR), *N*-nitrosodiethylamine (NDEA), *N*-nitrosomethylethylamine (NMEA), *N*-nitrosodi-*n*-propylamine (NDPA), and *N*-nitrosodi-*n*-butylamine (NDBA) to the Unregulated Contaminant Monitoring Rule 2 (UCMR 2) (*6*), and NDMA, NDEA, NDPA, NPYR and *N*-nitrosodiphenylamine (NDPhA) in the 3rd version of the Contaminant Candidate List (CCL 3) (*7*). The Ontario Ministry of the Environment has a maximum acceptable concentration (MAC) for NDMA of 9 ng/L while the California Department of Public Health has established a notification level of 10 ng/L for NDMA, NDEA, and NDPA, and a public health goal (PHG) of 3 ng/L for NDMA (*3*).

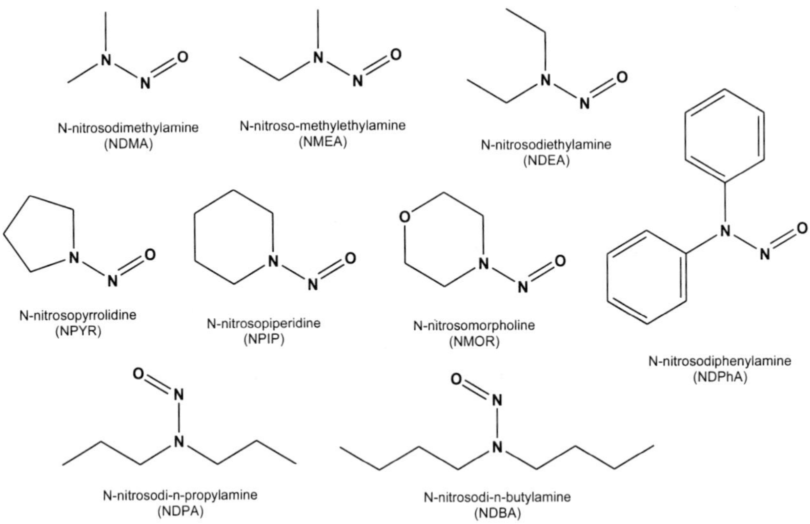

Figure 1. Molecular structures of N-nitrosamines

Table 1. Properties and Cancer Risks of Some Nitrosamines in Drinking Water. (Data from reference (*5*))

Compound	*Molecular Formula*	*Molecular Weight (g/mole)*	*10^{-6} Cancer Risk Concentration (ng/L)*
NDMA	$C_2H_6N_2O$	74.08	0.7
NMEA	$C_3H_8N_2O$	88.06	2
NPYR	$C_4H_8N_2O$	100.06	20
NDEA	$C_4H_{10}N_2O$	102.13	0.2
NPIP	$C_5H_{10}N_2O$	114.15	3.5
NMOR	$C_4H_8N_2O_2$	116.12	5
NDPA	$C_6H_{14}N_2O$	130.11	5
NDBA	$C_8H_{18}N_2O$	158.14	6
NDPhA	$C_{12}H_{10}N_2O$	198.23	7000

NDMA is the most commonly detected nitrosamine in drinking water at concentrations ranging from 0 to 180 ng/L (*3*, *4*). *N*-Nitrosomorpholine (NMOR), NPYR, *N*-nitrosopiperidine (NPIP), and NDPhA have also been detected in treated water at maximum concentrations near 3, 71, 118, and 2 ng/L, respectively (*3*, *4*). In the wastewater effluents, nitrosamine concentrations are generally 1-2 orders of magnitude higher than the levels found in drinking water (*2*, *8*). According to the results released from the UCMR2 survey, NDMA was found above the minimum reporting level (2 ng/L) in 10% of public water systems (PWS) samples collected nationwide (*9*). Future regulations of nitrosamines are likely to be implemented by the USEPA and have promoted widespread interest into minimizing the formation of nitrosamines within PWS.

Activated carbon (AC) materials are widely used in drinking water and wastewater treatment, primarily for removal of small molecular weight organic compounds. Liquid-phase applications of AC constitute 80% of the total AC demand in the U.S. (>500 million pounds), of which the removal of water contaminants represents the major market (55%) (*10*). Of the total U.S. water treatment market for AC adsorbents, about 50% is in drinking water, 40% is in wastewater, and the rest is in groundwater applications (*10*). Potable water is the largest end-use market for AC, accounting for more than 100 million pounds. Demand for AC in PWS is expected to grow at approximately 4.5 percent annually (*11*). Furthermore, AC plays an important role in analytical chemistry applications because AC is among the most common solid phase extraction (SPE) material used for analysis of nitrosamines as well as other organic compounds in water, food and beverage industries. AC is also frequently used as a catalyst or catalyst support in various industrial applications.

Because AC materials are widely used in analytical, environmental, and industrial applications, and since amine-containing organic compounds are commonly present in various aquatic environments, interactions between AC materials and amine-containing organics can be expected under environmentally relevant conditions. Moreover, recent studies by the authors (*12*, *13*) as well as some literature (*14*) have shown that AC materials can catalyze transformation of amine precursors to generate nitrosamines. Because of the significant health concerns of nitrosamine contaminants, the processes of AC-catalyzed nitrosamine formation have important environmental relevance and warrant a better understanding. The objective of this book chapter is to provide a critically-reviewed summary on the AC-catalyzed nitrosamine formation reactions, including reporting of new experimental results by the authors. In this chapter, background knowledge of nitrosamine formation chemistry and catalytic roles of ACs in chemical reactions is first summarized in the Background section. Following that, the Materials and Methods related to the experimental results are briefly described. Afterwards, the two major types (I & II) of AC-catalyzed nitrosamine formation reactions are discussed in detail with experimental data. Finally, conclusions are made, and recommendations for future research are proposed.

Background

Nitrosamine Formation Chemistry

It has long been known that nitrosamine can be formed from nitrosation of amine with nitrite. Nitrosation involves the formation of nitrosyl cation or similar nitrogen-containing species, such as dinitrogen trioxide (N_2O_3), during acidification of nitrite. The nitrosyl cation or N_2O_3 then reacts with an amine, such as dimethylamine (DMA), to form NDMA (*15*). The above *N*-nitrosation reaction occurs most rapidly at pH 3.4, reflecting a balance between the protonation of nitrite (pK_a of HNO_2 = 3.35) and the increased fraction of DMA in the reactive, deprotonated form with increasing pH (pK_a of $(CH_3)_2NH_2^+$ = 10.7). Due to the requirement of acidic conditions for the above reaction to occur, nitrosation is not considered a primary formation mechanism for NDMA in water treatment facilities which are operated at near circum-neutral pH (*2*). However, studies have reported that *N*-nitrosation can be catalyzed at circumneutral pH by photochemical reactions (*16*), and by formaldehyde (*17*), fulvic acid (*18*), or activated carbons (*14*). Additionally, formation of NDMA from the reaction of DMA and nitrite can be greatly enhanced by free chlorine at pH 7, owing to the formation of dinitrogen tetroxide (N_2O_4) nitrosating intermediate from oxidation of nitrite by free chlorine (*19*).

The other formation pathway of nitrosamines that has been recognized is during chloramination of amines in water and wastewater treatment. Choi and Valentine (*20*) and Mitch and Sedlak (*21*) proposed that NDMA could be preferably formed via an unsymmetrical dimethylhydrazine (UDMH) intermediate during the reaction of DMA and monochloramine instead of the nitrosation pathway. Later on, Schreiber and Mitch (*22*) refined the previous

UDMH oxidation pathway by revealing the important role of dichloramine and dissolved oxygen with the chlorinated UDMH intermediate rather than UDMH. The authors demonstrated that dichloramine plays a major role instead of monochloramine for NDMA formation. The chlorinated UDMH intermediate then undergoes further oxidation by dissolved oxygen to form NDMA, which is in competition with oxidation by dichloramine to form other products. The authors proposed that the low concentration of dichloramine solely contributes to NDMA formation via chlorine disproportionation even in typical chloramination conditions in which monochloramine exists as the dominant chloramine species, thereby possibly explaining the low yield of NDMA formation at the treatment plants.

Nitrosamines may also be formed via pathways other than the above two. Several studies have shown that other oxidants including ozone (O_3), chlorine dioxide (ClO_2) and permanganate (*23–29*) can oxidize amines to produce low levels of nitrosamines. Andrzejewski et al. (*24*, *27*) linked nitrosamine formation during ozonation to the formaldehyde-catalyzed nitrosation since formaldehyde is a common product of ozonation. In contrast, Yang et al. (*30*) proposed two other pathways: (i) secondary amine precursors reacting with hydroxylamine to form unsymmetrical dialkylhydrazine intermediates, which are further oxidized to their relevant nitrosamines; and (ii) a nitrosation pathway in which N_2O_4 acts as the nitrosating reagent, to be responsible. The former pathway was shown by Padhye et al. (*29*) to be more plausible.

Finally, while DMA is the most studied NDMA precursor, it accounts for only a portion of the NDMA precursors found in surface water and wastewater samples (*31–35*) Many other more complex NDMA precursors have been identified (*29*, *33*, *36*, *37*); however, more research is still needed in the identification and control of nitrosamine precursors in water sources.

Catalytic Role of Activated Carbons

While the principal use of ACs in water purification is for the adsorption of organic compounds, secondary reactions involving AC as a reductant or an oxidation catalyst has been recognized (*38*, *39*). Previous studies have reported AC-catalyzed oxidation of phenols under environmental conditions, leading to the so-called "irreversible adsorption" of phenols on AC surfaces (*40*). ACs are used as active catalysts in several industrial reactions such as production of phosgene (carbonyl chloride) by combination of carbon monoxide and chlorine over carbon (*41*) and oxidation of pollutant gases such as SO_2, NO, and H_2S in air pollution control (*42*). Comparatively, ACs are less frequently used as catalysts for the abatement of organic pollutants in the liquid phase. However, the use of ACs for catalytic wet air oxidation (CWAO) of organic pollutants in the treatment of wastewater is developing. CWAO of phenolic compounds has been studied in detail (*43*). ACs have also been shown as effective catalysts for CWAO of ammonia, aniline, DMA, methylamine, 1,3,6-naphthalene trisulphonic acid, hydrogen peroxide, cyclohexanone, and N-phosphonomethyliminodiactic acid (*43*, *44*).

The performance of a catalyst depends on the availability of suitable active sites, capable of chemisorbing the reactants and forming surface intermediates of adequate strength. In heterogeneous catalysis, the reaction occurs at the interface between the AC catalyst and the gas or liquid phase containing the reactants. While catalytic activity is expected to increase with carbon's specific surface area, more often the carbon's surface chemistry plays an even greater role (*45*). Hence, the activity of AC as a catalyst is determined by a number of factors including the nature, concentration, and accessibility of AC's active sites, the presence of functional groups of hetero-atoms (O, N, S), and trace metals (ash content) on the surface of AC (*46*, *47*). Oxygen and nitrogen functional groups, which are incorporated into the carbon materials during activation process or by a variety of chemical and thermal methods, play an important role in this context (*43*, *48–54*). Common oxygen containing surface functional groups observed on the surface of ACs include carboxyl, phenolic, quinonoid, lactone and anhydride groups (*55*). The influence of oxygen or nitrogen surface functional groups on AC's catalytic activity varies in different types of oxidation or reduction reactions (*48*, *51–54*, *56*, *57*). For example, CWAO and related oxidation studies have shown that basic carbons containing quinone groups are active as oxidation catalysts for organic wastewater pollutants probably due to their ability to form reactive free radicals (*43*).

Aguilar et al. (*58*) studied the AC-catalyzed CWAO of methylamine and DMA, and reported that the catalytic activity was related to AC's oxygenated surface functional groups. Methylamine and DMA were oxidized to nitrogen, water, and carbon dioxide under the employed conditions (195 °C and 16 atm). The authors proposed that the quinone groups (releasing CO by temperature programmed decomposition (TPD)) on surface of AC were responsible for the catalytic activity, while carboxylic, lactone and anhydride groups (releasing CO_2 in TPD), were responsible for a catalytic activity inhibition, despite that the latter groups appeared to exhibit stronger adsorption of methylamine and DMA. Gomes et al. (*44*) reported that the catalytic oxidation of aniline by AC led to a range of byproducts including nitrosobenzene and azobenzene. The oxidizing species are oxygen, either in the bulk or adsorbed phase, and hydroxyl radicals readily formed under CWAO's high temperature and pressure conditions. The authors reported that the catalytic removal of aniline was related to the mesoporous character of the AC materials and to the presence of specific oxygenated surface functional groups of quinone nature.

AC-Catalyzed Nitrosamine Formation Reactions

On the basis of research published thus far, AC materials can catalyze nitrosamine formation via two types of reactions: (I) AC surfaces may catalyze nitrosation of amine with nitrite to yield nitrosamine, and (II) AC may catalyze oxidation of amine to generate nitrosamine without the addition of nitrite, but in the presence of molecular oxygen and nitrogen (see Figure 2).

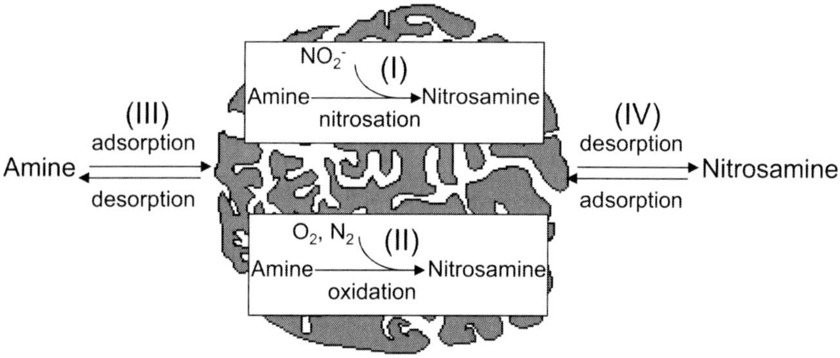

Figure 2. Possible reactions involved in activated carbon-catalyzed transformation of amine to nitrosamine. Adapted/reproduced from reference (59). Copyright (2013, Water Research Foundation).

We will refer to the Reaction I as "AC-catalyzed nitrosamine formation via nitrosation". Earlier, Dietrich et al. (*14*) reported AC-catalyzed *N*-nitroso-methylaniline (NNMA) formation from *N*-methylaniline and nitrite. The amount of NNMA formed in the presence of AC was 75 times more than that formed in the absence of AC under the same conditions. NNMA formation increased with increasing nitrite concentration, but did not increase with increasing AC mass. The amount of NNMA formed at pH 4 was 2 orders of magnitude greater than the corresponding treatment at pH 7. However, the authors did not identify the reaction mechanisms and no information is available regarding the impact of carbon or amine properties. Further research has been conducted by the authors to confirm this catalyzed reaction and will be discussed in detail in the later sections.

The Reaction II is a new discovery by the authors. Our work (*12*, *13*) identified that AC surfaces can react with molecular oxygen and nitrogen to generate reactive oxygen and nitrogen species (ROS and RNS), which then react with amine adsorbed on AC surfaces to yield nitrosamine. Thus, we will refer to the Reaction II as "AC-catalyzed nitrosamine formation via nitrogen fixation", with more details discussed later.

Lastly, as illustrated in Figure 2, since both Reactions I and II occur on AC's surface reactive sites, amine's adsorption to AC surfaces (Reaction III) is an important initial step for both reactions. Subsequently, the adsorption/desorption of nitrosamines with ACs (Reaction IV) affects the retention or leaching of nitrosamines from the AC surfaces. The Reactions III and IV, while relevant, are not the focus of this book chapter.

Materials and Methods

Chemicals, Reagents, and Activated Carbons

All chemicals and reagents were obtained at reagent grade or better from commercial suppliers and used without further purification. Details of the sources and purities of the chemicals and reagents were described previously (*12*, *13*). Trimethylmine (TMA), dimethylbenzylamine (DMBA) and benzalkonium chloride were obtained at >98% purity from Fisher Scientific Acros. TMA hydrochloride salt, ranitidine, 2-[(dimethylamino)methyl]phenol (2-DMP) and *N,N*-dimethylaniline (NNDA) were obtained from Sigma-Aldrich.

A variety of carbons were tested including coconut-shell based Siemens Aquacarb carbon (AqC) from Siemens Water Technologies (Warrendale, PA), coconut-shell based Prominent Systems carbon (PSC) from Prominent Systems Inc. (City of Industry, CA), a bituminous coal-based carbon, F400, from Calgon Corporation (Pittsburgh, PA), and synthetic Carboxen 572 (Ambersorb 572 equivalent) from Sigma-Aldrich (St. Louis, MO). Details of the AC materials tested and characterization were described previously (*12*, *13*).

Experimental Procedures

Batch experiments were conducted in deionized (DI) or real water matrices using amber borosilicate bottles with Teflon-lined caps. At first, AC particles (150-200 mg) were added to 100-125 mL of amine solution (at 200-222 μM) buffered at desired pH using 5-10 mM phosphate buffer. Afterwards, the amine+AC solution was shaken on a platform shaker at 200 or 250 rpm for 1–2 h at room temperature (22-23 °C) and filtered through a glass-fiber filter (Millipore, Jaffrey, NH). In evaluating Reaction I, nitrite was also amended to the solution and the solution was kept in the amber glass bottle with minimum headspace. NDMA-d_6 was spiked in the solution as a surrogate standard for analysis. After the reaction, the solution was quickly filtered through a glass-fiber filter (Millipore, Jaffrey, NH). The collected AC particles were immediately transferred into a microcentrifuge tube, desorbed by 1 mL dichloromethane (DCM), and centrifuged at 1000 rpm for 1 minute. The centrifugation separates the residual water from the particles, which forms a separate layer from the DCM layer. Then, the DCM layer was pipetted into a Gas Chromatography (GC) vial to analyze for nitrosamines by Gas Chromatography/Mass Spectrometry (GC/MS). In evaluating Reaction II, the collected AC particles were air dried for 1-6 hours, transferred into a GC vial, and desorbed by 0.5 mL DCM. Then, the DCM layer was analyzed for nitrosamines by GC/MS. The main difference between Reactions I and II was the spiking of nitrite and more limited air exposure of AC particles in Reaction I.

Chemical Analysis and Quantification

Analysis of nitrosamines and amines in the samples by GC/MS followed the methods described previously (*60*, *61*). The concentrations of NDMA and other nitrosamines were quantified by isotope dilution method using NDMA-d_6 which corrected for NDMA extraction efficiency. The amount of NDMA formed was also reported in mass (in nano-moles) without correction by NDMA-d_6 when investigating the Reaction II, since the formation of NDMA was found to be mainly on AC particles during the air-drying process.

Results and Discussion

AC-Catalyzed Nitrosamine Formation via Nitrosation (Reaction I)

Effect of Nitrite and Nitrate

Experiments were conducted by exposing DMA to nitrite or nitrate in the presence or absence of AC (AqC) particles to evaluate the impact on NDMA formation. Control experiments in which only DMA was exposed to AC were also conducted for comparison. The experimental results are shown in Figure 3.

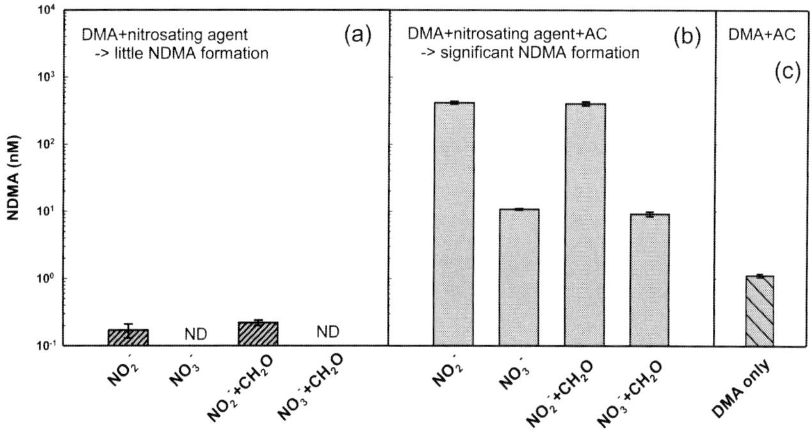

Figure 3. NDMA formation from pre-oxidation intermediates and DMA in the absence (a) and presence (b and c) of AC. Experiments were conducted at pH 7 (10 mM phosphate buffer) and 22 °C with 200 µM DMA, 200 µM nitrite or nitrate, and 200 mg AqC carbon for 7 days. Formaldehyde was added at the same concentration as nitrite or nitrate. NDMA concentration was determined by NDMA-d_6 isotope dilution. ND = non-detectable. Adapted/reproduced from reference (59). Copyright (2013, Water Research Foundation).

The results (Figure 3) clearly show that AC significantly catalyzed formation of NDMA from DMA in the presence of nitrite and nitrate. Without AC, only a very small amount of NDMA was formed from DMA and nitrite, whereas no

NDMA was detected from DMA with nitrate (Figure 3a). With AC, a significantly greater amount of NDMA was formed when DMA was present together with either nitrite or nitrate (Figure 3b). The above observations are consistent with the study by Dietrich et al. (*14*), even though a lower amine concentration and a lower nitrite concentration were employed in our study. Based on the results in Figure 3, the yield of AC-catalyzed NDMA formation from DMA via nitrosation is about 0.2% and 0.005% of the initial aqueous DMA concentration for nitrite and nitrate, respectively. Since nitrite and nitrate are both nitrosating agents commonly found in wastewaters and less pristine surface waters, the above results suggest that the AC-catalyzed nitrosation may be relevant in such waters.

Experiments were also conducted to investigate formaldehyde in combination with either nitrite or nitrate because literature has reported that formaldehyde can promote nitrosation of amines by nitrite at neutral pH conditions (*17*). However, the results showed that formaldehyde had negligible promoting effect on the NDMA formation from DMA with nitrite or nitrate whether in the presence or absence of AC (Figure 3a&b).

It should be noted that DMA alone with AC also yielded NDMA (i.e., without addition of nitrite or nitrate) (Figure 3c), but at a level considerably lower than those when nitrite or nitrate was present together with DMA (Figure 3b). The formation of NDMA presented in Figure 3c was caused by the Reaction II and will be discussed later.

Effect of Different Amines

A range of different amines including three secondary amines (DMA, pyrrolidine (PYR), and di-*n*-butylamine (DBA)) and three tertiary amines (TMA, DMBA, and NNDA) (structures shown in Figure 4) were evaluated in similar experiments as described in the previous section to assess the impact of amine structure on the AC-catalyzed nitrosamine formation via nitrosation. In Figure 5, Figures 5b and 5c show the AC-catalyzed nitrosamine formation from the secondary amines by either nitrate or nitrite, respectively via nitrosation (i.e., DMA formed NDMA, PYR formed NPYR, and DBA formed NDBA). In comparison, Figure 5a shows the catalytic effect of AC to promote nitrosamine formation from the secondary amines without addition of nitrite or nitrate. When comparing Figures 5b and 5c against Figure 5a, it is evident that AC can significantly enhance formation of nitrosamines from their corresponding secondary amines via the nitrosation pathway. The yield of nitrosamine formation was much higher with nitrite than with nitrate, consistent with expectation that nitrite is a better nitrosating reagent than nitrate. Also notably, the promoting effect was especially stronger for DBA among the three secondary amines; this may be partly due to the stronger hydrophobicity of DBA, which allows its greater adsorption to the AC surfaces and thus more catalyzed reaction.

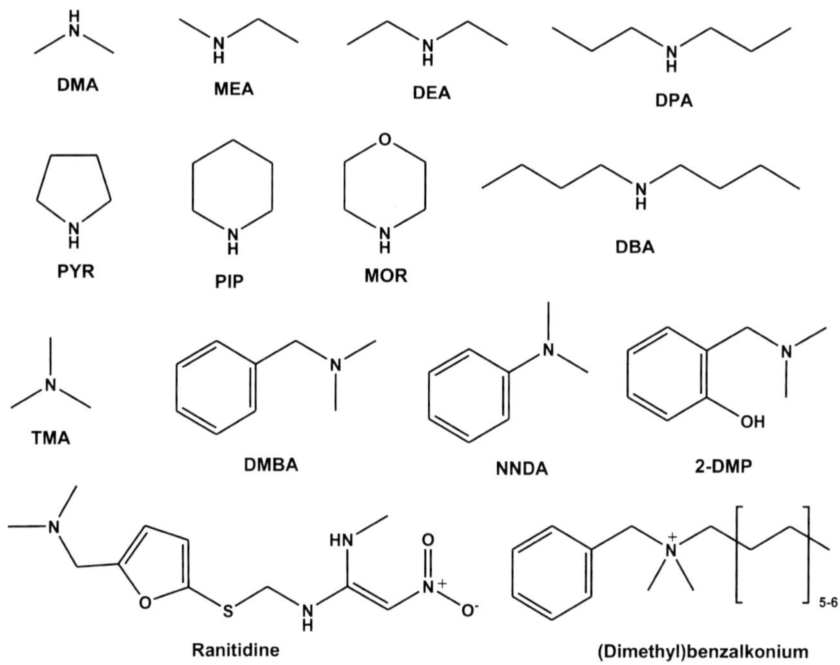

Figure 4. Secondary, tertiary, and quaternary amines investigated in this study.

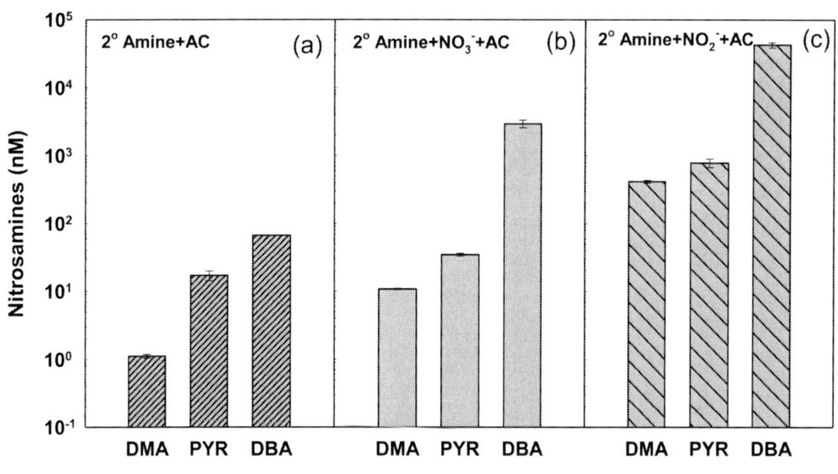

Figure 5. Nitrosamines formation from secondary amines in AC-catalyzed nitrosation. Experimental conditions were similar to those in Figure 3. NDMA concentration was determined by NDMA-d_6 isotope dilution. ND = non-detectable. Adapted/reproduced from reference (59). Copyright (2013, Water Research Foundation).

The experimental results with tertiary amines are shown in Figure 6. Comparing Figures 6b and 6c against Figure 6a, it is quite interesting to observe that AC can promote the formation of NDMA from only some tertiary amines (TMA and DMBA, but not NNDA) with the presence of nitrate and nitrite. The catalyzed transformation is likely via nitrosation pathway in which AC promotes cleavage of TMA and DMBA to DMA via nitrosative cleavage by nitrous acid (*62*), followed by nitrosation of DMA to form NDMA that can also be catalyzed by AC surfaces. Accordig to Smith and Loeppky (*62*), tertiary amines may react with aqueous nitrous acid to undergo dealkylation to form a carbonyl compound, a secondary nitrosamine, and nitrous oxide. The reaction mechanism involves formation of an N-nitrosoammonium ion and *cis* elimination of nitroxyl to form a ternary immonium ion, $R_2N^{+}{=}CR_2$, which then undergoes hydrolysis and further nitrosation (Scheme 1). Such a pathway, however, was apparently not possible with NNDA, because it lacks the α-carbon hydrogen. Moreover, the NDMA formation via the AC-catalyzed nitrosation was much greater from DMBA than from TMA; thus, the structure of tertiary amine is a critical factor for the AC-catalyzed nitrosation to nitrosamine reaction. It is also noteworthy that AC surfaces can catalyze formation of NDMA from DMBA by nitrogen fixation (i.e., without addition of nitrite or nitrate) (Figure 6a).

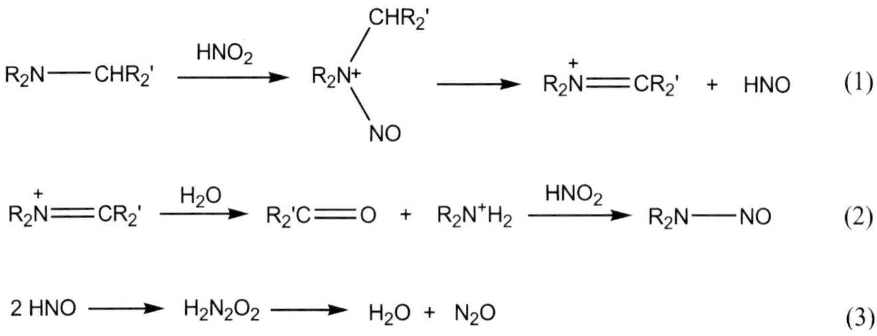

Scheme 1. Nitrosative cleavage of tertiary amines proposed by reference (62).

AC-Catalyzed Nitrosamine Formation via Nitrogen Fixation (Reaction II)

Effect of Different Carbons

The results reported by Padhye et al. (*12*) and presented above in Figures 3c, 5a, and 6a clearly demonstrate that secondary amines can transform to nitrosamines in the presence of AC particles even without the presence of nitrite or nitrate (Reaction II). Padhye et al. (*12, 13*) investigated a range of ACs commonly used in water treatment or water analysis for their effects to catalyze NDMA formation from DMA (without nitrite or nitrate) and the results are shown in Table 2. The tested ACs included: (i) synthetic Ambersorb 572 (Rohm and Haas) and Carboxen 572 (Sigma-Aldrich) particles, (ii) coconut shell-based activated carbons: PSC (Prominent Systems Inc.), AqC (Siemens Water Technologies), UCT activated carbon (UCT), and OLC (Calgon), (iii) a bituminous coal-based carbon, F400 (Calgon), (iv) a lignite coal-based carbon, HD4000, (Norit Americas), (v) three phenol formaldehyde-based activated carbon fibers (ACF10, ACF15, ACF20H) (American Kynol), and (vi) a granular activated carbon (GAC) (manufacturer unknown) from a local wastewater treatment plant. A relatively high amine concentration was employed in the experiments in order to obtain substantial differences in nitrosamine formation from different AC particles. Control experiments confirmed that there was no NDMA contamination in the DMA stocks or in carbon particles alone (*12*). Results in Table 2 show that many ACs are capable of promoting transformation of DMA to generate NDMA but the yield varies significantly, likely related to the different properties of the AC materials. Among the different ACs, the NDMA yield at pH 7.5 ranged from 0.001% to 0.01% of the initial aqueous DMA concentration, but at 0.05-0.29% of the amount of adsorbed DMA by AC (*12*).

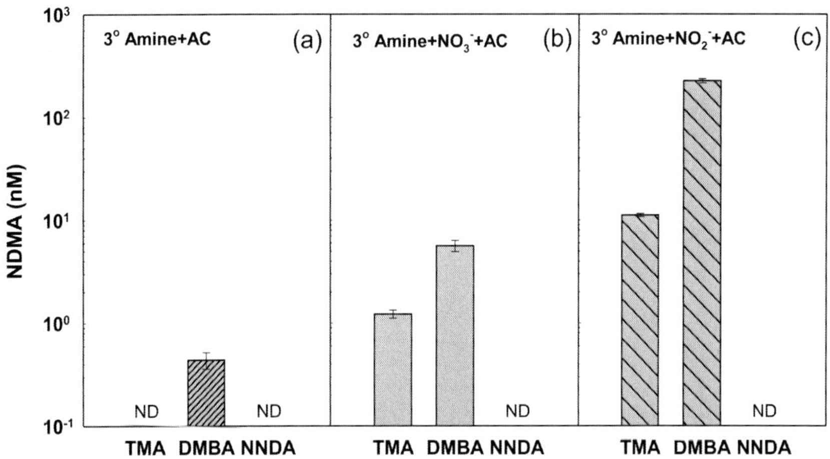

Figure 6. NDMA formation from tertiary amines in AC-catalyzed nitrosation. Experimental conditions were similar to those in Figure 3. NDMA concentration was determined by NDMA-d_6 isotope dilution. ND = non-detectable. Adapted/reproduced from reference (59). Copyright (2013, Water Research Foundation).

Table 2. NDMA Formation from DMA in the Presence of Activated Carbon Materials.[a] (Data from references (*12*, *13*)).

Carbon Particles/ Carbon Fibers	*Specific Area S_{BET} (m^2/g)*	*pH_{zpc}*	*Test Cond.*	*NDMA (nano moles)*	*NDMA (ng/L)*	*NDMA (ng/L) by NDMA-d_6 Isotope Dilution*
Ambersorb 572 AC particles	1020	7.0	a	0.45 ± 0.33	333 ± 244	972 ± 702
Siemens Aquacarb AC particles (AqC)	1202	9.6	a c	2.14 ± 0.18 1.65 ± 0.10	1584 ± 133 1221 ± 74	11337 ± 954 -
Prominent Systems AC particles (PSC)	1114	9.7	a c	1.37 ± 0.09 0.56 ± 0.07	422 ± 52 414 ± 52	4971 ± 618 -
UCT AC particles	NA	NA	a	1.68 ± 0.09	1243 ± 67	4463± 249
GAC particles	819	7.4	a	0.28 ± 0.03	207 ± 22	2316 ± 258
F400 (AC particles)	1044	9.2	b c	0.61 ± 0.08 0.54 ± 0.06	451 ± 59 400 ± 44	3773 ± 509 -
OLC (AC particles)	983	9.3	b	0.86 ± 0.02	636 ± 15	5794 ± 141
HD4000 (AC particles)	706	6.6	b	0.60 ± 0.01	444 ± 7	4459 ± 74
Carboxen 572 particles	858	7.0	c	0.07 ± 0.02	52 ± 15	-
ACF 10 (AC fibers)	972	7.3	b	0.17 ± 0.01	126 ± 7	1275 ± 73
ACF15 (AC fibers)	1520	8.9	b	0.21 ± 0.01	155 ± 7	1558 ± 71
ACF 20H (AC fibers)	1740	9.5	b	0.16 ± 0.02	118 ± 15	627 ± 80

[a] (a) 222 μM DMA, 200 mg carbon, 2 h suspension shaking time, 3 h carbon drying time; (b) 200 μM DMA, 150 mg carbon, 1.5 h suspension shaking time, 6 h carbon drying time; (c) 222 μM DMA, 150 mg carbon, 1.5 h suspension shaking time, 3 h carbon drying time. All reaction suspensions were buffered at pH 7.5 by 10 mM phosphate buffer. NA = not available; pH_{zpc}: zero point of charge.

Further research was conducted to assess how the AC's surface properties might affect the AC-catalyzed NDMA formation from DMA. It was found that there were no specific correlations between NDMA formation and the specific surface areas of ACs. On the other hand, AC's surface chemistry played a greater role. ACs treated with HNO_3 (i.e., oxidation treatment) showed decreased NDMA formation despite an increase in DMA adsorption to ACs was observed. Similarly, ozone pre-treated ACs also showed a decrease in NDMA formation in comparison to untreated ACs. In contrast, for ACs treated with H_2 or baked at high temperature (i.e., reduction treatment), an increase in NDMA formation was observed compared to untreated ACs (*12*). Further characterization of AC's surface functional groups was conducted by Padhye et al. (*13*), in which the contents of acidic (i.e., carboxyl, lactone, phenolic hydroxyl, and carbonyl) and basic (e.g., pyrone-type and chromene-type structures) oxygen-containing surface functional groups were determined. The results showed that there was a nearly linear correlation ($R^2 = 0.97$) between the carbonyl group content on the surfaces of ACs and the amount of NDMA formed (Figure 7). The higher the carbonyl group content on the AC surfaces, the higher was the NDMA formation potential. Overall, it was concluded that AC's properties play a crucial role in the catalyzed nitrosamine formation. More nitrosamine formation was associated with carbon surfaces with higher surface area, more surface defects, reduced surface properties, higher O_2 uptake capacity, and higher carbonyl group content (*13*).

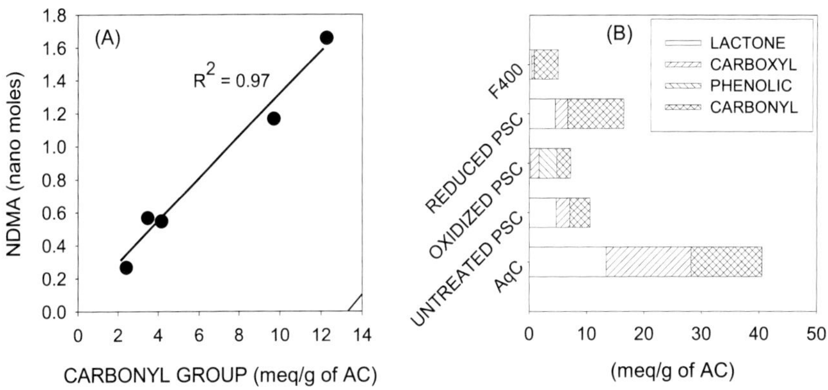

Figure 7. (A) Relationship between the amount of NDMA formation and the surface carbonyl group content of different carbons; (B) Surface acidic functional group contents of differnet carbons. Adapted/reproduced from reference (13). Copyright (2011, American Chemical Society).

Effect of Different Amines

Experiments were conducted to evaluate the impact of amine properties on the catalyzed transformation of amines to nitrosamines on the AC surfaces without the addition of nitrite or nitrate. A range of secondary amines including

DMA, methyletheyl amine (MEA), diethylamine (DEA), di-*n*-propylamine (DPA), DBA, PYR, morpholine (MOR) and piperidine (PIP) (structures shown in Figure 4) were investigated. As shown in Figure 8, ACs can enhance oxidation of all of the secondary amines to form their corresponding nitrosamines. The structural properties of amine affect how the amine molecules adsorb to the AC surfaces and the tendency of the amine to undergo oxidation to yield nitrosamine. Longer-chain and larger molecular-weight secondary amines showed a higher yield of nitrosamine formation; this is likely due to their increased adsorption on the carbon surface because of their greater hydrophobicity.

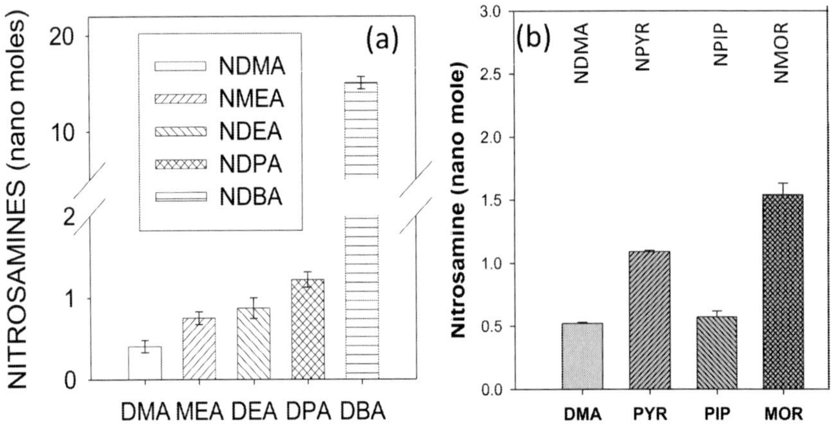

Figure 8. Formation of nitrosamines from corresponding secondary amines. Experimental conditions: 222 μM secondary amine, 200 mg PSC (a) or AqC (b) carbon, 2 h suspension shaking time and 3 h carbon drying time. Reaction pH was maintained at 7.5 by 10 mM phosphate buffer. Adapted/reproduced from reference (59). Copyright (2013, Water Research Foundation).

Several tertiary amines and one quaternary amine were also investigated for their susceptibility to the AC-catalyzed nitrosamine formation via nitrogen fixation. The amines investigated included TMA, DMBA, 2-DMP, ranitidine and (dimethyl)benzalkonium chloride (structures shown in Figure 4). These tertiary and quaternary amines were selected because they have been shown to be NDMA precursors during chlorination or chloramination. Among them, ranitidine is a popularly prescribed gastrointestinal pharmaceutical drug. Previous studies have reported a higher NDMA percentage yield from DMBA and ranitidine than from DMA during chlorination/chloramination (*63–65*). Evidently, the tendency of tertiary and quaternary amines to yield nitrosamines under chlorination or chloramination is strongly influenced by the structure of amine.

Our investigation found that ranitidine was the only tertiary amine that could be converted to NDMA by AC surfaces, while the other tertiary amines (TMA, DMBA and 2-DMP) did not transform to NDMA (Table 3). Exposing (dimethyl)benzalkonium chloride to either AqC or F400 carbon also did not generate any NDMA (data not shown). On the basis of the same molar concentration of amine precursors, the AC-catalyzed NDMA formation from ranitidine was about two times more than that from DMA for AqC (Table 3). Evidently, our study shows that not all the tertiary amines that are shown as precursors of NDMA during chlorination/chloramination can be transformed to yield NDMA on the surfaces of ACs.

Table 3. Formation of NDMA from Tertiary Amines vs. from DMA on ACs.[a] Adapted/reproduced from reference (*59*). Copyright (2013, Water Research Foundation).

Chemical	*AC*	*NDMA (nano moles)*	*NDMA (ng/L)*	*AC*	*NDMA (nano moles)*	*NDMA (ng/L)*
DMA	AqC	1.10 ± 0.13	812 ± 93	F400	0.77 ± 0.07	568 ± 49
TMA	AqC	ND	ND	F400	ND	ND
DMBA	AqC	ND	ND	F400	ND	ND
2-DMP	AqC	ND	ND	F400	ND	ND
Ranitidine	AqC	2.58 ± 0.24	1908 ± 178	F400	3.25 ± 0.15	2401 ± 114

[a] Note: Experiments were conducted using particle method in 100 mL suspension at pH 7.5 with 10 mM phosphate buffer, 222 μM amine, 200 mg activated carbon, 2 h shaking time, and 3 h particle drying time; Mean ± standard deviation (n = 2); ND = not detected.

Reaction Mechanisms

Few studies have thus far investigated the AC-catalyzed nitrosation of amine with nitrite (or nitrate) to yield nitrosamine (Reaction I) except for our study and the earlier work by Dietrich et al. (*14*). It may be hypothesized that AC surfaces catalyze the reaction by concentrating the reactants (amine precursor and nitrite) on surface active sites, as well as promoting nitrosation via surface reactive functional groups. AC's surface functional groups may provide locally acidic conditions that facilitate formation of the more favorable nitrosyl cation or dinitrogen trioxide nitrosating species from nitrite. Furthermore, the experimental results, showing that some tertiary amines were susceptible to the AC-catalyzed nitrosation in the presence of nitrite to transform into nitrosamine (Figure 6), indicate that cleavage of a N-C bond occurred during the reaction. Previously, Smith and Loeppky (*62*) suggested transformation of tertiary amine to secondary amine via nitrosative cleavage by nitrous acid. More research is needed to elucidate the kinetics and mechanisms involved in the AC-catalyzed nitrosation of amine to yield nitrosamine.

For the AC-catalyzed nitrosamine formation from amine without the presence of nitrite (or nitrate) (Reaction II), Padhye et al. (*13*) devised a range of experimental methods and techniques to probe the reaction mechanism. First, the authors conducted experiments to confirm that trace amounts of background nitrite (if present at all) in the AC could not contribute to the amount of nitrosamine observed. Spiking ^{15}N-labeled nitrite at the level as high as 7 times of the method detection limit (70 nM) for nitrite to the reaction did not yield appreciable levels of ^{15}N-labeled NDMA. Note that in the investigation for Reaction I described above, different reaction conditions with a higher nitrite concentration and longer reaction time were employed. Second, experiments conducted using ^{15}N-labeled DMA did not yield regular NDMA or ^{15}N,^{15}N-NDMA but only ^{15}N-NDMA, indicating that DMA is responsible for yielding only one nitrogen in NDMA formation, and the source of second nitrogen of the nitroso group of NDMA is either atmospheric, dissolved, or through carbon. Third, experiments showed that the AC-catalyzed nitrosamine formation from amine without the presence of nitrite required exposure to both molecular oxygen and nitrogen (gases or dissolved). When the reaction was restricted in co-exposure to molecular oxygen and nitrogen, catalyzed NDMA formation was significantly decreased. The above findings strongly suggested that, under Reaction II, atmospheric and dissolved nitrogen contributes to one nitrogen atom for NDMA formation from DMA.

The involvement of molecular nitrogen in the AC-catalyzed transformation of DMA to NDMA implies that AC's active surface sites are capable of reacting with molecular nitrogen to generate reactive nitrogen species (RNS), posssibly with involvement of reactive oxygen species (ROS) generated from oxygen on AC surfaces. The ability of AC surfaces to react with oxygen to generate ROS has been well recognized and is related to surface defects (*66*). ROS generated or carbene centers created by reactions on carbon surfaces could be responsible for nitrogen fixation by reacting with dissolved N_2 to form RNS like N_2O or diazo intermediates (*67*, *68*). Reductive transformation promoted by AC surfaces may convert N_2O to yield NH-containing species such as H_2N_2O, NH_2OH and ammonia (*69*). Using x-ray photoelectron spectroscopy (XPS), an increase in nitrogen peak (~400 eV) for AC particles that were exposed to DI water, compared to AC particles that were not in contact with water, was observed, confirming the occurrence of "nitrogen fixation" on the AC surfaces (*13*). Furthermore, experiments conducted with addition of *tert*-butanol (TBA), a hydroxyl radical scavenger, yielded less NDMA, confirming the importance of radical species. Experiments also confirmed that the presence of intermediates such as dinitrogen oxide (N_2O) and hydroxylamine (NH_2OH) significantly increased the AC-catalyzed nitrosamine formation via nitrogen fixation (*13*).

Overall, Figure 9 depicts the proposed reaction mechanism for the AC-catalyzed nitrosamine formation via nitrogen fixation (Reaction II) (*13*). Reactive sites on the AC surfaces react with molecular oxygen to form ROS, particularly hydroxyl radicals. The ROS generated promote fixation of molecular nitrogen to form RNS such as N_2O, H_2N_2O and NH_2OH. The subsequent reaction of NH_2OH with DMA can be catalyzed by AC's surface sites or ROS to yield NDMA. For carbon surface that is more reduced and with more defects and available surface sites, formation of ROS is more favorable, so is the formation of N_2O and conversion of N_2O to NH_2OH, and thus more NDMA can be formed. AC's properties play a crucial role as more nitrosamine formation is associated with carbon surfaces with higher surface area, more surface defects, reduced surface properties, higher O_2 uptake capacity, and higher carbonyl group content. Although demonstrated only for NDMA in Figure 8, catalyzed formation of other

N-nitrosamines from corresponding secondary amines on AC surfaces is expected to follow the same reaction mechanism.

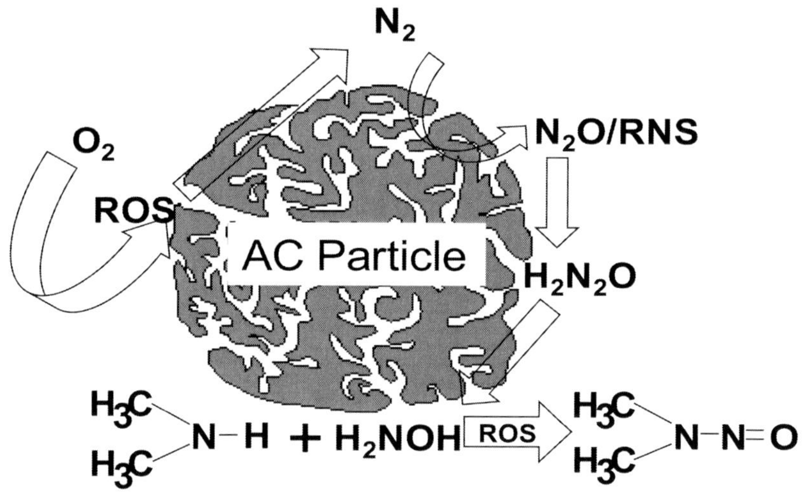

Figure 9. Proposed mechanism for formation of N-nitrosamine from secondary amine on the surface of AC particles. ROS = reactive oxygen species; RNS = reactive nitrogen species. Adapted/reproduced from reference (13). Copyright (2011, American Chemical Society.

Conclusions and Environmental Significance

The phenomena that AC materials are capable of catalyzing transformation of amine to yield nitrosamine have been discovered by recent studies. The AC-catalyzed nitrosamine formation from amine can be categorized into two types: (I) in the presence of nitrosating reagents such as nitrite or nitrate; and (II) without the presence of nitrite or nitrate. Reaction I may be similar to the classical nitrosation of amine by nitrite but with catalysis by AC surface sites (i.e., AC-catalyzed nitrosamine formation via nitrosation). In contrast, Reaction II has been identified to be a radical-involved "nitrogen fixation" mechanism in which AC's surface sites react with molecular oxygen, molecular nitrogen, and amine to generate nitrosamine (i.e., AC-catalyzed nitrosamine via nitrogen fixation). Many commercial AC materials were found to exhibit the above catalytic influence, but the nitrosamine yield varied significantly. The catalyzed nitrosamine formation is significantly influenced by AC properties, amine properties, and environmental conditions. All secondary amines and select tertiary amines tested have been shown to be susceptible to the above AC-catalyzed nitrosamine formation.

Since co-existence of AC and amine precursors occurs in many environmental and industrial applications, the AC-catalyzed nitrosamine formation is highly relevant, and a better understanding of such phenomena will help minimize undesirable nitrosamine formation. Results of this study are clearly helpful to avoid nitrosamine analytical errors when AC is applied to analyze amine-containing samples such as wastewater or wastewater-impacted surface water. For example, selection of ACs with fewer attributes that catalyze nitrosamine formation (i.e., higher surface area, more surface defects, reduced surface properties, higher O_2 uptake capacity, and higher carbonyl group content) or minimization of co-exposure to molecular oxygen and nitrogen should be exercised to minimize analytical artifact. More research is needed to assess the impact of carbon properties and reaction conditions on Reaction I, thus allowing development of strategies to minimize AC-catalyzed nitrosamine formation in systems, such as wastewater, where significant amounts of amine precursors and nitrosating species (nitrite or nitrate) are both present. Lastly, research in this topic may have far-reaching implications in many other AC-involved processes and should be extended beyond water and wastewater analysis/treatment conditions to derive more insight and applications.

Acknowledgments

This study was funded by a grant from the Water Research Foundation (Project 4343). The comments and views detailed herein may not necessarily reflect the views of the Water Research Foundation, its officers, directors, affiliates, or agents. L.P.P. thanks financial support from the Amirtharajah Fellowship. Y.L.W. thanks financial support from the Taiwan government scholarship program.

References

1. Richardson, S. D. *Anal. Chem.* **2007**, *79*, 4295–4323.
2. Mitch, W. A.; Sharp, J. O.; Trussell, R. R.; Valentine, R. L.; Alvarez-Cohen, L.; Sedlak, D. L. *Environ. Eng. Sci.* **2003**, *20*, 389–404.
3. California Department of Public Health. California Drinking Water: NDMA-related activities. http://www.cdph.ca.gov/certlic/drinkingwater/Pages/NDMA.aspx (accessed March 31, 2012).
4. Charrois, J. W. A.; Boyd, J. M.; Froese, K. L.; Hrudey, S. E. *J. Environ. Eng. Sci.* **2007**, *6*, 103–114.
5. USEPA. United States Environmental Protection Agency IRIS database. http://www.epa.gov/ncea/iris (accessed 07/09/2013).
6. USEPA. United States Environmental Protection Agency unregulated contaminant monitoring regulation for public water systems. http://www.epa.gov/fedrgstr/EPA-WATER/2007/January/Day-04/w22123.htm (accessed 07/09/2013).
7. USEPA. United States Environmental Protection Agency Contaminant Candidate List 3. http://water.epa.gov/scitech/drinkingwater/dws/ccl/ccl3.cfm (accessed 07/09/2013).
8. Krauss, M.; Hollender, J. *Anal. Chem.* **2008**, *80*, 834–842.
9. Russell, C. G.; Blute, N. K.; Via, S.; Wu, X. Y.; Chowdhury, Z. *J. Am. Water Works Assoc.* **2012**, *104*, 57–58.
10. Marsh, H.; Rodríguez-Reinoso, F. *Activated Carbon*; Elsevier: Amsterdam, Boston, 2006.

11. The Innovation Group. *Activated Carbon, Chemical Market Reporter*; Schnell Publishing Company: New York, NY, 2002.
12. Padhye, L.; Wang, P.; Karanfil, T.; Huang, C.-H. *Environ. Sci. Technol.* **2010**, *44*, 4161–4168.
13. Padhye, L. P.; Hertzberg, B.; Yushin, G.; Huang, C.-H. *Environ. Sci. Technol.* **2011**, *45*, 8368–8376.
14. Dietrich, A. M.; Gallagher, D. L.; DeRosa, P. M.; Millington, D. S.; DiGlano, F. A. *Environ. Sci. Technol.* **1986**, *20*, 1050–1055.
15. Mirvish, S. S. *Toxicol. Appl. Pharmacol.* **1975**, *31*, 325–351.
16. Lee, C.; Yoon, J. *J. Photochem. Photobiol., A: Chem.* **2007**, *189*, 128–134.
17. Keefer, L. K.; Roller, P. P. *Science* **1973**, *181*, 1245–1247.
18. Weerasooriya, S. V. R.; Dissanayake, C. B. *Toxicol. Environ. Chem.* **1989**, *25*, 57–62.
19. Choi, J. H.; Valentine, R. L. *Environ. Sci. Technol.* **2003**, *37*, 4871–4876.
20. Choi, J.; Valentine, R. L. *Water Res.* **2002**, *36*, 817–824.
21. Mitch, W. A.; Sedlak, D. L. *Environ. Sci. Technol.* **2002**, *36*, 588–595.
22. Schreiber, I. M.; Mitch, W. A. *Environ. Sci. Technol.* **2006**, *40*, 6007–6014.
23. Andrzejewski, P.; Kasprzyk-Hordern, B.; Nawrocki, J. *Desalination* **2005**, *176*, 37–45.
24. Andrzejewski, P.; Kasprzyk-Hordern, B.; Nawrocki, J. *Water Res.* **2008**, *42*, 863–870.
25. Schmidt, C. K.; Brauch, H.-J. *Environ. Sci. Technol.* **2008**, *42*, 6340–6346.
26. Oya, M.; Kosaka, K.; Asami, M.; Kunikane, S. *Chemosphere* **2008**, *73*, 1724–1730.
27. Andrzejewski, P.; Nawrocki, J. *Water Sci. Technol.* **2007**, *56*, 125–131.
28. Park, S. H.; Piyachaturawat, P.; Taylor, A. E.; Huang, C.-H. *Water Sci. Technol.: Water Supply* **2009**, *9*, 279–288.
29. Padhye, L. P.; Luzinova, Y.; Cho, M.; Mizaikoff, B.; Kim, J. H.; Huang, C.-H. *Environ. Sci. Technol.* **2011**, *45*, 4353–4359.
30. Yang, L.; Chen, Z. L.; Shen, J. M.; Xu, Z. Z.; Liang, H.; Tian, J. Y.; Ben, Y.; Zhai, X.; Shi, W. X.; Li, G. B. *Environ. Sci. Technol.* **2009**, *43*, 5481–5487.
31. Mitch, W. A.; Gerecke, A. C.; Sedlak, D. L. *Water Res.* **2003**, *37*, 3733–3741.
32. Chen, Z.; Valentine, R. L. *Environ. Sci. Technol.* **2007**, *41*, 6059–6065.
33. Mitch, W. A.; Sedlak, D. L. *Environ. Sci. Technol.* **2004**, *38*, 1445–1454.
34. Padhye, L.; Tezel, U.; Mitch, W. A.; Pavlostathis, S. G.; Huang, C.-H. *Environ. Sci. Technol.* **2009**, *43*, 3087–3093.
35. Padhye, L.; Kim, J.-H.; Huang, C.-H. *Water Res.* **2012**, *47*, 725–736.
36. Lee, C.; Schmidt, C.; Yoon, J.; von Gunten, U. *Environ. Sci. Technol.* **2007**, *41*, 2056–2063.
37. Park, S. H.; Wei, S.; Mizaikoff, B.; Taylor, A. E.; Favero, C.; Huang, C.-H. *Environ. Sci. Technol.* **2009**, *43*, 1360–1366.
38. Voudrias, E. A.; Larson, R. A.; Snoeyink, V. L. *Carbon* **1987**, *25*, 503–515.
39. Snoeyink, V. L. In *Activated Carbon for Water Treatment*; Sontheimer, H., Crittenden, J. C., Summers, R. S., Eds.; DVGW-Forschungsstelle: 1988.
40. Dabrowski, A.; Podkoscielny, P.; Hubicki, Z.; Barczak, M. *Chemosphere* **2005**, *58*, 1049–1070.
41. Abotsi, G. M. K.; Scaroni, A. W. *Fuel Proc. Technol.* **1989**, *22*, 107–133.

42. Radovic, L. R.; RodriguezReinoso, F. *Chem. Phys. Carbon* **1997**, *25*, 243–358.
43. Stuber, F.; Font, J.; Fortuny, A.; Bengoa, C.; Eftaxias, A.; Fabregat, A. *Top. Catal.* **2005**, *33*, 3–50.
44. Gomes, H. T.; Machado, B. F.; Ribeiro, A.; Moreira, I.; Rosario, M.; Silva, A. M. T.; Figueiredo, J. L.; Faria, J. L. *J. Hazard. Mater.* **2008**, *159*, 420–426.
45. Rodriguez-Reinoso, F. *Carbon* **1998**, *36*, 159–175.
46. Stuber, F.; Font, J.; Fortuny, A.; Bengoa, C.; Eftaxias, A.; Fabregat, A. *Top. Catal.* **2005**, *33*, 3–50.
47. Serp, P.; Figueiredo, J. L. s. *Carbon Materials for Catalysis*; John Wiley & Sons: Hoboken, NJ, 2009.
48. Szymanski, G. S.; Grzybek, T.; Papp, H. *Catal. Today* **2004**, *90*, 51–59.
49. Lisovskii, A. E.; Aharoni, C. *Catal. Rev.: Sci. Eng.* **1994**, *36*, 25–74.
50. Petrosius, S. C.; Drago, R. S.; Young, V.; Grunewald, G. C. *J. Am. Chem. Soc.* **1993**, *115*, 6131–6137.
51. Szymanski, G. S.; Rychlicki, G. *Carbon* **1991**, *29*, 489–498.
52. Garten, V. A.; Weiss, D. E. *Austr. J. Chem.* **1957**, *10*, 309–328.
53. Bente, P. F.; Walton, J. H. *J. Phys. Chem.* **1943**, *47*, 133–148.
54. Firth, J. B.; Watson, F. S. *J. Chem. Soc.* **1923**, *123*, 1219–1222.
55. Mattson, J. S.; Mark, H. B. *Activated Carbon: Surface Chemistry and Adsorption from Solution*; M. Dekker: New York, 1971.
56. Bansal, R. C.; Goyal, M. *Activated Carbon Adsorption*; Taylor & Francis: Boca Raton, 2005.
57. Pereira, M. F. R.; Orfao, J. J. M.; Figueiredo, J. L. *Appl. Catal., A* **2001**, *218*, 307–318.
58. Aguilar, C.; Garcia, R.; Soto-Garrido, G.; Arraigada, R. *Top. Catal.* **2005**, *33*, 201–206.
59. Huang, C.-H. *Impact of Activated Carbon on Nitrosamine Formation*; Draft Report; Water Research Foundation: 2013.
60. Padhye, L.; Tezel, U.; Mitch, W. A.; Pavlostathis, S. G.; Huang, C.-H. *Environ. Sci. Technol.* **2009**, *43*, 3087–3093.
61. Park, S.-H.; Wei, S.; Mizaikoff, B.; Taylor, A. E.; Favero, C.; Huang, C.-H. *Environ. Sci. Technol.* **2009**, *43*, 1360–1366.
62. Smith, P. A. S.; Loeppky, R. N. *J. Am. Chem. Soc.* **1967**, *89*, 1147–1157.
63. Kemper Jerome, M.; Walse Spencer, S.; Mitch William, A. *Environ. Sci. Technol.* **2010**, *44*, 1224–31.
64. Le Roux, J.; Gallard, H.; Croué, J.-P. *Water Res.* **2011**, *45*, 3164–3174.
65. Le Roux, J.; Gallard, H.; Croué, J.-P. *Environ. Sci. Technol.* **2012**.
66. Boehm, H. P. *Carbon* **1994**, *32*, 759–69.
67. Pichugina, D. A.; Ibragimova, R. I.; Emel'yanova, N. S.; Koryagina, N. L.; Shestakov, A. F.; Vorob'ev-Desyatovskii, N. V. *Russ. J. Gen. Chem.* **2008**, *78*, 557–566.
68. Vorob'ev-Desyatovskii, N. V.; Ibragimova, R. I.; Gordeev, S. K.; Nikolaev, B. P. *Russ. J. Gen. Chem.* **2006**, *76*, 946–954.
69. Shestakov, A. F.; Shilov, A. E. *Russ. Chem. Bull.* **2001**, *50*, 2054–2059.

Chapter 6

Insights into Interactions of Propranolol with Nano TiO_2

Jingjing Du and Chuanyong Jing*

State Key Laboratory of Environmental Chemistry and Ecotoxicology, Research Center for Eco-Environmental Sciences, Chinese Academy of Sciences, Beijing 100085, China

***Tel.: +86 10 6284 9523; Fax: +86 10 6284 9523; E-mail: cyjing@rcees.ac.cn**

Beta blockers in the environment are acknowledged to cause health risk towards humans. Insights from molecular-level mechanisms of propranolol adsorption can further our understanding of the fate and transport of beta blockers in the environment. The motivation of our study is to explore the dynamic adsorption process of propranolol at the TiO_2/aqueous interface on the molecular scale. Complementary techniques including macroscopic adsorption experiments, UV analysis, surface enhanced Raman scattering (SERS), flow-cell ATR-FTIR measurement, XPS, and quantum chemical calculations were used to study the interaction between propranolol and TiO_2. The results show that propranolol adsorption on TiO_2 increased from 0.3 to 2.3 μmol/g in the pH range 5 to 9. The ATR-FTIR and XPS analysis indicated that the hydroxyl and amino groups of propranolol strongly interacted with the TiO_2 surface. The DFT calculations suggested the molecular structure of surface complexes with hydrogen bonding and the charge transfer from propranolol to TiO_2 surface upon adsorption. The TiO_2 doped nanomaterial shows high performance SERS ability together with photocatalytic property towards propranolol.

Introduction

The presence of pharmaceuticals in the environment has been repeatedly reported for years (*1–3*). Pharmaceuticals can cause adverse effects due to their biological potency toward ecosystem and humans. Conventional urban wastewater treatment plants are not able to completely degrade these chemicals. Therefore, these compounds are considered as emerging contaminants, and has been the target of many recent studies (*4–6*).

Among pharmaceuticals, propranolol is the first successful beta blocker developed to treat high blood pressure and cardiovascular diseases (*7*). The ubiquitous occurrence of propranolol in aquatic environment (*8*) and potential harm to organisms (*9–11*) motivate extensive study in its adsorption behaviors (*8, 12–14*). Propranolol exists as a positively charged species in natural environments with a pK_a of 9.53 (Figure 1), and behaves like a cation at the aquatic-mineral interface. Previous study reports that hydrophobic, electrostatic, and chemical specific interactions play important roles in the uptake of propranolol by sediment/soil (*12*). However, different propranolol surface configurations are proposed. The uncharged surface complexes are suggested in the adsorption of three beta blockers including propranolol on iron-containing natural geosorbents (*13*). In contrast, positively charged surface species are assumed to dominate the adsorption of nine beta blockers on two sediments (*8*). The difference in explaining macroscopic adsorption behaviors highlights the need to investigate the adsorption mechanism of propranolol on the molecular level.

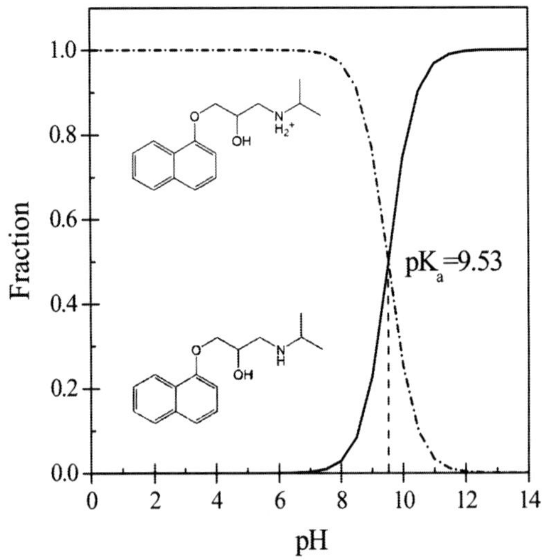

Figure 1. Structure and property of propranolol.

TiO_2 is one of the most investigated and understood metal oxides (*15, 16*). Compared with other metal oxides in sediments, TiO_2 is chemically more stable and has a large number of active sorption sites due to its nanometer-scale particle size. Because of its long-term thermodynamic stability and strong oxidizing capability, TiO_2 has been considered as one of the best photocatalytic materials. Adsorption and photodegradation mechanism of organic molecules on TiO_2 has been investigated using spectroscopic studies and theoretical calculations (*17–19*). Spectroscopic measurements are sensitive to surface reactions, and complementary quantum chemical calculations using periodical Density Functional Theory (DFT) can assist in interpreting the interactions between adsorbate and surface (*18*).

The photo-degradation of propranolol in the presence of TiO_2 is of great importance in assessing its toxicity and environmental fate (*4*). Most previous photocatalytic reactions were conducted using TiO_2 nanoparticles in suspension (*6*, *20*). Therefore, the separation of TiO_2 nanoparticles from the solution is a major bottleneck limiting the application of the photocatalytic processes in water treatment (*21*, *22*). Separation of TiO_2 from the solution is important not only from the regenerable perspective, but also to avoid the adverse ecological effects of the semiconductor TiO_2 (*21*). The need for facile separation of nanopartilces motivated us to fabricate a recycable nanomaterial taking advantage of magnatic Fe_3O_4, Ag, and TiO_2 for propranolol monitoring and degradation.

In the first section of this chapter, we investigated the adsorption mechanism of propranolol on TiO_2 on the molecule level. Multiple complementary techniques including batch adsorption experiments, UV analysis, Raman spectroscopy, *in situ* flow-cell attenuated total reflectance Fourier transform infrared (ATR-FTIR) measurements, X-ray photoelectron spectroscopy (XPS), .surface complexation modeling, and periodic DFT calculations were used to gain insights into the adsorption process. In the second part, the photodegradation of propranolol in the presence of TiO_2 was studied using batch tests and spectroscopic analysis. We synthesized a TiO_2 loaded magnetic nanomaterial that can both capture and photocatalytically degradate propranolol in water. The functionalized nanoparticles can be magnetically separated from the dispersion and hence be regenerated and reused. The results should further our understanding of the fate and transport of propranolol and similar pharmaceuticals on the molecular level in aquatic systems.

Results and Discussion

The Adsorption Behavior of Propranolol on TiO_2 Surface

Adsorption Kinetics

Adsorption experiments were first performed to assess the kinetics. Figure 2 indicated that the rate of propranolol adsorption on TiO_2 was rapid during the initial 4 h and the propranolol concentration kept relatively constant thereafter. A pseudo-second-order kinetics model was used in this study as follows:

$$t/q_t = t/q_e + 1/k \cdot q_e^2$$

where q_t and q_e are the amounts (μmol/g) of propranolol adsorbed at time t and at equilibrium respectively, and k (g/μmol·h) is the rate constant of pseudo-second-order kinetics. The model parameters as inserted in Figure 2 were obtained by linear regression of the integrated rate equation. Kinetics following the pseudo-second-order model has been widely used to describe organics adsorption on metal oxides (*23*), indicating the adsorption of propranolol on TiO_2 might be a chemisorption process.

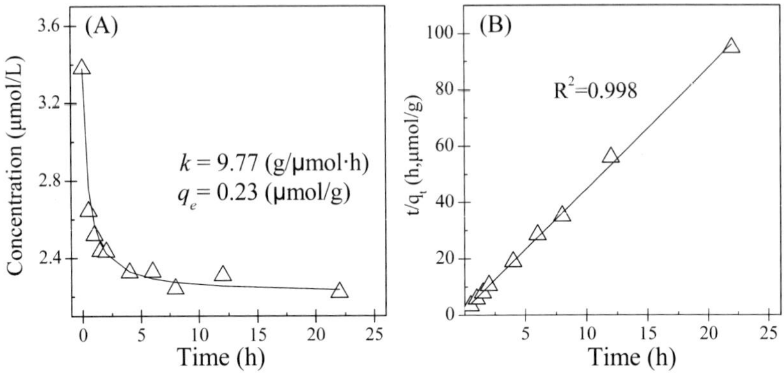

Figure 2. (A) Adsorption kinetics of 3.38e-6 M (1 mg/L) propranolol on 5 g/L TiO_2 at pH 7 in 0.01 M NaCl solution. The adsorption followed pseudo-second-order kinetics (B), and the k= 9.77 (g/µmol·h), q_e= 0.23 (µmol/g).

Adsorption kinetics of propranolol was faster on minerals than on natural sediments. The uptake of propranolol by Dausenau sediment sample took 6 h to reach apparent equilibrium (*8*). Yang et al. (*5*) reported the adsorption of propranolol on TiO_2 reached equilibrium in about 20 min, which was much faster than that in our study. The difference may be ascribed to the substantially high propranolol concentrations used in Yang's research (100 µM) compared with this study (3.38 µM). The high concentration of propranolol may enable its rapid access to TiO_2 surface and resulted in a short equilibration time.

Batch Adsorption Experiments

The adsorption of propranolol on TiO_2 was highly pH-dependent in the pH range 4 to 11 (Figure 3). The increased adsorption as a function of pH can be attributed to the pH-dependent change of propranolol speciation (Figure 1) and TiO_2 surface charge. At pH < pH_{pzc} (5.8), both the TiO_2 surface and propranolol were positively charged, and the electrostatic repulsion resulted in a low adsorption capacity. At pH_{pzc} < pH < pKa (9.53), the TiO_2 surface was negatively charged and propranolol was positively charged. The electrostatic attraction led to an enhanced adsorption affinity. At pH > pKa, propranolol exists mainly in an uncharged species, resulting in a reduced adsorption. Therefore, electrostatic interactions play an important role in the adsorption of propranolol on TiO_2.

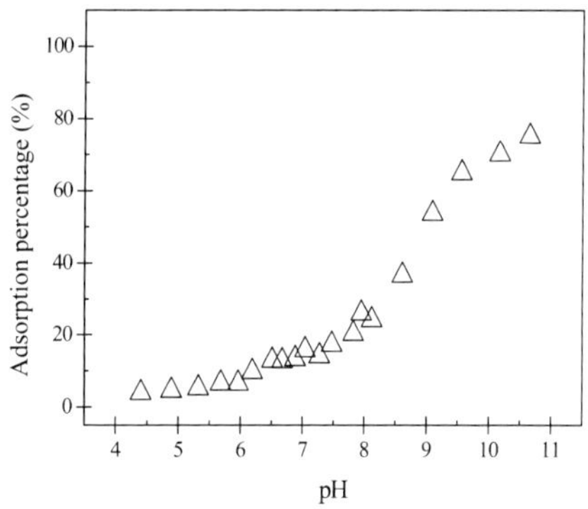

Figure 3. Experimental results of 2.03e-5 M (6 mg/L) propranolol adsorption on 5 g/L TiO_2 as a function of pH in 0.01 M NaCl solution.

However, electrostatic interactions alone cannot explain the observed adsorption in the pH range 4 to 5.8, where both surface and propranolol were positively charged, and at pH > 9.53, where propranolol mainly exist as a neutral form. Interactions other than electrostatic force would be involved in propranolol adsorption on TiO_2.

Adsorption isotherms of propranolol on TiO_2 at pH 5, 7, and 9 in the presence of 0.01 and 0.1 M NaCl followed Freundlich model (Figure 4):

$$q = K_d * C^{1/n}$$

where q (μmol/g) is the amount of adsorption, C (μmol/L) is the aqueous propranolol concentration, K_d is the distribution coefficient, and $1/n$ is a correction factor, the deviation of which from 1 indicates the extent of nonlinearity.

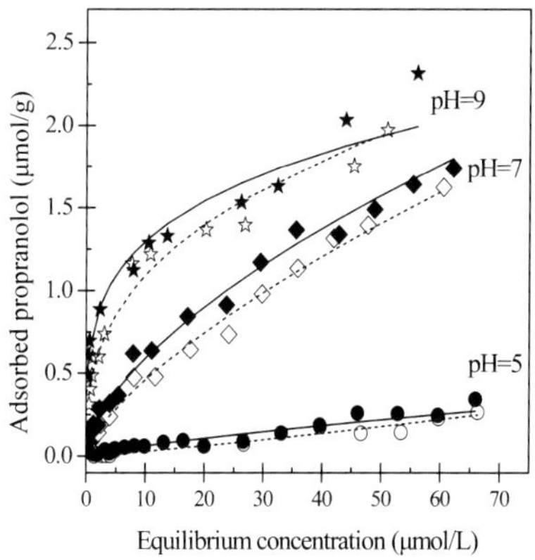

Figure 4. Adsorption isotherms of propranolol on 5 g/L TiO_2 at pH 5 (○), pH 7 (◇), and pH 9 (☆) in 0.01 M (closed symbols) and 0.1 M NaCl (open symbols) solution. Solid (dashed) lines are Freundlich models in the presence of 0.01 (0.1) M NaCl.

As shown in Figure 4, the maximum adsorption capacity was a function of pH, which was consistent with the pH edge results. The $1/n$ values were in the range 0.25 to 1.17, indicating that the adsorption of propranolol was nonlinear. The nonlinearity in adsorption isotherms may be partly attributed to heterogeneous adsorption sites and sorbate-sorbate interactions including electrostatic repulsion (*24*). In this study, the repulsive electrostatic interactions among cationic propranolol may exist due to its high surface loading (0.07 molecule/nm^2). Moreover, the large molecular size of propranolol (about 1.4 nm) may lead to the steric exclusion or obstruction. In addition, ionic strength had a negligible effect on the adsorption of propranolol (Figure 4), suggesting the involved adsorption driving force was stronger than electrostatic interaction. Generally, electrostatic interaction is significantly influenced by ionic strength, whereas inner-sphere surface complexes should not vary as a function of ionic strength (*25*).

The Spectroscopic Investigations of Propranolol on TiO_2 Surface

UV-Vis Absorption Spectroscopy

UV-vis absorption spectra of TiO_2 film and propranolol/TiO_2 are presented in the range of 200-600 nm (Figure 5-A). The adsorption of propranolol on TiO_2 increased the UV-vis absorbance compared with the virgin TiO_2. Subtracting the spectrum of TiO_2 film from the spectrum of propranolol/TiO_2, the difference absorption spectrum was obtained (Figure 5-B). The absorption peaks of propranolol were at 215 and 290 nm. The incorporation of propranolol on TiO_2 film showed two pronounced peaks at 213 and 237 nm, while the latter peak was shoulder-like in the spectrum of propranolol. In the range of 200-400 nm, the absorption of propranolol is attributed to the π-π^* electron transition of the aromatic moiety (*26*). Therefore, the change observed in the spectra indicated that TiO_2 has interacted with propranolol and affected the electron transition of naphthalene ring. Similar results have been obtained by Mizutani (*27*).

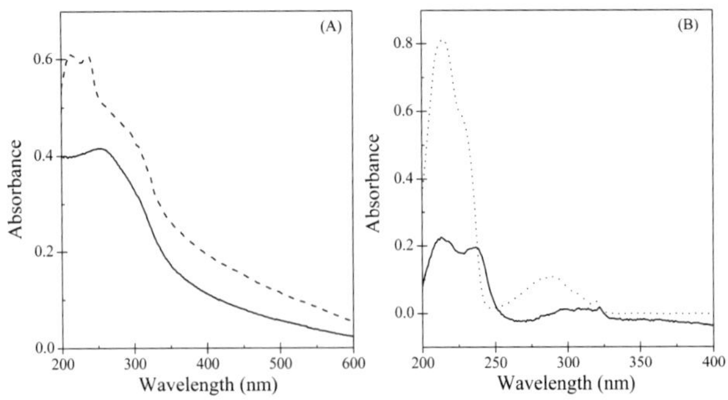

Figure 5. (A) UV-vis absorption spectra of TiO_2 film (solid line) and propranolol/TiO_2 (dashed line); (B) Difference absorption spectra of propranolol on TiO_2 film (solid line), the UV-vis absorption spectra of propranolol in water (dotted line).

The Raman spectrum was acquired on propranolol hydrochloride powder as shown in Figure 6, and the assignments were summarized in Table I.

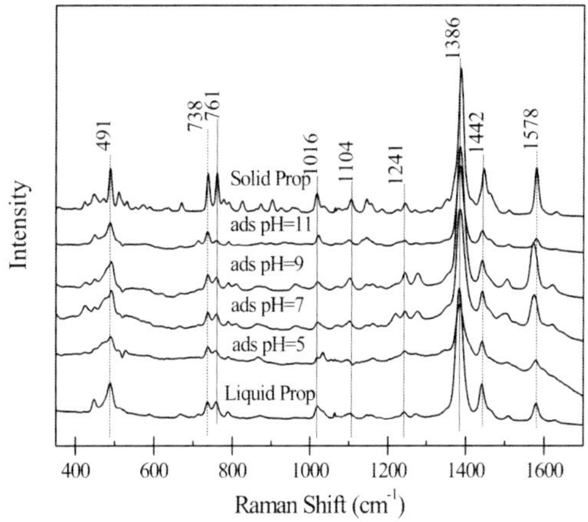

Figure 6. Raman spectra of solid and aqueous propranolol, and adsorbed propranolol at pH 5, 7, 9, and 11.

Table I. Assignments of Raman spectra for propranolol.[a]

Raman shift (cm^{-1})	*Assignments*
491	Symmetric longitudinal stretching of the naphthalene ring
738	Naphthalene ring breathing
761	γ(NH) + δ(CN) + tυ(CC)
1016	γ(CC) +δ(CH)
1104	υ(CO) + δ(CCC) + tδ(CH)
1241	δ(CH) + δ(OH)
1386	υ(CC) (naphthalene) + δ(CH)
1442	δ(CH) (naphthalene)
1578	υ(CC) + δ(CH) (naphthalene)

[a] γ: out-of-plane bending, δ: bending, υ: stretching

Intense Raman bands can be observed at 738, 1386, 1442, and 1578 cm^{-1}, all these bands are attributed to different modes of naphthalene ring vibrations as shown in Table I (*28*, *29*). The Raman spectra of propranolol doped TiO_2 film in the 5-11 pH range are also plotted in Figure 6. The samples exhibit the main spectral features of propranolol, indicating the succeccful adsoption of propranolol on TiO_2 surface. Moreover, as shown in Figure 6, the peak position of Raman shifts for propranolol were not affected by the adsorption, suggesting the weak interactions between propranolol and TiO_2.

It should be noted that the Raman response recorded in pH 5-9 increased as a function of pH. In the 5-9 pH range only the protonated propranolol molecules are adsorbed on the TiO_2 surface. The hydrogen bonding between the $-NH_2^+$-group of propranolol and the TiO_2 surface would increase from pH 5 to 9 due to the decreasing positive charge of TiO_2 surface (pH_{pzc}=5.8). For the pH value at 11, the weak Raman response can be attributed to the existence of propranolol in its neutral form. The molecules are less easily adsorbed on TiO_2 surface and consequently weak spectral signatures was obtained.

ATR-FTIR Analysis

To further study the interactions of propranolol with TiO_2 on the molecule level, *in situ* flow-cell ATR-FTIR spectra were collected at pH 5, 7, and 9 as a function of time and the results are shown in Figure 7. The peak positions and the assignments are listed in Table II.

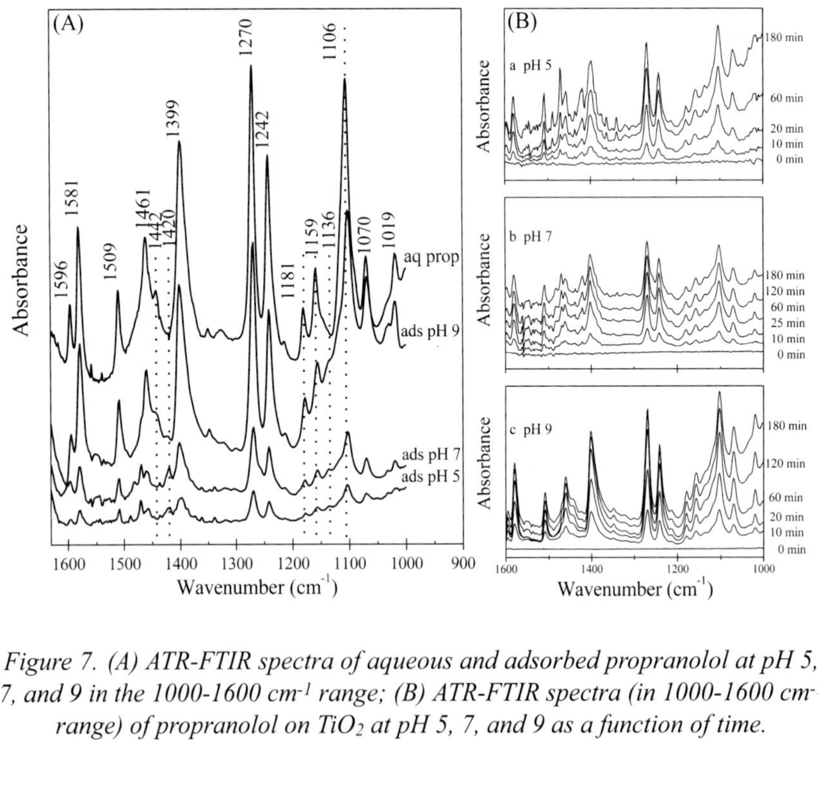

Figure 7. (A) ATR-FTIR spectra of aqueous and adsorbed propranolol at pH 5, 7, and 9 in the 1000-1600 cm-1 range; (B) ATR-FTIR spectra (in 1000-1600 cm-1 range) of propranolol on TiO_2 at pH 5, 7, and 9 as a function of time.

Table II. ATR-FTIR peak position of standard propranolol and the assignment of main peaks.[a]

wavenumber (cm^{-1})	*assignment*
1019	ν(C=C)
1070	ν(C-O)
1106	ν(C-O)
1159	ν(C-O)
1181	ν(C-N)
1242	ν(C-O)
1270	ν(C-O)
1399	δ(C-H)
1442	δ(O-H)
1461	δ(C-H)ring
1509	ν(C=C)
1581	ν(C=C)
1596	ν(C=C)

[a] ν: stretching, δ: bending

The peak intensity increased as a function of time and pH, while the shape of the spectra did not change. The spectrum of adsorbed propranolol at pH 9 exhibited a strong similarity with that of soluble propranolol (Figure 7-A). Meanwhile, peak shifts were observed for C-N and C-O bands. The C-N stretching band at 1181 cm^{-1} in propranolol solution (*30*, *31*) was shifted to 1178 cm^{-1} upon adsorption. The magnitudes of the shift were considerably lower than that in the formation of inner-sphere surface complex (*32*). This peak shift may be attributed to the hydrogen bonding between the $-NH_2^+-$ group in propranolol and the negative TiO_2 surface sites at pH 9. The C-O stretching vibrations at about 1159, 1106, and 1147 cm^{-1} in propranolol (*33*) were shifted to 1156, 1102, and 1136 cm^{-1}, respectively, upon adsorption. Corresponding to the C-O band shift, the sharpness of the O-H bending vibration (*34*) at 1442 cm^{-1} was decreased. The changes in C-O stretching and O-H bending vibrations were attributed to the hydrogen bonding of hydroxyl group in propranolol on TiO_2 surfaces (*32*, *35*). The changes in peak positions upon adsorption at pH 9 were also observed at pH 5 and 7 during the whole process of adsorption (Figure 5). The O-H bending vibration at 1442 cm^{-1} in propranolol standard was shifted to 1420 cm^{-1} at pH 5 and 7. Similar to pH 9, the bands at 1159, 1147, and 1106 cm^{-1} due to C-O stretching were shifted to 1156, 1136, and 1102 cm^{-1}, respectively. The shift of O-H bending and C-O stretching vibrations indicated a strong hydrogen bonding between the -OH of propranolol and TiO_2 surface. In addition, the C-N stretching band shift was observed at pH 5 and 7, confirming the hydrogen bonding between the $-NH_2^+-$ group and the surface.

The FTIR results indicated that the hydroxyl and amino group of propranolol interacted with TiO_2 surface at pH 5 to 9. Meanwhile, a pronounced shift in O-H bending vibration suggested that interactions between hydroxyl group of propranolol and the surface may be the dominant adsorption mechanism at pH 5 to 7. Moreover, the peak shift due to hydrogen bonding between hydroxyl group of propranolol and the TiO_2 surface was about 10-20 cm^{-1}, suggesting strong interactions comparable with inner-sphere surface complexes. In fact, hydrogen bond is partly covalent (*36*) and could exhibit covalent bond characters (*37*, *38*). Besides, the amino group and hydroxyl group of propranolol interacted with TiO_2 surface simultaneously, which enhanced the affinity of propranolol to TiO_2. The ATR-FTIR results indicated the interaction of $-NH_2^+-$ and -OH of propranolol with the TiO_2 surface, and detailed information regarding chemical and electronic state of N, O, and Ti at the interface merit the XPS study.

XPS Analysis

The binding energies of Ti $2p_{1/2}$ and Ti $2p_{3/2}$ for pristine TiO_2 were observed at 464.4 and 458.7 eV, respectively. No striking shift in the Ti peaks was detected upon adsorption at pH 6.5 and pH 10, respectively (Figure 8A). The spin-orbit splitting (SOS) of 5.7 eV between Ti $2p_{1/2}$ and $2p_{3/2}$ was in good agreement with previous study (*39*).

The peak shifts in the O 1s spectra implied that the hydroxyl group of propranolol was interacted with the surface (Figure 8B). The peak at 533.2 eV and 532.3 eV (Figure 8B-a, Table III) was attributed to the O atom bonding to naphthalene ring (*40*) and the O atom in hydroxyl group (*41*) of propranolol, respectively. Two O 1s peaks were observed for TiO_2 (Figure 8B-b) where the peak at 529.6 eV was attributed to lattice O atom (O^{2-}) and the peak at 531.4 eV was assigned to the O atom in surface hydroxyl group (*42*). Upon adsorption at pH 6.5 and 10 (Figure 8B-c, d), the peak for the O associating naphthalene ring (533.2 eV) showed no change, whereas the O peak for the hydroxyl group in propranolol decreased from 532.3 eV to about 531.7 eV (pH 6.5) and 531.6 eV (pH 10). The decrease in binding energy suggested that the O atom in the hydroxyl group of propranolol accepted electrons upon adsorption on TiO_2.

Analysis of the N 1s spectra supported the contribution of $-NH_2^+-$ group to the propranolol adsorption. The N 1s peak of propranolol was observed at 401.7 eV (Figure 8C-a), which was consistent with the reported value of $-NH_2^+-$ (*43*). Upon adsorption at pH 6.5 and 10, the binding energy of N 1s decreased to 400.2 and 399.7 eV, respectively (Figure 8C-b, c). The decrease in the N 1s binding energy indicated that the nitrogen atom of propranolol accepted electrons and led to a less electron deficient environment upon adsorption. This binding energy change was attributed to the hydrogen bonding between the amino group of propranolol and TiO_2 surface, in agreement with our ATR-FTIR results.

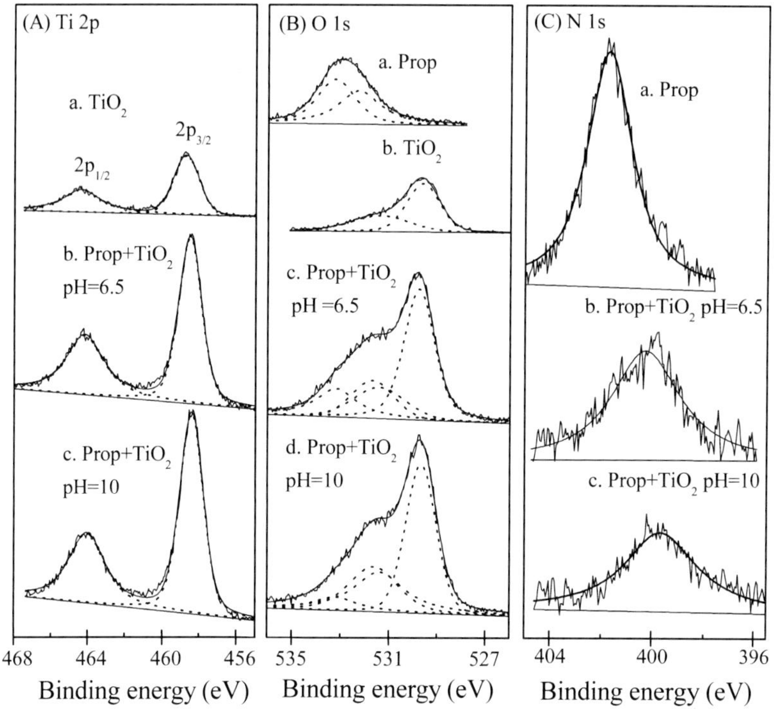

Figure 8. XPS spectra of Ti 2p (A), O 1s (B), and N 1s (C) for propranolol, TiO_2, and propranolol on TiO_2 at pH 6.5 and 10.

Table III. Ti 2p, O 1s, and N 1s binding energy (eV) obtained from XPS analysis.

	Ti 2p		*O 1s*				*N 1s*
	$2p_{1/2}$	$2p_{3/2}$	*C-O-C*	*C-OH*	*OH*	O^{2-}	NH_2^+
prop[a]			533.2	532.3			401.7
TiO_2	464.4	458.7			531.4	529.6	
Ads[b] (pH 6.5)	464.2	458.5	533.2	531.7	531.4	529.8	400.2
Ads (pH 10)	464.1	458.4	533.2	531.6	531.4	529.7	399.7

[a] propranolol; [b] adsorbed propranolol on TiO_2

Quantum Chemical Calculations

The formation of propranolol-TiO_2 complex was confirmed using DFT calculations (*44*). Figure 9 presents the geometric optimized structure of propranolol on TiO_2 surface. The hydrogen end of amino and hydroxyl group in propranolol molecule was oriented to the bridging oxygen atoms on the TiO_2 surface. The distance between the hydrogen atom in hydroxyl group and the surface oxygen atom was 1.6 Å, which is smaller than the sum of the van der Waals radius of hydrogen and oxygen atoms (2.6 Å), indicating the formation of hydrogen bond. The distance between the hydrogen atom in amino group and the surface oxygen atom was 2.2 Å, suggesting a weak hydrogen-bond interaction between amino group of propranolol and the TiO_2 surface. This adsorption model was in agreement with our ATR-FTIR results.

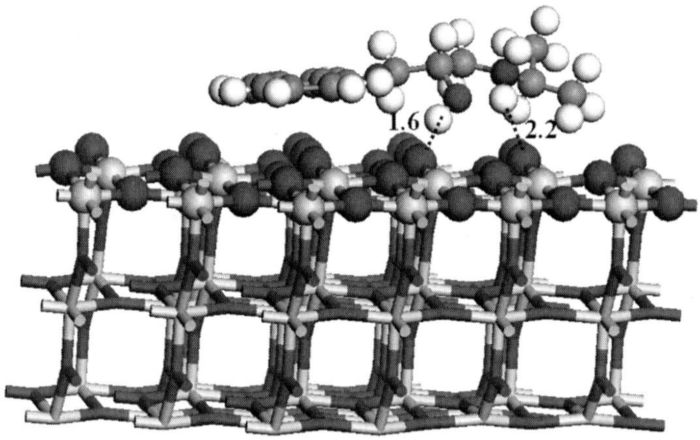

Figure 9. Geometric optimized structure of propranolol adsorption on TiO_2 (101) surfaces. Atoms are presented as dark grey (carbon, C), white (H), red (O), blue (N), and light grey (Ti). Distance in Å.

The calculated E_{ads} was -3.767 eV for the adsorption of propranolol on TiO_2 surface. This large value of adsorption energy indicated strong interactions between adsorbent and adsorbate. Notably, the calculated E_{ads} was much larger than the sum of two hydrogen bonding energy (4-15 kcal/mol) (*37*), which may be explained from two aspects. First, the simplified model without considering solvation effect between adsorbent and adsorbate could lead to high E_{ads}. Water molecules may inhibit propranolol adsorption by trapping propranolol or capping the adsorption sites (*45*). Second, the slab established in the model may not be large enough to eliminate potential interactions between the propranolol molecules, although the size of the slab (16.734 Å×15.191 Å×29.775 Å) was comparable to reported values (*16*, *45*, *46*). Nevertheless, the results indicated that the hydrogen bonding of propranolol on the TiO_2 surface is energetically favorable and a stable inner-sphere surface complex could be formed.

The interaction between propranolol and TiO_2 induced the change in electron density of the surface and the adsorbate. In the electron density difference map (Figure 10), the yellow and blue represents the charge accumulation and depletion, respectively. The charge accumulation occurred between the hydroxyl group of propranolol and the bridging-O (O1) atom of TiO_2. An increased electron density was also observed between the amino group and bridging-O (O_2) atom. The charge depletion occurred near the H atoms in hydroxyl and amino group bound to the surface. Moreover, the O atom in hydroxyl group and the N atom in amino group were in a state of electron density accumulation, which was in accordance with the decreased binding energy of the O 1s and N 1s in XPS results (Figure 8 and Table III). Therefore, charge transfer possibly occurred from surface oxygen atoms to the hydroxyl and amino group in propranolol. Interestingly, charge transfer between the naphthyl group and TiO_2 surface was expected (Figure 10) where the naphthyl group acted as an electron donor and TiO_2 surface as an electron acceptor. The UV-vis absorption spectra (Figure 5) showed the change in peak intensity of propranolol upon adsorption, indicating that the TiO_2 has interacted with propranolol and induced the electron transition of naphthalene ring. However, the ATR-FTIR spectra did not exhibit relative changes and we considered the interaction between the naphthyl group and TiO_2 surface play a minor role in the adsorption of propranolol.

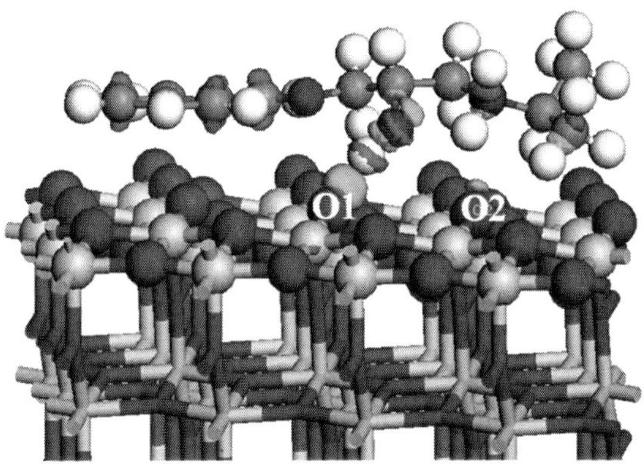

Figure 10. Electron density difference for propranolol on TiO_2 (101) surface with the isovalues of $\Delta\rho$ =+0.02 e (in purple) and -0.02 e (in green). Green represents charge accumulation, and purple represents charge depletion.

The density of states (DOS) for propranolol adsorption on the TiO_2 (101) surface was calculated to characterize the interactions. Figure 8 shows the partial density of states (PDOS) of adsorbed propranolol and TiO_2 (101) surface. The atomic orbitals of hydroxyl group (Figure 11-B) of adsorbed propranolol overlapped with the orbitals of the bridging-O (O1) atom (Figure 11-A). This overlap demonstrated the formation of the bond between the hydroxyl group in adsorbed propranolol and TiO_2 surface. Similarly, overlaps between the atomic orbitals of the amino group (Figure 11-D) of adsorbed propranolol and that of the bridging-O (O_2) atom of the TiO_2 surface (Figure 11-C) were observed with a less extent than that of hydroxyl group and surface O atom orbitals. The analysis of PDOS was consistent with the result of electron density difference and demonstrated the specific interactions between propranolol and TiO_2.

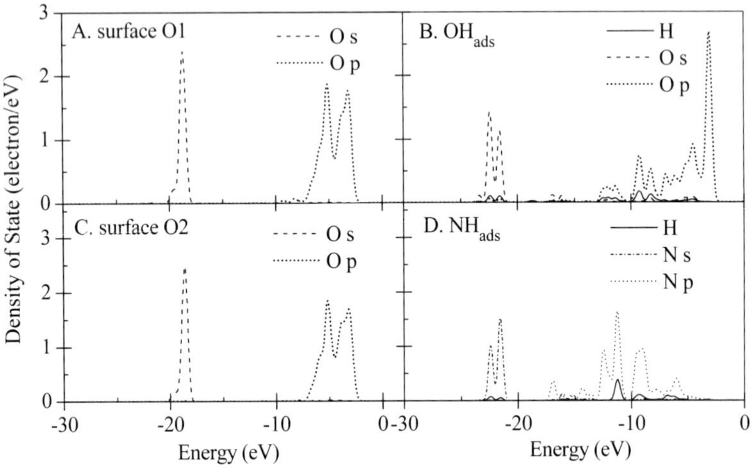

Figure 11. Partial density of states of propranolol on TiO_2 (101) surfaces. (A) Surface O1, (B) hydroxyl group of propranolol, (C) surface O2, (D) amino group of propranolol. Surface O1 and O2 atoms are marked in Figure 7.

Photocatalytic Degradation of Propranolol in the Presence of TiO_2

In this section, we developed a recycable nanomaterial with Fe_3O_4, Ag, and TiO_2 constituents, which can be used as an active photocatalyst for propranolol degradation. The interaction of propranolol with Ag and TiO_2 allows its SERS analysis, concomitantly providing molecular information of degradation process.

Characterization

The EDX analysis served as important evidence for the existence of Ag and Ti on the Fe_3O_4 surface (Figure 12-A). The structure of the $Fe_3O_4/TiO_2/Ag$ nanoparticles was further inspected by the XRD patterns. As shown in Figure 12-B, all the prominent peaks at 2θ values of 30.2°, 35.69°, 43.29°, 53.68°, 57.18°, and 62.79° agree well with the JCPDS data of Fe_3O_4 (PDF, file No.19-0629). Peaks located at 38.09°, 44.23°, 64.5°, 77.6° and 81.6° representing the 111, 200, 220, 311 and 222 Bragg's reflections of metallic silver (*47*). From XRD patterns it is also inferred that anatase is doped on the substrate surface, since the nanocomposites exhibited the characteristic reflections of anatase at 25.3°, 47.6°, and 54.8° (*48*). According to several previous reports (*6, 49*), anatase is more active than rutile, which may be ascribed to the higher aptitude of anatase to photoadsorb oxygen due to higher density of superficial hydroxyl groups, and the lower relative electron/hole recombination rate in anatase.

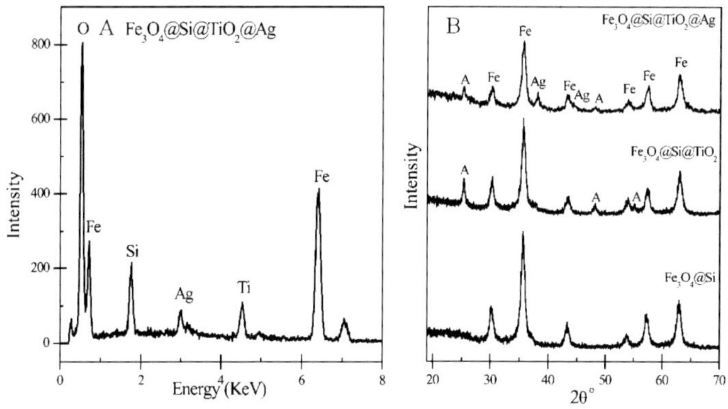

Figure 12. (A) EDX spectra of $Fe_3O_4/SiO_2/TiO_2/Ag$ NPs; (B) XRD survey of (a) Fe_3O_4/SiO_2; (b) $Fe_3O_4/SiO_2/Ag$; (c) $Fe_3O_4/SiO_2/TiO_2$; (d) $Fe_3O_4/SiO_2/TiO_2/Ag$ NPs.

Photocatalytic Kinetics

Generally, the degradation of propranolol takes place through two parallel processes: photolysis and photocatalysis. As shown in the degradation pathways (Figure 13), the photo catalytic processes can decompose propranolol to CO_2, and H_2O (*4*, *5*, *20*). The evolution of the corresponding C/C_0 with exposure time is presented in Figure 14. Under 2 h UV irradiation, about 40% propranolol was degradated. In the presence of 0.25 g/L of $Fe_3O_4/TiO_2/Ag$ as the catalyst, the degradation efficiency of propranolol reach 100% within 2 h.

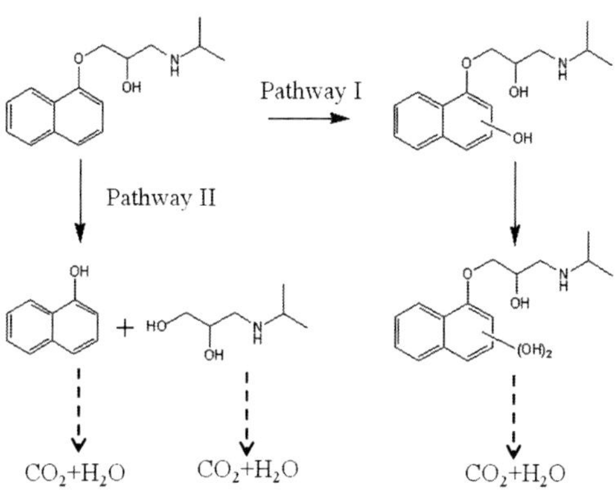

Figure 13. Scheme of propranolol degradation pathways.

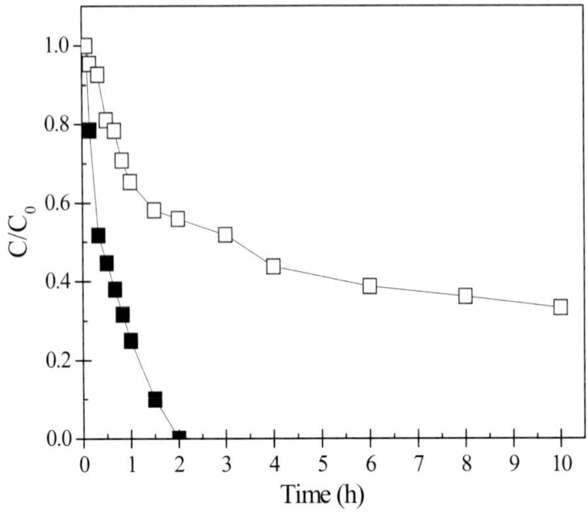

Figure 14. Evolution of propranolol in the presence (▪) and absence (▫) of catalyst during UV irradiation (propranolol 3 mg/L, $Fe_3O_4/TiO_2/Ag$ 0.25 g/L).

SERS Activity

SERS shows surface selectivity for certain functional groups, and it could be useful for identification of β-blocker drugs (*28*, *50*). In our study, the recycable photocatalyst with Fe_3O_4, Ag, and anatase constituents can also be used as active SERS substrate for propranolol recognition. The SERS spectrum of propranolol on $Fe_3O_4/TiO_2/Ag$ substrate shows several strong peaks at 448, 491, 738, 760, 1016, 1386, 1442, and 1578 cm^{-1} (Figure 15), which are characteristic Raman bands of propranolol (*28*). The peak at 570 cm^{-1} was attributed to the T_{2g} mode of Fe_3O_4, which is consistent with our previous study (*51*). The peak intensity was pronounced in the presence of Ag film on Fe_3O_4. The intense bands can be observed at 1386 and 1442 cm^{-1}, and were clearly observable at 10^{-6} M concentration level. The small change between the Raman and SERS shifts indicated the weak interaction of propranolol with the substrate.

Degradation of Propranolol on $Fe_3O_4/TiO_2/Ag$ Surface

Under the UV irradiation, TiO_2 is excited and generate highly active oxidative species on TiO_2 or Ag surfaces due to the charge-transfer from TiO_2 to Ag. These active oxidative species can lead to the decomposition of propranolol adsorbed on the surfaces of $Fe_3O_4/TiO_2/Ag$ (*52*). The photocatalytic property of $Fe_3O_4/TiO_2/Ag$ was evaluated by monitoring the degradation process under the UV irradiation. As shown in Figure 16-A, propranolol (5×10^{-4} M) can be photodegradated within 60 min. To further justify the catalytic mechenism, a reference material that composed of Fe_3O_4/Ag nanoparticles was prepared.

This reference material did not show any photocatalytic activity under the same experimental condition (Figure 16-B). The results demonstrated that the $Fe_3O_4/TiO_2/Ag$ can be used as a recyclable SERS substrate and a renewable photocatalyst.

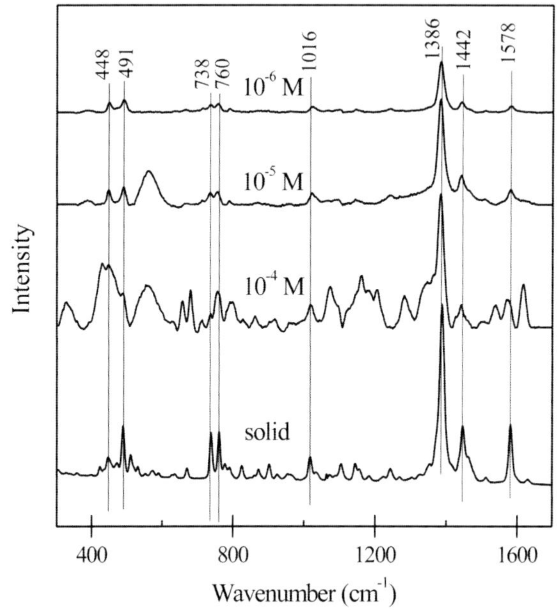

Figure 15. SERS spectra of propranolol on $Fe_3O_4/TiO_2/Ag$ surface.

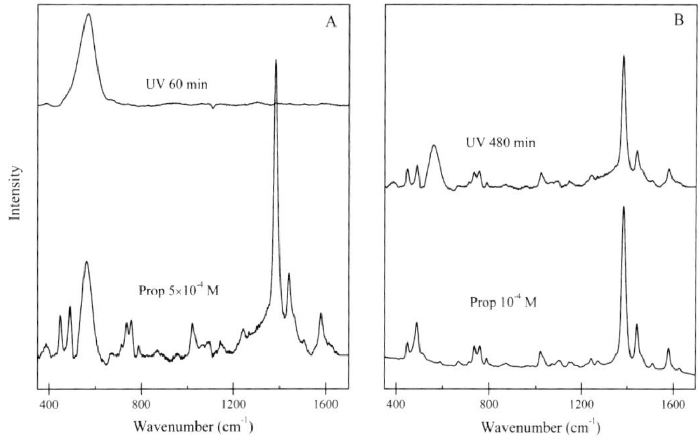

Figure 16. (A) Photodegradation of propranolol in the presence of $Fe_3O_4/TiO_2/Ag$, the concentration of propranolol was 5×10^{-4} M; (B) Photodegradation of propranolol in the presence of Fe_3O_4/Ag, concentration of propranolol was 10^{-4} M.

Conclusions

In this study, the adsorption capacity of propranolol on TiO_2 increases from 0.3 to 2.3 μmol/g in the pH range 5 to 9. The DFT calculations agree well with our ATR-FTIR and XPS analysis and suggest the hydroxyl and amino groups of adsorbed propranolol strongly interact with the TiO_2 surface. This strong affinity can slow the migration of propranolol and reduce the potential for exposure. The interaction mechanisms at water-solid interfaces for propranolol may be safely extrapolated to other pharmaceuticals with same functional groups. The degradation process of propranolol is monitored by SERS analysis. The unique structure enabled the Fe_3O_4/ TiO_2/Ag to serve as an easily recyclable and highly efficient SERS substrate. This work should encourage the development of synthetic approaches to other novel nanocomposite materials with dual functions. The aquatic-mineral interface model established in this study can improve our understanding of the fate and risk assessment of propranolol and analogous compounds.

References

1. Halling-Sorensen, B.; Nielsen, S. N.; Lanzky, P. F.; Ingerslev, F.; Lutzhoft, H. C. H.; Jorgensen, S. E. *Chemosphere* **1998**, *36*, 357–394.
2. Daughton, C. G.; Ternes, T. A. *Environ. Health Perspect.* **1999**, *107*, 907–938.
3. Lee, J.; Hong, S.; Mackeyev, Y.; Lee, C.; Chung, E.; Wilson, L. J.; Kim, J. H.; Alvarez, P. J. J. *Environ. Sci. Technol.* **2011**, *45*, 10598–10604.
4. Santiago-Morales, J.; Aguera, A.; Gomez, M. D.; Fernandez-Alba, A. R.; Gimenez, J.; Esplugas, S.; Rosal, R. *Appl. Catal., B* **2013**, *129*, 13–29.
5. Yang, H.; An, T. C.; Li, G. Y.; Song, W. H.; Cooper, W. J.; Luo, H. Y.; Guo, X. D. *J. Hazard. Mater.* **2010**, *179*, 834–839.
6. Ioannou, L. A.; Hapeshi, E.; Vasquez, M. I.; Mantzavinos, D.; Fatta-Kassinos, D. *Sol. Energy* **2011**, *85*, 1915–1926.
7. Stapleton, M. P. *Texas Heart Inst. J.* **1997**, *24*, 336–342.
8. Ramil, M.; El Aref, T.; Fink, G.; Scheurer, M.; Ternes, T. A. *Environ. Sci. Technol.* **2010**, *44*, 962–970.
9. Franzellitti, S.; Buratti, S.; Valbonesi, P.; Capuzzo, A.; Fabbri, E. *Aquat. Toxicol.* **2011**, *101*, 299–308.
10. Giltrow, E.; Eccles, P. D.; Winter, M. J.; McCormack, P. J.; Rand-Weaver, M.; Hutchinson, T. H.; Sumpter, J. P. *Aquat. Toxicol.* **2009**, *95*, 195–202.
11. Huggett, D. B.; Brooks, B. W.; Peterson, B.; Foran, C. M.; Schlenk, D. *Arch. Environ. Contam. Toxicol.* **2002**, *43*, 229–235.
12. Yamamoto, H.; Nakamura, Y.; Moriguchi, S.; Nakamura, Y.; Honda, Y.; Tamura, I.; Hirata, Y.; Hayashi, A.; Sekizawa, J. *Water Res.* **2009**, *43*, 351–362.
13. Kibbey, T. C. G.; Paruchuri, R.; Sabatini, D. A.; Chen, L. X. *Environ. Sci. Technol.* **2007**, *41*, 5349–5356.

14. Lin, A. Y. C.; Lin, C. A.; Tung, H. H.; Chary, N. S. *J. Hazard. Mater.* **2010**, *183*, 242–250.
15. Mudunkotuwa, I. A.; Grassian, V. H. *J. Am. Chem. Soc.* **2010**, *132*, 14986–14994.
16. Zhao, Z. Y.; Li, Z. S.; Zou, Z. G. *Phys. Lett. A* **2011**, *375*, 2939–2945.
17. Li, C.; Monti, S.; Carravetta, V. *J. Phys. Chem. C* **2012**, *116*, 18318–18326.
18. Almeida, A. R.; Calatayud, M.; Tielens, F.; Moulijn, J. A.; Mul, G. *J. Phys. Chem. C* **2011**, *115*, 14164–14172.
19. Li, H. Q.; Hong, W. S.; Cui, Y. M.; Fan, S. H.; Zhu, L. J. *J. Alloys Compd.* **2013**, *569*, 45–51.
20. De la Cruz, N.; Dantas, R. F.; Gimenez, J.; Esplugas, S. *Appl. Catal., B* **2013**, *130*, 249–256.
21. Chalasani, R.; Vasudevan, S. *ACS Nano* **2013**, *7*, 4093–4104.
22. Jing, J.; Li, J.; Feng, J.; Li, W.; Yu, W. W. *Chem. Eng. J* **2013**, *219*, 355–360.
23. Tanis, E.; Hanna, K.; Emmanuel, E. *Colloids Surf., A* **2008**, *327*, 57–63.
24. Higgins, C. P.; Luthy, R. G. *Environ. Sci. Technol.* **2006**, *40*, 7251–7256.
25. O'day, P. A. *Rev. Geophys.* **1999**, *37*, 249–274.
26. Govindarajan, M.; Karabacak, M. *Spectrochim. Acta, Part A* **2012**, *85*, 251–260.
27. Mizutani, N.; Yang, D.-H.; Selyanchyn, R.; Korposh, S.; Lee, S.-W.; Kunitake, T. *Anal. Chim. Acta* **2011**, *694*, 142–150.
28. Stiufiuc, R.; Iacovita, C.; Lucaciu, C. M.; Stiufiuc, G.; Nicoara, R.; Oltean, M.; Chis, V.; Bodoki, E. *J. Mol. Struct.* **2013**, *1031*, 201–206.
29. Ruperez, A.; Laserna, J. J. *Anal. Chim. Acta* **1996**, *335*, 87–94.
30. Castro, R. A. E.; Canotilho, J.; Nunes, S. C. C.; Eusebio, M. E. S.; Redinha, J. S. *Spectrochim. Acta, Part A* **2009**, *72*, 819–826.
31. Canotilho, J.; Castro, R. A. E. *Spectrochim. Acta, Part A* **2010**, *76*, 395–400.
32. Johnson, S. B.; Yoon, T. H.; Kocar, B. D.; Brown, G. E. *Langmuir* **2004**, *20*, 4996–5006.
33. Farooqi, Z. H.; AboulEnein, H. Y. *Biospectroscopy* **1996**, *2*, 131–141.
34. Sidiras, D.; Batzias, F.; Schroeder, E.; Ranjan, R.; Tsapatsis, M. *Chem. Eng. J.* **2011**, *171*, 883–896.
35. Morrison, W. H. *J. Colloid Interface Sci.* **1984**, *100*, 121–127.
36. Isaacs, E. D.; Shukla, A.; Platzman, P. M.; Hamann, D. R.; Barbiellini, B.; Tulk, C. A. *Phys. Rev. Lett.* **1999**, *83*, 4445–4445.
37. Parthasarathi, R.; Subramanian, V.; Sathyamurthy, N. *J. Phys. Chem. A* **2006**, *110*, 3349–3351.
38. Emsley, J. *Chem. Soc. Rev.* **1980**, *9*, 91–124.
39. Fierro, J. L. G.; Arrua, L. A.; Nieto, J. M. L.; Kremenic, G. *Appl. Catal.* **1988**, *37*, 323–338.
40. delaPuente, G.; Pis, J. J.; Menendez, J. A.; Grange, P. *J. Anal. Appl. Pyrolysis* **1997**, *43*, 125–138.
41. Gleason, N.; Guevremont, J.; Zaera, F. *J. Phys. Chem. B* **2003**, *107*, 11133–11141.
42. Simmons, G. W.; Beard, B. C. *J. Phys. Chem.* **1987**, *91*, 1143–1148.
43. Li, P.; Lin, J. Y.; Tan, K. L.; Lee, J. Y. *Electrochim. Acta* **1997**, *42*, 605–615.

44. Ye, C.; Hu, S.; Yan, W.; Duan, J.; Jing, C. *J. Phys. Chem. C* **2013**, *117*, 5785–5791.
45. Guo, Y. N.; Lu, X.; Zhang, H. P.; Weng, J.; Watari, F.; Leng, Y. *J. Phys. Chem. C* **2011**, *115*, 18572–18581.
46. Ojamae, L.; Aulin, C.; Pedersen, H.; Kall, P. O. *J. Colloid Interface Sci.* **2006**, *296*, 71–78.
47. Abdulla-Al-Mamun, M.; Kusumoto, Y.; Islam, M. S. *J. Mater. Chem.* **2012**, *22*, 5460–5469.
48. Papoulis, D.; Komarneni, S.; Panagiotaras, D.; Stathatos, E.; Toli, D.; Christoforidis, K. C.; Fernandez-Garcia, M.; Li, H. H.; Yin, S.; Sato, T.; Katsuki, H. *Appl. Catal., B* **2013**, *132*, 416–422.
49. Carp, O.; Huisman, C. L.; Reller, A. *Prog. Solid State Chem.* **2004**, *32*, 33–177.
50. Levene, C.; Correa, E.; Blanch, E. W.; Goodacre, R. *Anal. Chem.* **2012**, *84*, 7899–7905.
51. Du, J.; Jing, C. *J. Colloid Interface Sci.* **2011**, *358*, 54–61.
52. Zhou, Y.; Chen, J.; Zhang, L.; Yang, L. B. *Eur. J. Inorg. Chem.* **2012**, 3176–3182.

Chapter 7

Photocatalytic Degradation of Bisphenol A Using TiO_2/CNTs Nanocomposites under UV Irradiation

Chung-Hsuang Hung,[1] Bo-Chao Chuang,[2] Hsing-Lung Lien,[2] and Ching Yuan*,[2]

[1]Department of Safety, Health and Environmental Engineering, National Kaohsiung First University of Science and Technology, 824 Kaohsiung, Taiwan
[2]Department of Civil and Environmental Engineering, National University of Kaohsiung, 811 Kaohsiung, Taiwan
***E-mail: caroline@nuk.edu.tw**

Bisphenol A (BPA) is an endocrine disruptor and a emerging contaminant that is widely distributed in the environment. In this work, a novel nanocomposite prepared by coating TiO_2 on the surface of multi-wall carbon nanotubes (TiO_2/CNTs) was used to photocatalytically degrade BPA under UV irradiation. Two types of TiO_2/CNTs nanocomposites were synthesized by sol-gel hydrolysis precipitation of titanium tetraisopropoxide onto CNTs, followed by calcination or hydrothermal treatment. The precursor ratio of CNT to Ti was set at 1/99 (w/w). Mesoporous crystalline anatase TiO_2 was observed for both types of TiO_2/CNTs. The BET specific surface area of calcined TiO_2/CNTs and hydrothermal treated TiO_2/CNTs was 113.7 and 131.8 m^2/g, respectively. Generally, relatively smaller particle sizes, higher surface areas and larger pore sizes were found in the hydrothermal treated TiO_2/CNTs as compared to the calcined TiO_2/CNTs. This work indicated a faster BPA photocatalytic degradation rate for TiO_2/CNTs nanocomposites in comparison to TiO_2 nanoparticles alone, indicating the photocatalytic activity of the anatase nanocrystalline TiO_2 is enhanced by the interaction with CNTs. Both solution pH and phosphate concentrations showed a large influence on the photocatalytic activity of TiO_2/CNTs nanocomposites for BPA degradation.

Keywords: emerging contaminant; nanotechnology; carbon nanotube; titanium dioxide; bisphenol A

Introduction

Bisphenol A (BPA) is an important industrial chemical produced in large quantities for widely use in the production of polycarbonate plastics, epoxy resins and flame retardants (*1*). It has a wide variety of applications in many consumer products such as plastic bottles, beverage cans, household electronics, and medical devices. The global production of BPA was 3.2 million metric tons in 2003 and it was predicted to exceed 5.5 million metric tons for global consumption in 2011 (*2*). Thus, BPA is inevitably released into environment through many routes at least in part due to the widespread usage. Of many possible BPA contamination sources, the primary route is effluent from wastewater treatment plants and landfill sites. In the United States, BPA was detected in the surface water samples with a median concentration of 0.14 μg/L (*3*) while concentrations ranging from below detection limit to 4.23 μg/L were found in a river in Taiwan (*4*). BPA is not a recalcitrant organic compound. The half-life of biodegradation was reported between 2 to 7 days (*4*). However, BPA is an endocrine-disrupting chemical that may cause reproductive damage to aquatic organisms. BPA can mimic estrogen because it has some structural similarity to human hormone, estradiol and diethylstilbestrol. Because of the health concern, many countries have banned BPA-based plastic bottles for infant use. In Taiwan, BPA has been classified as "Class 4 toxic chemical substances" for which there is concern of pollution of the environment or the endangerment of human health since 2009 while Canada declared BPA a toxic substance in 2010 (*5*).

Titanium dioxide (TiO_2) is a well-known nanoscale photocatalyst that has extensively been examined for the degradation of a wide array of contaminants including dyes, surfactants, pesticides, halogenated organics (*6*) and emerging contaminants such as acetaminophen, caffeine, ibuprofen and sulfamethoxazole (*7*). In general, the photocatalytic degradation involves several steps occurring on the surface of TiO_2. First, the generation of electron-hole pairs by exciting the semiconductor with ultraviolet (UV) light. Second, separation of electrons and holes by traps available on the TiO_2 surface and third, production of reactive oxygen species (e.g., OH•) through a redox process induced by the separated electrons and holes with water (*6*). As a result, to effectively separate electrons and holes, and to prevent the electron-hole recombination should enhance the photocatalytic activity of TiO_2. In other words, the electron-hole recombination is detrimental to the photocatalytic activity. Many methods have been proposed to retard the electron-hole recombination including the surface modification by metal doping and immobilized TiO_2 with a support.

Recently, immobilization of TiO_2 on carbon nanotubes (CNTs) has been examined for the possibility of retarding electron-hole recombination by the electron transfer from TiO_2 to CNTs (*8*). CNTs have unique physical, mechanical, electrical and thermal properties and are good support materials because of their large surface area and their ability to disperse catalytic metal nanoparticles (*9*). The metallic and semiconducting properties of CNTs offer a possible process for photoexcited electrons from the TiO_2 surface to be transferred to the underlying CNTs. In addition, phosphate is known to exhibit a strong affinity to adsorb on the surface of TiO_2. Surface modification of TiO_2 by phosphate was found capable of enhancing the production of hydroxyl radicals (*10*). Other researchers have observed the photocatalytic degradation of BPA by Zr-doped TiO_2 in the presence of phosphate (*11*) and by supported TiO_2 (*12*). In this study, we present novel TiO_2/CNTs nanocomposites for the BPA degradation under illumination of UV light. Two types of TiO_2/CNTs nanocomposites were prepared by sol-gel hydrolysis precipitation of titanium tetraisopropoxide onto CNTs, followed by calcination or hydrothermal treatment. The surface characterization of TiO_2/CNTs was examined using X-ray diffraction (XRD), transmission electron microscopy (TEM) and Brunauer-Emmett-Teller (BET) surface analysis. The effects of pH and phosphate concentration on the photocatalytic activity of both TiO_2/CNTs nanocomposites for the BPA degradation were investigated.

Experimental Section

Chemicals

Bisphenol A and titanium tetraisopropoxide (TTIP) were obtained from Aldrich. Multi-wall carbon nanotubes (MWCNTs) used in this study was obtained from Centron Biochemistry Technology Co. Ltd., Taiwan. Isopropanol, sulfuric acid, nitric acid and phosphoric acid were purchased by Wako Pure Chemical Industries, Ltd. Osaka, Japan.

Functionalization of MWCNTs

The diameter and length of MWCNTs were about 10 nm and 5-15 μm, respectively. The specific surface area was measured to be about 145 m^2/g. The pristine MWCNTs synthesized by chemical vapor deposition (CVD) are hydrophobic and inevitably contain carbonaceous impurities and metal catalyst particles. Thus, they are required to be functionalized with hydrophilic organic groups to offer attractive interaction with the titanium sol. Purification of MWCNTs was conducted in a reflux system with nitric acid (68%) according to the process provided by Yu et al. (*13*). Carboxyl modification of the purified MWCNTs was carried out by chemical oxidation in 98% concentrated H_2SO_4 and 68% HNO_3 in a 3:1 ratio followed by sonication for 8 h in a fume cupboard. The suspension containing COOH-functionalized MWCNTs was cooled to room temperature and the COOH-functionalized MWCNTs were recovered by vacuum filtration and washed thoroughly with deionized water to pH about 6-7. The COOH-functionalized MWCNTs were then dried in the vacuum oven at 100°C overnight.

Preparation of TiO_2/CNTs Nanocomposites

TiO_2/CNTs nanocomposites were prepared by sol-gel hydrolysis precipitation of titanium tetraisopropoxide (TTIP) onto MWCNTs, followed by calcination or hydrothermal treatment. A specific amount of TTIP and MWCNTs was mixed in an isopropanol solution containing nitric acid, followed by sonication for 15 min and then rigorous stirring at room temperature for 3 h. Our preliminary studies have indicated the precursor ratio of CNT/Ti = 1/99 (w/w) exhibited the best photocatalytic activity among others. Thus, the precursor ratio of CNT to Ti was set at 1/99 (w/w) in this work unless indicated otherwise. The amorphous precipitates of TiO_2/CNTs were then treated by either calcination or hydrothermal method. In the former, the resulting materials were filtered, dried and then calcined at 400°C for 1 h. In the latter, the resulting materials were placed in a stainless steel autoclave and aged at 200°C for 10 h. The hydrothermal treated TiO_2/CNTs were washed with deionized water and dried in the vacuum oven at 60°C.

Surface Characterization of TiO_2/CNTs Nanocomposites

The surface characterization of TiO_2/CNTs was conducted by using XRD, TEM and BET surface analysis. XRD measurements were performed on a X-ray diffractometer (Siemens D5000) using Cu K_α radiation producing X-ray with a wavelength of 1.54056 Å. Samples were scanned from 20° to 80° (2θ) at a rate of 2° 2θ/minute. XRD is used to identify crystalline species in a material and can also be used to determine strain, crystallographic structure, preferred orientation, and grain size of crystalline materials. TEM analysis was carried out using a field emission transmission electron microscope (Hitachi Model HF-2000) at 120 kV. It uses a high energy electron beam transmitted through an ultra-thin specimen to image and analyze the microstructure of materials with atomic scale resolution. Specific surface areas and pore size of TiO_2/CNTs were determined by BET-N_2 method using a COULTER SA 3100 surface area analyzer (Coulter Co.).

Photocatalytic Tests

A schematic of the laboratory-scale quartz photoreactor surrounded by four UV lamps (F10 TBLB, fluorescent black light, GE Company, 10 W, λ = 360 nm) is shown in Figure 1. The design is to ensure a homogenous radiation field inside the reactor. Complete mixing was achieved by a magnetic stirrer at 150 rpm to ensure a complete suspension of catalyst particles. The UV intensity in the reactor was about 0.34 mW/cm^2 determined by a radiometer (UVX Digital Radiometer, UVX-36, USA).

For a typical experiment, the initial BPA concentration and the catalyst concentration were 10 and 50 mg/L in 1000 mL aqueous solution, respectively. The concentration of BPA was determined by HPLC-UV (Hitachi, L-2000 series, Japan) coupled with column of Hypersil APS-2NH_2 (250 × 4.6 mm, 5 μm) and detected at a wavelength of 225 nm. The mobile phase was methanol at a flow rate of 1 mL/min. The pH of the solution was adjusted by adding HCl or NaOH solution as per requirement.

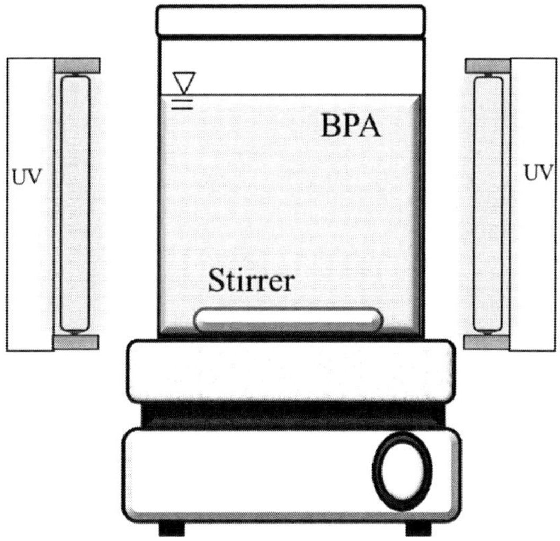

Figure 1. A schematic of the laboratory-scale quartz photoreactor.

A pseudo first-order kinetic model was applied to the photocatalytic degradation of BPA in this work.

$$r = -\frac{dC}{dt} = k_a C \quad (1)$$

where r is the reaction rate; C is the concentration of BPA (mg/L); k_a is the apparent pseudo-first order rate constant (1/min) and t is time (min).

Results and Discussion

Characterization of TiO_2/CNTs Nanocomposites

White precipitates produced from the sol-gel hydrolysis precipitation of TTIP were amorphous. During the hydrothermal aging process or calcination, the amorphous hydrous gels/CNTs were gradually developed into crystalline titania/CNTs. The TEM images of TiO_2/CNTs prepared by calcination and hydrothermal treatment is shown in Figure 2. As shown in Figure 2(a) and (b), the external surface of CNTs was fully loaded with TiO_2 nanoparticles in both calcined TiO_2/CNTs and hydrothermal treated TiO_2/CNTs at the precursor ratio of CNT/Ti = 1/99 (w/w). However, it should be noted that there is no evidence to indicate whether the internal surface of CNTs is coated with TiO_2 nanoparticles or not at the current stage. Increasing the precursor ratio of CNT/Ti to 10/90 (w/w), CNTs were clearly observed with TiO_2 nanoparticles coated for calcined TiO_2/CNTs (Figure 2(c)). The average particle size of the TiO_2 was about 10 nm.

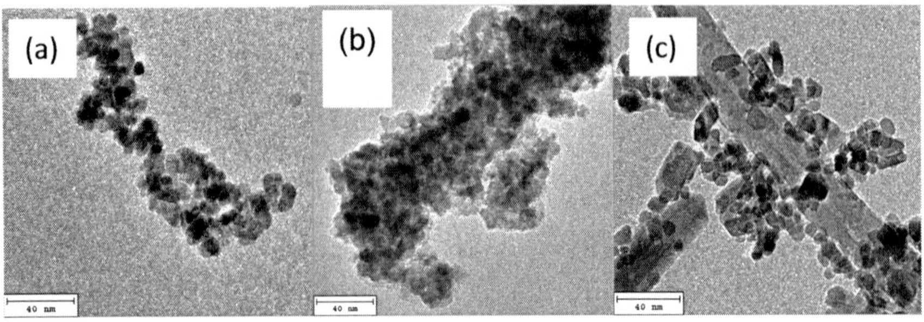

Figure 2. TEM micrographs of (a) calcined TiO_2/CNTs (b) hydrothermal treated TiO_2/CNTs at the precursor ratio of CNT/Ti = 1/99 (w/w) and (c) calcined TiO_2/CNTs at the precursor ratio of CNT/Ti = 10/90 (w/w).

The XRD analysis for hydrothermal treated TiO_2/CNTs and calcined TiO_2/CNTs is shown in Figure 3. The XRD patterns of TiO_2 nanoparticles alone prepared by hydrothermal and calcination processes are also included. The XRD pattern exhibited diffraction peaks of (101), (004), (200), (105) and (204), which are consisted with The Joint Committee on Powder Diffraction Standards (JCPDS) reference pattern of anatase crystalline phase. This indicated that TiO_2 prepared by the synthetic methods provided in this work is an anatase phase, which is preferred to the photocatalysis.

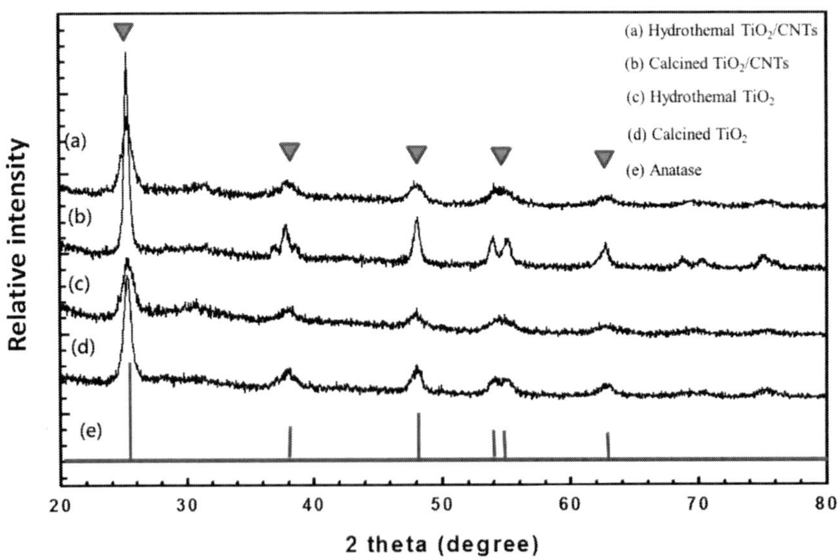

Figure 3. XRD patterns of (a) hydrothermal treated TiO_2/CNTs, (b) calcined TiO_2/CNTs, (c) hydrothermal treated TiO_2 and (d) calcined TiO_2.

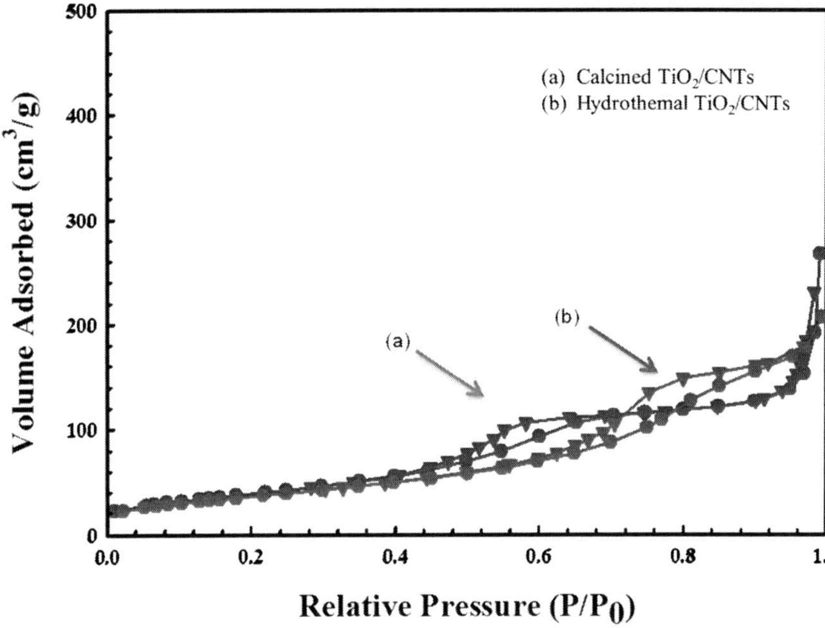

Figure 4. Nitrogen adsorption-desorption isotherms of (a) calcined TiO_2/CNTs and (b) hydrothermal treated TiO_2/CNTs.

The grain size of TiO_2 immobilized on the surface of CNTs was estimated from the broadening of the anatase (101) XRD peak using the Scherrer's equation (*14*):

$$d = \frac{0.89 \times \lambda}{B \times \cos \theta_B} \quad (2)$$

where d is the grain size (Å), λ is 1.54056 Å, B is the full width at half-maximum (FWHM) and θ_B is Bragg angle. Based on the XRD patterns shown in Figure 3, the grain size of calcined TiO_2/CNTs and hydrothermal treated TiO_2/CNTs was calculated to be about 15.5 nm and 10.8 nm, respectively. This is consistent with the result observed in the TEM analysis.

The specific surface area of calcined TiO_2/CNTs and hydrothermal treated TiO_2/CNTs determined by the BET analysis was 113.7 and 131.8 m^2/g, respectively. The nitrogen adsorption-desorption isotherms of the calcined TiO_2/CNTs and hydrothermal treated TiO_2/CNTs are shown in Figure 4. Both TiO_2/CNTs exhibited isotherms of type IV (BDDT classification) indicating the presence of mesopores. In the case of calcined TiO_2/CNTs, two hysteresis loops were observed. The primary loop appearing at low relative pressures between 0.4 and 0.7 is of a type H2, associated with ink-bottle pores while a minor loop appeared at high relative pressure between 0.9 and 1.0. On the other hand, only a hysteresis loop was found for hydrothermal treated TiO_2/CNTs. According to pore size distributions using Barrett-Joyner-Halanda (BJH) methodology, the pore diameter of calcined TiO_2/CNTs and hydrothermal treated TiO_2/CNTs was about 2.39 and 3.52 nm, respectively. Both are in the mesopore range, which is consistent with the results of the nitrogen adsorption-desorption isotherms. It is worth noting that the type IV isotherm with two hysteresis loops indicating bimodal pore size distributions was often observed in TiO_2 nanoparticles prepared by the hydrothermal treatment (*15*). In general, the mesopore structure is favorable to the chemical transport process by alleviating diffusion limitations. We believe that the mesopore structure may be beneficial to BPA degradation.

In general, the hydrothermal treatment requires a mild operating temperature allowing the generation of a more defined structure with less surface defects (*16*). The calcination process, on the other hand, involves rapid purification and thermal transformation of hydrous gels that had not yet fully reacted (*16*). As a result, a relatively smaller particle size, higher surface area and larger pore size were found in the hydrothermal treated TiO_2/CNTs nanocomposite, compared to the calcined TiO_2/CNTs in this study.

Photocatalytic Activity of TiO_2/CNTs Nanocomposite

Photocatalytic degradation of BPA by TiO_2/CNTs was carried out using the laboratory-scale quartz photoreactor. TiO_2 nanoparticles alone synthesized by either calcination or hydrothermal treatment were used for comparison. In the control experiment, the BPA disappearance was negligible in the absence of catalysts under UV irradiation while surface adsorption accounted for about 16% and 17% of the BPA disappearance in the presence of calcined TiO_2/CNTs and hydrothermal treated TiO_2/CNTs, respectively without UV irradiation. Figure 5 depicts the photocatalytic degradation of BPA by various catalysts under near neutral pH at 25°C. The BPA degradation efficiency was 48%, 59%, 61%, and 62% for calcined TiO_2, hydrothermal treated TiO_2, calcined TiO_2/CNTs, and hydrothermal treated TiO_2/CNTs, respectively. The apparent first-order reaction rate constant was therefore calculated to be 4.84×10^{-3}, 6.85×10^{-3}, 7.1×10^{-3} and 7.7×10^{-3} 1/min for calcined TiO_2, hydrothermal treated TiO_2, calcined TiO_2/CNTs, and hydrothermal treated TiO_2/CNTs, respectively. The TiO_2/CNTs exhibited a faster BPA degradation rate in comparison to the case of TiO_2 nanoparticles alone, indicating the photocatalytic activity of the anatase nanocrystalline TiO_2 is enhanced by the interaction with MWCNTs.

Though the product analysis was not carried out in this study, it has been demonstrated that TiO_2 is capable of photocatalytically mineralizing BPA to carbon dioxide based on TOC analysis (*11*, *12*). Intermediates by decomposition of BPA with TiO_2, such as hydroxylated-BPA, carboxylic intermediates (e.g., 4-OH-benzoic acid), phenolic intermediates (e.g., 4-OH-acetophenone and 1,4-dihydroxy-benzene), and other intermediates (e.g., glycolic acid) have been identified (*17*).

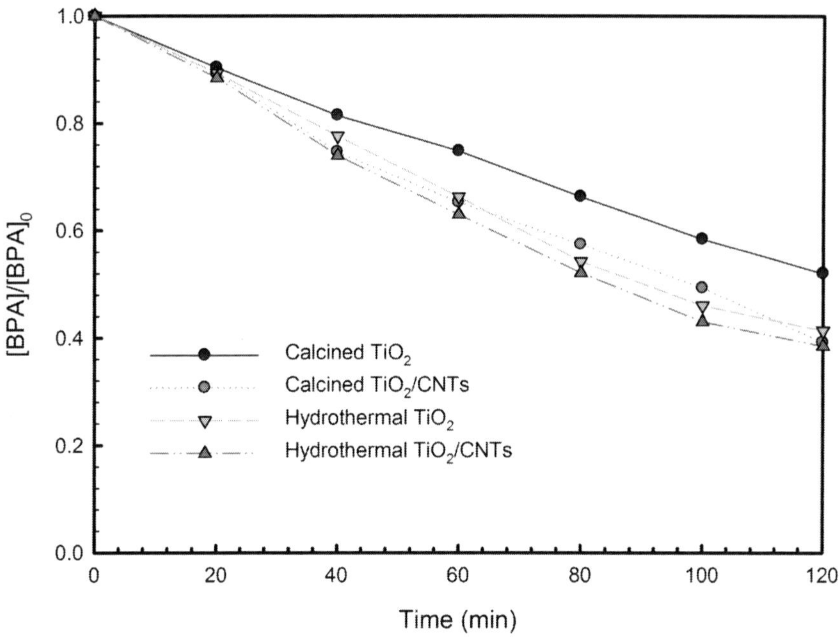

Figure 5. Comparison of the photocatalytic degradation of BPA for various TiO_2/CNTs and controls.

Effects of pH on the Photocatalytic Activity

Figure 6 shows the effect of initial pH on the photocatalytic degradation of BPA with TiO_2/CNTs. The hydrothermally treated TiO_2/CNTs exhibited a relatively higher degree of BPA degradation than was the case with calcined TiO_2/CNTs. Yet, the same trend of pH effects was found for both types of TiO_2/CNTs; that is, the degradation efficiency increased with increasing pH and to a relative maximum at pH 7 before declining under alkaline conditions. The amphoteric behavior was quite commonly observed in the TiO_2 photocatalytic system (*12*, *15*). Generally, a higher concentration of hydroxide ions tends to facilitate the formation of hydroxyl radicals, which are a strong oxidizing species subjected to the BPA degradation (*15*). On the other hand, BPA dissociates into its conjugate base, which is negatively charged under alkaline conditions (pk_{a1} = 9.52, pK_{a2} = 10.2) (*1*). The zero point of charge (pH_{zpc}) of the both types of TiO_2/CNTs was measured to be about pH 5.5. As a result, the surface of TiO_2/CNTs tends to be negatively charged at pH greater than 5.5. The development of repulsive forces between TiO_2/CNTs surface and the bisphenolate anions under alkaline conditions leads to the decrease in efficiency of the BPA degradation (*12*, *15*). The optimum pH determined at neutral pH is likely a compensation effect between the nature of BPA and TiO_2/CNTs nanocomposites.

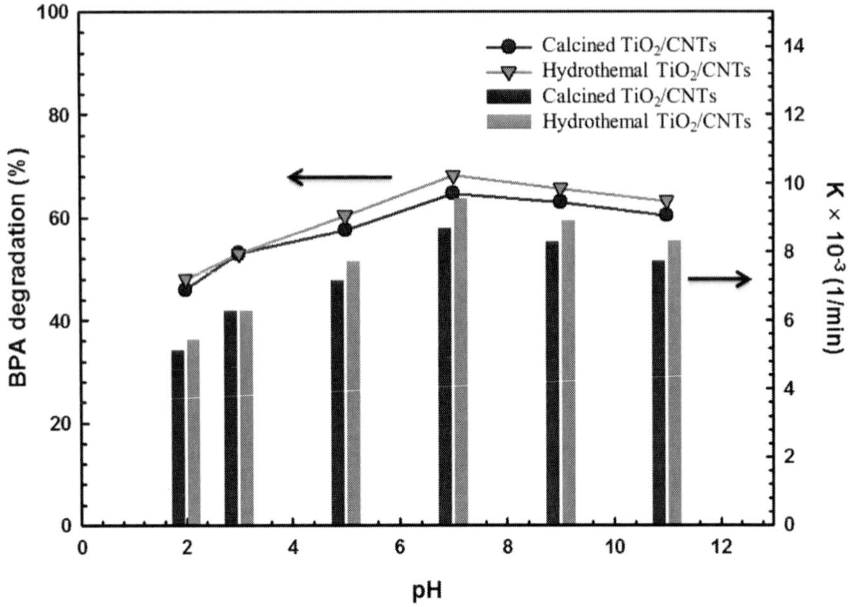

Figure 6. Effect of initial pH on the photocatalytic degradation of BPA using two types of TiO_2/CNTs nanocomposites.

It should be pointed out that both calcined TiO_2/CNTs and hydrothermally treated TiO_2/CNTs showed only slightly difference for BPA degradation efficiency. For example, the BPA degradation efficiency was 61% and 62% for calcined TiO_2/CNTs and hydrothermally treated TiO_2/CNTs, respectively at pH 7. In deed, the surface characteristics of two types of nanocomposites such as pore diameters and specific surface areas are quite similar. The insignificant difference at BPA degradation efficiency may be attributed to the similarity of these TiO_2/CNTs nanocomposites.

Effects of Phosphate on the Photocatalytic Activity

Phosphate anions have been shown to adsorb strongly on the surface of TiO_2 by a bidentate chelation, which may largely influence the interfacial and surface chemistry of TiO_2 (*18*). Phosphate anions are also a common inorganic ion in wastewater and are considered an important species for the wastewater treatment using TiO_2 photocatalytic process. Thus, we examined the effect of phosphate on the photocatalytic activity of TiO_2/CNTs nanocomposites in the study.

The effect of phosphate on the photocatalytic activity of TiO_2/CNTs nanocomposites at pH 9 is shown in Figure 7. The calcined and hydrothermal treated TiO_2/CNTs showed only slightly difference in the BPA degradation efficiency. An increase in phosphate concentrations resulted in an increase in the BPA degradation efficiency. The highest BPA degradation efficiency was found at the phosphate concentration of 10^{-4} mM where the BPA degradation efficiency was about 81.3% and 82.9% for calcined TiO_2/CNTs and hydrothermal treated TiO_2/CNTs, respectively, which correspond to the apparent pseudo-first order rate constant of 14×10^{-3} and 14.7×10^{-3} (1/min). Phosphate modified TiO_2 was found capable of enhancing the photocatalytic activity by generating a larger quantity of hydroxyl radicals (*10*). The enhancement was attributed to the surface-bound phosphate anion through the surface modification of phosphate. It was found that the photocatalytic degradation of BPA in the presence of phosphate anions resulted in an increasing in degradation rates using Zr-doped TiO_2 as a catalyst (*11*), which is in a good agreement with our study. The promotional effect caused by phosphate was attributed to hinder the electron-hole recombination via a proper separation of electrons and holes (*11*). An anion-induced negative electrostatic field and electron-transfer pathway were proposed for the phosphate modified TiO_2 under UV irradiation (*10*). The phosphate ions have strong interaction with water through hydrogen bond that facilitates the charge transfer between the hole and water to form hydroxyl radicals (*10*). On the contrary, excessive amounts of phosphate were detrimental to the photocatalytic activity of TiO_2/CNTs. A decrease of the BPA degradation efficiency was found at the phosphate concentration of 10^{-3} mM. The presence of phosphate anions reduced the photocatalytic rates was documented (*19*, *20*). It is suggested that phosphate anions block the active sites on the surface of TiO_2 that limits the electron transfer during the photocatalytic process (*20*).

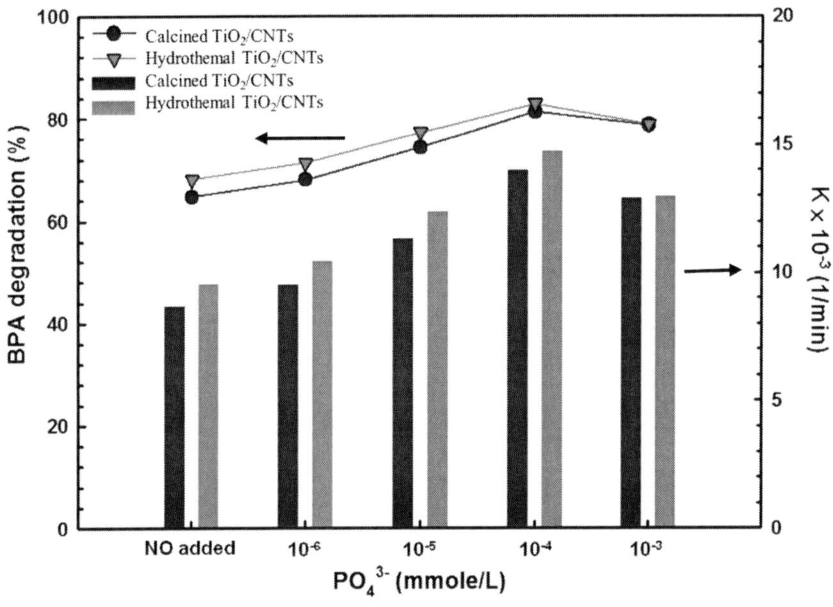

Figure 7. Effect of phosphate concentrations on the photocatalytic degradation of BPA using two types of TiO_2/CNTs nanocomposites.

Conclusion

A novel nanocomposite prepared by coating TiO_2 on the surface of multi-wall carbon nanotubes (TiO_2/CNTs) was used to photocatalytically degrade BPA under UV irradiation in this work. Two types of TiO_2/CNTs nanocomposites were synthesized by sol-gel hydrolysis precipitation of titanium tetraisopropoxide onto CNTs, followed by calcination or hydrothermal treatment. The precursor ratio of CNT to Ti was set at 1/99 (w/w). Mesoporous crystalline anatase TiO_2 was observed for both types of TiO_2/CNTs. The specific surface area of calcined TiO_2/CNTs and hydrothermal treated TiO_2/CNTs determined by the BET analysis was 113.7 and 131.8 m^2/g, respectively.

From this study, the conclusions can be drawn as follows:

- The hydrothermal treated TiO_2/CNTs showed a relatively smaller particle size, higher surface area and larger pore size, compared to the calcined TiO_2/CNTs.
- Studies of BPA photocatalytic degradation indicated a faster degradation rate for TiO_2/CNTs nanocomposites than occurred with TiO_2 nanoparticles alone, indicating the photocatalytic activity of the anatase nanocrystalline TiO_2 is enhanced by the interaction with CNTs.
- Regardless of pH (range 2-11), the hydrothermally treated TiO_2/CNTs exhibited a relatively higher efficiency of BPA degradation than the calcined TiO_2/CNTs. The same trend of pH effects was observed for both types of TiO_2/CNTs; that is, the degradation efficiency increased in acidic solution as the pH increased, while it researched a maximum at pH 7 before declining under alkaline conditions. The BPA degradation efficiency was 61% and 62% for calcined TiO_2/CNTs and hydrothermally treated TiO_2/CNTs, respectively at pH 7.
- The calcined and hydrothermal treated TiO_2/CNTs showed only slightly difference in BPA degradation efficiency in the presence of phosphate. Increasing in phosphate concentrations resulted in an increase of BPA degradation efficiency while excessive amounts of phosphate were detrimental to the photocatalytic activity of TiO_2/CNTs. The highest BPA degradation efficiency was found at the phosphate concentration of 10^{-4} mM where BPA degradation efficiency was about 81.3% and 82.9% for calcined TiO_2/CNTs and hydrothermal treated TiO_2/CNTs, respectively, which correspond to the apparent pseudo-first order rate constant of 14×10^{-3} and 14.7×10^{-3} (1/min).

While still early in our research, this study shows promising results for TiO_2/CNTs nanocomposites that may be potentially useful for various environmental applications, including treatments of hazardous pollutants in both water and air streams, which have also been suggested by Woan et al. (*21*). Still, further studies are needed including the durability and reusability of TiO_2/CNTs nanocomposites, the underlying mechanisms of the photocatalytic enhancement by TiO_2-CNTs system, and the technology for mass production of TiO_2/CNTs nanocomposites.

Acknowledgments

The authors would like to thank the National Science Council (NSC), Taiwan, R.O.C. for the financial support under Grant no. NSC 97-2221-E-390-007.

References

1. Staples, C. A.; Dorn, P. B.; Klecka, G. M.; O'Block, S. T.; Harris, L. R. A review of the environmental fate, effects, and exposures of bisphenol A. *Chemosphere* **1998**, *36*, 2149–2173.
2. Flint, S.; Markle, T.; Thompson, S.; Wallace, E. Bisphenol A exposure, effects, and policy: A wildlife perspective. *J. Environ. Manage.* **2012**, *104*, 19–34.
3. Kolpin, D. W.; Furlong, E. T.; Meyer, M. T.; Thurman, E. M.; Zaugg, S. D.; Barber, L. B. Pharmaceuticals, hormones, and other organic wastewater contaminants in U.S. streams, 1999-2000: A national reconnaissance. *Environ. Sci. Technol.* **2002**, *36*, 1202–1211.
4. Chen, T. C.; Shue, M. F.; Yeh, Y. L.; Kao, T. J. Bisphenol A occurred in Kao-Pin River and its tributaries in Taiwan. *Environ. Monit. Assess.* **2010**, *161*, 135–145.
5. Vandenberg, L. N. Exposure to bisphenol A in Canada: invoking the precautionary principle. *Can. Med. Assoc. J.* **2011**, *183*, 1265–1270.
6. Hoffmann, M. R.; Martin, S. T.; Choi, W.; Bahnemann, D. W. Environmental applications of semiconductor photocatalysis. *Chem. Rev.* **1995**, *95*, 69–96.
7. Miranda-García, N.; Maldonado, M. I.; Coronado, J. M.; Malato, S. Degradation study of 15 emerging contaminants at low concentration by immobilized TiO_2 in a pilot plant. *Catal. Today* **2010**, *151*, 107–113.
8. Yao, Y.; Li, G.; Ciston, S.; Lueptow, R. M.; Gray, K. A. Photoreactive TiO_2/carbon nanotube composites: synthesis and reactivity. *Environ. Sci. Technol.* **2008**, *42*, 4952–4957.
9. Serp, P.; Corrias, M.; Kalck, P. Carbon nanotubes and nanofibers in catalysis. *Appl. Catal., A* **2003**, *253*, 337–358.
10. Zhao, D.; Chen, C.; Wang, Y.; Ji, H.; Ma, W.; Zang, L.; Zhao, J. Surface modification of TiO_2 by phosphate: Effect on photocatalytic activity and mechanism implication. *J. Phys. Chem. C* **2008**, *112*, 5993–6001.
11. Gao, B.; Lim, T. M.; Subagio, D. P.; Lim, T. T. Zr-doped TiO_2 for enhanced photocatalytic degradation of bisphenol A. *Appl. Catal., A* **2010**, *375*, 107–115.
12. Wang, R.; Ren, D.; Xia, S.; Zhang, Y.; Zhao, J. Photocatalytic degradation of Bisphenol A (BPA) using immobilized TiO_2 and UV illumination in a horizontal circulating bed photocatalytic reactor (HCBPR). *J. Hazard. Mater.* **2009**, *169*, 926–932.
13. Yu, Y.; Yu, J. C.; Chan, C. Y.; Che, Y. K.; Zhao, J. C.; Ding, L.; Ge, W. K.; Wong, P. K. Enhancement of adsorption and photocatalytic activity of TiO_2 by using carbon nanotubes for the treatment of azo dye. *Appl. Catal. B* **2005**, *61*, 1–11.

14. Cullity, B. D.; Stock, S. R. *Elements of X-ray Diffraction*, 3rd ed.; Prentice-Hall: New York, 2001.
15. Yu, J.; Wang, G.; Cheng, B.; Zhou, M. Effects of hydrothermal temperature and time on the photocatalytic activity and microstructures of bimodal mesoporous TiO_2 powders. *Appl. Catal., B* **2007**, *69*, 171–180.
16. Wang, C. C.; Ying, J. Y. Sol-gel synthesis and hydrothermal processing of anatase and rutile titania nanocrystals. *Chem. Mater.* **1999**, *11*, 3113–3120.
17. Nomiyama, K.; Tanizaki, T.; Koga, T.; Arizono, K.; Shinohara, R. Oxidative degradation of BPA using TiO_2 in water, and transition of estrogenic activity in the degradation pathways. *Arch. Environ. Contam. Toxicol.* **2007**, *52*, 8–15.
18. Connor, P. A.; McQuillan, A. J. Phosphate adsorption onto TiO_2 from aqueous solutions: an in situ internal reflection infrared spectroscopic study. *Langmuir* **1999**, *15*, 2916–2921.
19. Abdullah, M.; Low, G. K. C.; Matthews, R. W. Effects of common inorganic anions on rates of photocatalytic oxidation of organic carbon over illuminated titanium dioxide. *J. Phys. Chem.* **1990**, *94*, 6820–6825.
20. Chen, F.; Zhao, J.; Hidaka, H. Adsorption factor and photocatalytic degradation of dye-constituent aromatics on the surface of TiO_2 in the presence of phosphate anions. *Res. Chem. Intermed.* **2003**, *29*, 733–748.
21. Woan, K.; Pyrgiotakis, G.; Sigmund, W. Photocatalytic carbon-nanotube–TiO_2 composites. *Adv. Mater.* **2009**, *21*, 2233–2239.

Chapter 8

Iron-Based Nanomaterials for the Treatment of Emerging Environmental Contaminants

Bangxing Ren,[1] Changseok Han,[1] Abdulaziz H. Al Anazi,[1] Mallikarjuna N. Nadagouda,[2] and Dionysios D. Dionysiou*,[1,3]

[1]Environmental Engineering and Science Program, University of Cincinnati, Cincinnati, Ohio 45221-0012, USA
[2]7723 Alexandra Dr., Mason, Ohio 45040, USA
[3]Nireas-International Water Research Centre, University of Cyprus, 20537 Nicosia, Cyprus
***E-mail: dionysios.d.dionysiou@uc.edu**

The occurrence of pharmaceutical and personal care products (PPCPs), endocrine disrupting chemicals (EDCs), and pesticides in drinking water sources has drawn public concerns in recent years. These emerging contaminants, including PPCPs and EDCs, may have significant adverse effects on environment and human health even at a very low concentration range. Iron-based nanomaterials became a promising candidate to solve these environmental problems due to their unique properties. This book chapter focuses on the use of ferrate, ferrite, and TiO_2-composite magnetic iron oxides for the removal of contaminants of emerging concern, including PPCPs, EDCs, and pesticides. Some of the preparation techniques for iron-based nanomaterials are also briefly presented.

Introduction

Iron is one of the most abundant materials in the Earth. Ferrous iron (Fe(II)) and ferric iron (Fe(III)) are typical valence states of iron in the environment (*1–3*). However, metallic iron has been widely used for environmental remediation. Ever since the use of metallic iron (zero-valent iron, ZVI) in permeable reactive barrier systems was reported (*4–6*), ZVI-based materials have been widely used for environmental remediation such as groundwater treatment, remediation of contaminated soil, and drinking water treatment (*1*, *3*, *7*). ZVI materials can be used as sorbents and reducing reagents for treatment of various environmental contaminants. ZVI can efficiently adsorb Cr, Cu, As, and Zn in contaminated soils and reduce the mobility and bioavailability of these contaminants (*8*, *9*). Heavy metals, including Zn and Cd, were removed by adsorption on ZVI from water (*10*, *11*). It was also demonstrated that ZVI was also effective in the reduction of heavy metals and the dechlorination of chlorinated organic solvents, organochlorine pesticides, and polychlorinated biphenyls, which resulted in the removal of these contaminants from water (*3*, *7*).

In addition to ZVI, ferrate (Fe(VI)) has attracted much attention in the field of environmental remediation (*2*, *11*). Due to the unique properties of Fe(VI), including a high redox potential of 2.20 V (under acidic conditions) and selective reactivity, ferrate can be used as an oxidant, coagulant, and/or disinfectant for water treatment (*12*, *13*). Ferrate can decompose various water contaminants, including pharmaceuticals and toxins, as well as inactivate harmful bacteria and viruses in water due to its higher redox potential compared to that of ozone under acidic conditions (2.08 V). Moreover, Fe(VI) can remove particulate matter (PM) from water because Fe(VI) effectively destabilizes PM in water due to its high valence state (+6), which results in coagulation (*13*).

Except for Fe(VI), ferrites (MFe_2O_4, M=Ni, Zn, Ti, etc.) have also been widely studied in applications for environmental remediation (*14–16*). Ferrites can be used as sorbents as well as reagents for Fenton-like reactions to decompose water contaminants. During a Fenton-like reaction, metal oxides and hydrogen peroxide (H_2O_2) are used to produce hydroxyl radicals ($^{\bullet}OH$), which can attack organic contaminants, in a process similar to the traditional Fenton reaction (i.e., combination of Fe^{2+} and H_2O_2) which is also employed to generate $^{\bullet}OH$ for decomposing organic contaminants in water (*14*). However, in the Fenton-like reactions, there is more flexibility to perform the reactions at neutral or circum-neutral pH which is encounted in most types of waters.

Even though ferrate and ferrites have been directly used to remove water contaminants, certain forms of iron oxide materials have also gained much attention in the field of environmental remediation due to their magnetic properties. For example, iron oxide can be used as magnetic core material combined with titanium dioxide (titania, TiO_2) photocatalysts for easy separation in water treatment and soil remediation (*17–19*).

In addition to the use of micro-size iron-based materials as catalysts, sorbents, and magnetic core materials, nanostructured iron-based materials have also attracted much attention in various applications (*3*, *7*, *17*, *20*). Nanostructured iron materials demonstrated higher adsorption capacity and catalytic activity for the removal of contaminants from the environment. In particular, significantly increased surface area of nanostructured iron materials can play an important role for the interaction between nanomaterials and environmental contaminants. Large surface area is associated with large active surface/(nano)interface. Their surface interaction, adsorption, and reaction with other molecules as well as their mobility in different media can be enhanced due to the particle size (*3*, *7*, *20*). This chapter will discuss the application of different iron-based nanomaterials for environmental remediation, such as water treatment, and groundwater and soil decontamination. In particular, the use of ferrate, ferrites, and TiO_2-composite magnetic iron oxides will be discussed for the removal of contaminants of emerging concern.

Ferrates

Ferrate (Fe(VI)) has drawn increasing attention as an emerging water-treatment chemical. Fe(VI)-based materials are known with the molecular formula M_2FeO_4 (M = Li, Na, K, Rb, Cs, and Ag) and $M'FeO_4$ (M′ = Ca, Ba, and Sr) (*21*). Fe(VI) does not react with bromide, thus the carcinogenic disinfection by-product bromate would not be produced during the treatment of bromide-containing water (*22*). During the oxidation process, Fe (VI) can oxidize pollutants and then be reduced to ferric ions or ferric hydroxide, which simultaneously generates a coagulant in a single unit process (*23*). Compared to the required dose of common coagulants, much less dose was needed for Fe (VI) to achieve similar treatment performance (*24*). In recent years, the application of Fe(VI) to treat emerging contaminants has been extensively studied due to these superior advantages of Fe(VI) (*21–33*).

Studies on the reactivity of Fe(VI) with a number of EDCs have been reported. Among EDCs of environmental importance, three synthetic EDCs (nonylphenol (NP), bisphenol A (BPA), and 17α-ethynylestradiol (EE)) and three natural EDCs (estrone (E1), 17β-estradiol (E2), and estriol (E3)) were chosen to investigate their reaction kinetics with Fe(VI) (*21*, *22*, *25–27*). All six EDCs are phenolic substances. In the experiments, high purity potassium ferrate (K_2FeO_4) was prepared by the oxidation of ferric nitrate with hypochlorite. The reactivity tests were conducted in the pH range of 8.2-12, and with different molar ratios of the reactants. The reactions were observed to follow an overall second-order kinetic model and the rate constants generally decreased with increase in pH. The mono-protonated Fe(VI), $HFeO_4^-$, was found to have higher oxidizing power than the non-protonated Fe(VI), FeO_42^- (*25*) and the fraction of $HFeO_4^-$ increased with decreasing pH, which contributed to the greater degradation performance of EDCs by Fe(VI) at lower pH. Among the six EDCs, the Fe(VI) oxidation of the four steroid estrogens exhibited higher reaction rates than those of BPA and NP (*22*, *25*, *26*). Generally, potassium Fe(VI) removed EDCs in a great degree within seconds (*21*). Liquid chromatography/mass spectrometry–mass spectrometry (LC/MS–MS) and gas chromatography/mass spectrometry–mass spectrometry (GC/MS–MS) were used to analyze the intermediate reaction products formed in the reaction of BPA with Fe(VI). Nine specific compounds were identified, including *p*-isopropylphenol, *p*-isopropenylphenol, 4-isopropanolphenol, and dicarboxylic acids (e.g., oxalic acid) (*25*).

Other studies reported the degradation of pharmaceuticals in water using Fe(VI). Pharmaceuticals studied include carbamazepine, sulfonamide, trimethoprim, tetracycline, ibuprofen, ciprofloxacin, enrofloxacin, diclofenac, triclosan, and propranolol (*21*, *28–32*). The pH range of the tests was from 6 to

10 and oxidation reactions of pharmaceuticals by Fe(VI) were found to follow first-order kinetics for each reactant.

The study on the triclosan showed that it reacted rapidly with Fe(VI). At a stoichiometric ratio of 10: 1 ([Fe(VI)]: [triclosan]), triclosan was completely degraded. Oxidation of triclosan by Fe (VI) involves the break of ether bond and phenoxy radical reaction. Coupling reaction may also occur during Fe(VI) treatment. According to the growth inhibition tests with alga *P. subcapitata*, the oxidation products of triclosan by Fe(VI) possess less toxicity than triclosan (*29*).

The second-order rate constants for the reaction of Fe(VI) with sulfonamide decreased non-linearly with increasing pH (*30*). The oxidation of sulfamethoxazole (SMX) by Fe(VI) was found to follow a molar stoichiometry of 4:1 ([Fe(VI)]: [SMX]). According to the products of SMX, the isoxazole ring was disrupted and the amino group was converted into a nitro/nitroso group during the oxidation. It was found that Fe(VI) removed SMX in water, while it produced by-products that are less toxic than the original compound (*31*).

The reaction between Fe(VI) and propranolol (PPL), a β-blocker, followed second-order kinetics. A molar stoichiometry of 6:1 ([Fe(VI)]: [PPL]) was determined for the complete removal of PPL. The formed oxidation products of the reaction could be further oxidized by Fe(VI) and thus resulted in a non-linear relationship between degradation of PPL and the added amount of Fe(VI) (*32*).

Besides, removal of sucralose by Fe(VI) was also studied recently (*33*). Sucralose is a chlorinated derivative of the disaccharide sucrose and is being used as an artificial noncaloric sweetener in many food and beverage products. It is also considered to be an emerging contaminant due to its persistence in the environment, with a half-life of up to several years. The oxidation of sucralose by Fe(VI) was found to follow second-order kinetics and the rate constants increased with a decrease in pH. Comparing the reactivity of Fe(VI) with other oxidants, it suggests that $^{\bullet}OH$ has much higher reactivity than Fe(VI) towards carbohydrates (*33*).

Nanostructured Ferrite

Our environment is rich with iron as it exists in two valence states, one of them being the water soluble state Fe(II) and the other is the highly water insoluble state Fe(III) which is also the stable form of iron (*1*). Due to their unique structure and magnetic properties, ferrite-based catalysts have attracted much attention to decompose emerging contaminants of concern, including PPCPs, and organic dyes (*17*, *34*).

One unique property of iron is being an electron donor, which can be utilized in endorsing a Fenton reaction in which $^{\bullet}OH$ radicals are generated by reducing hydrogen peroxide similar to what happens in Fe^{2+} here (*1*),

$$Fe^{2+} + H_2O_2 \rightarrow Fe^{3+} + {}^{\bullet}OH + OH^-$$

A good example is the degredation of pharmaceutical ibuprofen present in effluents of wastewater treatment plants by using Fe^{2+} and H_2O_2 (*35*). Furthermore, in many cases, researchers combine ferrite with transition elements in order to enhance their structure which can lead to better lasting magnetic properties since ferrite alone might lose its magnetic properties during the process of degrading water contaminants. Many researchers have synthesized $M_xFe_{2-x}O_4$ where M is iron, cobalt, copper or another transition element and x = 0.1, 0.3, 0.7, 1.0 or any desired element ratios as a combination formula which is effective in degrading contaminants while retaining their magnetic properties for sufficient time (*36*).

Depending on the types of contaminants, ferrite-based technologies can be competitive or have advantages over some other advanced oxidation processes (AOPs) to degrade emeging contaminants of concern in water and soil. This efficiency is considered to be due to the generation of the $^{\bullet}OH$ radicals such as in the case of homogenous Fenton, heterogeneous Fenton, Fenton-like, photo-Fenton and ozonation which are among the most commonly used processes (*37*, *38*). In the case of Fenton-like reactions based on ferrite catalysts, hydroxyl radicals are generated due to the presence of a metal oxide instead of Fe^{2+} dissolved in solution. The kinetics of such heterogeneous Fe (hydr)oxides catalytic process are mainly depended on the particle size, surface area, morphology, isomorphic substitution, pH and temperature (*38*). Moreover, ferrite-based catalysts are capable of removing nanomaterial contaminants, including natural, incidental, and engineered, from water by adsorption due to the distinguished spinel ferrite structure and the chemical stability because of the lattice bonding between heavy metal ions and ferric ion (*1*).

Most of ferrite-based catalysts are environmental friendly since they have the magnetic properties which allow them to be recovered magnetically from treated solutions. As there are many ferrite-based catalysts preparation techniques, the self propagating high temperature synthesis (SHS), also called composition synthesis, is one of the most practical techniques due to the ability of minimizing or eliminating the experimental variations (*39*). The SHS synthesis process is very versatile, rapid, and energetically favorable process where the energy needed to sustain the process results from the chemical energy of the exothermic reaction mix. Ferrites and the desired metal nitrates are mixed with varying polyvinyl alcohol (PVA) concentrations and ignited in muffle furnace set at desired temperature for determined period of time. This cause the PVA to self-ignite and drive the resulting chemical reaction.

Tin, cobalt, manganese, zinc, copper, vanadium, titanium and palladium, are among the major elements researchers have studied as a composition with iron. Some of these composition can have photocatalytic effect (*34*, *40*). As researchers have shown the photocatalytic activity of titanium, vanadium, zinc and tin, a lot more efforts still needed to be made to study photocatalytic properties of ferrite-based catalysts. In the case of vanadium, researchers were able to utilize visible light as an energy source along with nanosized iron-based catalysts to obtain the Fenton-like reaction system necessary for the oxidation of many organic and inorganic compounds by H_2O_2 using such heterogeneous Fenton methods (*34*).

Iron-Based Magnetic TiO_2 Photocatalysts

Photocatalysis using titanium dioxide (TiO_2) has been intensively studied for environmental remediation, including drinking water and wastewater treatment, air purification, and soil and groundwater remediation due to the low cost, non-toxicity, high chemical and thermal stability, and relatively high degradation activity of TiO_2 photocatalysts (*41–43*). UV-induced TiO_2 photocatalysis can result in the generation of highly reactive oxygen species (ROS), including $^{\bullet}OH$, superoxide anion radical ($O_2^{\bullet -}$), and singlet oxygen ($^1O_2$), which promptly attack and break down most organic contaminants in the environment. However, most researchers have used slurry-type TiO_2 photocatalysts for water treatment. Thus, the difficulty of separation and reusability of the catalysts have become a significant challenge for their applications for water treatment processes (*17*, *44*). Recently, magnetic separation has attracted much attention in the water treatment process using slurry-type TiO_2 photocatalysts due to their easy separation and reusability (*17–19*, *44–47*). Magnetic core materials, such as magnetite (Fe_3O_4), maghemite (γ-Fe_2O_4), or ferrite (e.g., $ZnFe_2O_4$ and $NiFe_2O_4$), have been employed for magnetic separation (*17*, *48*, *49*). For photocatalytic water treatment and soil remediation, TiO_2 photocatalysts were immobilized on the magnetic core materials (*19*, *45*, *47*, *48*). Chalasani and Vasudevan (*19*) prepared cyclodextrin-functionalized Fe_3O_4@TiO_2 core-shell magnetic nanoparticles (CMCD-Fe_3O_4@TiO_2) to decompose EDCs, such as BPA and dibutyl phthalate (DBP), in water supplies. The light absorption edge was blue-shifted (bandgap of 3.44 eV) due to the combination of particle size, surface groups, and the presence of Fe as a dopant. The stability of dispersions resulted in very high resistance to separation from the solution by ultracentrifugation due to the exposed hydroxyl groups on the rims of the cyclodextrin cavities. However, the suspended particles were easily separated by a magnet (modest magnetic field < 0.4 T), and the saturation magnetization value (M_s) of CMCD-Fe_3O_4@TiO_2 was 28 emu g^{-1} at 300 K. The adsorption of both compounds by CMCD-Fe_3O_4@TiO_2 was studied. The photocatalytic degradation of BPA and DBP by CMCD-Fe_3O_4@TiO_2 and Fe_3O_4@TiO_2 were also studied. The adsorption data for BPA and DBP was fitted well to the Langmuir isotherm and the maximum uptake of BPA was higher than that of DBP. For the UV-assisted photocatalytic degradation of BPA and DBP, both compounds of 20 mg L^{-1} were added in 1 g L^{-1} catalyst solution. After 60 min of UV illumination, both BPA and DBP were completely degraded by using CMCD-Fe_3O_4@TiO_2, but they were not completely removed by using Fe_3O_4@TiO_2 after 180 min of UV irradiation. In addition, the effective pH range for the degradation of BPA and DBP was 7-9 and 7-10, respectively. The water dispersibility of the particles and the presence of the hydrophobic cyclodextrin cavities played an important role to enhance the photocatalytic degradation of BPA and DBP in their research. Moreover, the photocatalytic degradation efficiency of the materials was kept over 90% during the 10 cycle degradation experiments, which indicated that CMCD-Fe_3O_4@TiO_2 possess highly stable photocatalytic activity for decomposing EDCs in water.

Other studies reported the degradation of fish pathogens, phenol, and chlorophenol in water using TiO_2-Fe_3O_4 or $ZnFe_2O_4$ composites (*45*, *47*, *48*). Choi et al. (*45*) reported the efficient decomposition of 2,4,6-trichlorophenol (2,4,6-TCP) using Fe_3O_4@TiO_2 under UV illumination. The synthesized particles were very uniform with a diameter of 250 nm and their M_s was 63.2 emu g^{-1}. For the photocatalytic degradation of 2,4,6-TCP, the concentrations of the contaminant and catalyst were 7.5×10^{-5} M and 5 g L^{-1}, respectively. After 100 min of UV irradiation, 93% of 2,4,6-TCP was decomposed, and the degradation efficiency was maintained during the reusability test (seven-cycle consecutive experiments). Cheng et al. (*47*) reported on the bactericidal effect and photocatalytic degradation of indigo carmine dye of TiO_2/Fe_3O_4 particles under blue LED illumination (λ_{max}=475 nm) and the effect of different mole ratios of TiO_2/Fe_3O_4 on the decolorization of the dye. In particular, the antibacterial effects and the photocatalytic activity of the samples were investigated in seawater of different salinities. The mole ratio of 400:1 for TiO_2/Fe_3O_4 demonstrated the highest photocatalytic activity to decolorize the dye, however, the sample containing only TiO_2 did not remove the dye after 180 min of light illumination. The incorporation of Fe^{3+}, which is disassociated from Fe_3O_4, into the lattice of TiO_2 resulted in the activation of TiO_2/Fe_3O_4 by blue LED for the degradation of the dye. In addition, the photocatalytic activity of the samples decreased at higher salinity of the seawater resulting from the depletion of the electrons, which are generated during TiO_2 photocatalysis, due to the ions in the seawater. Similar results that the lower antibacterial activity in the seawater of higher salinity were also obtained for the antibacterial activity of the particles in the seawater. In addition, photocatalytic degradation of phenol by using TiO_2-$ZnFe_2O_4$ under UV, UV-visible, and visible light illumination was reported by Srinivasan et al. (*48*). For this study, different mole fractions of TiO_2-$ZnFe_2O_4$ were applied from 0.01 to 0.2. Even though the light absorption edge was significantly red-shifted at the ratio of 0.2, the photocatalytic degradation activity depended on the type of light sources. The highest photocatalytic activity was observed at the ratio of 0.01 and 0.1 under UV and UV-visible, and visible light illumination, respectively. Figure 1 shows the energy levels of different nanomaterials.

However, significant drawbacks of direct-junction between TiO_2 and magnetic materials (e.g., magnetite and ferrite) have been reported (*17*, *18*, *43*, *44*, *46*, *56*). The magnetic materials can absorb strong UV light, resulting in the decreased photocatalytic activity of TiO_2 (*57*). Also, the unfavorable heterojunction of TiO_2 and magnetic materials increases the recombination of electrons and holes as well as the photodissolution of iron (*58*, *59*). Therefore, to overcome the drawbacks, an inner layer between TiO_2 and a magnetic core has been proposed. SiO_2 was commonly employed as an inner layer material due to its stability for surface modification and protection of core materials (*58*, *60*). Figure 2 shows the photoexcitation mechanism of conventional TiO_2 photocatalysts, the electron-hole recombination due to the direct-junction between TiO_2 and magnetic materials, and the application of an inner layer to prevent the electron-hole recombination for the magnetic separatable TiO_2 photocatalyts.

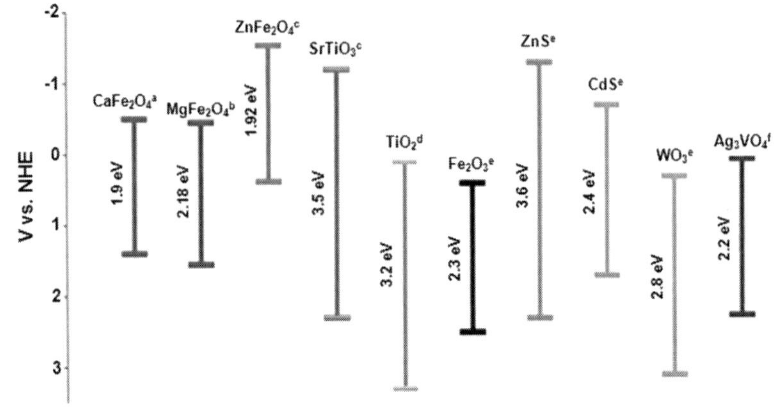

Figure 1. The energy levels of different nanomaterials; (a) (49), (b) (50), (c) (51), (d) (52), (e) (53), and (f) (54). (Reproduced with permission from reference (55). Copyright 2012 Elsevier.)

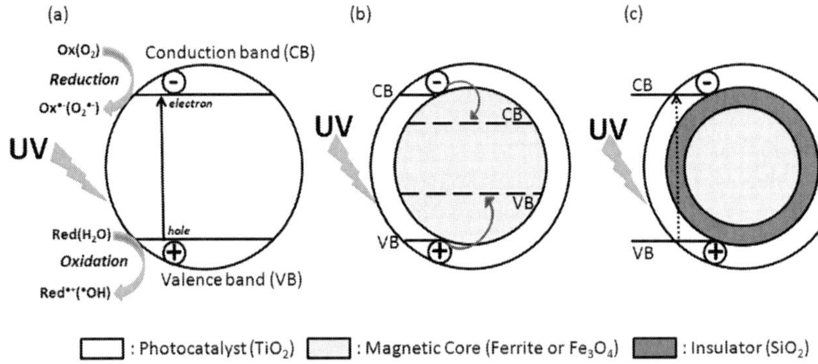

Figure 2. The diagrams of photoexcitation mechanism of TiO_2 photocatalyst (a), electron-hole recombination (b), and the application of an inner layer between TiO_2 photocatalyst shell and magnetic core.

The use of magnetic $Fe_3O_4/SiO_2/TiO_2$ composites for the degradation of water contaminants, including dyes, pharmaceuticals and personal care products (PPCPs), herbicides, and EDCs, has been reported (*17*, *18*, *44*, *46*). Álvarez et al. (*17*) prepared $Fe_3O_4/SiO_2/TiO_2$ nanoparticles to decompose several PPCPs, such as antipyrine, caffeine, metoprolol, and BPA. They reported the M_s and the BET surface area decreased when SiO_2 and TiO_2 layers were formed on the core Fe_3O_4. The M_s and BET surface area of the particles were 39.3 emu g^{-1} and 22 m^2 g^{-1}, respectively. The dissolution and photodissolution of iron from the particles were significantly decreased when SiO_2 inner layer was added. The photodissolution of iron decreased from 4.4% to 0.15% at pH 3. As a result, the photocatalytic activity and reusability of $Fe_3O_4/SiO_2/TiO_2$ particles was enhanced compared to that of Fe_3O_4/TiO_2 samples for degradation of these PPCPs due to the inhibition of electron-hole recombination by the SiO_2 inner layer. During the four-cycle reusability test, the photocatalytic degradation efficiency was very stable with high first-order rate constants. Chang and Man (*44*) also reported the photocatalytic degradation of an EDC, dimethyl phthalate (DMP), using $Fe_3O_4/SiO_2/TiO_2$ under UV irradiation. To maximize the deposition of TiO_2 on SiO_2/Fe_3O_4 core, the ratio of water to titanium (WTR) was varied from 60 to 250. The maximal deposition of TiO_2 was obtained at WTR = 130 and its bandgap, pH_{PZC}, and M_s were 3.2 eV, 5.6, and 8.9 emu g^{-1}, respectively. Its photocatalytic activity was the highest for the degradation of DMP at the experimental conditions among the samples obtained at different WTR. In addition, the optimal loading of catalyst was 1.2 g L^{-1} for the photocatalytic degradation due to the agglomeration of samples, which is associated with active surface area and turbidity of solution for light penetration. The adsorption of DMP on the samples was dependant on the pH of the solution, which was inversely proportional to the pH. The surface charge of the particles and DMP at different pH played an important role in the DMP adsorption because the positively charged surface of the particles is favorable to attract the oxygens of DMP chains and electron-rich aromatic nuclei at lower pH than pH_{PZC}. Different studies also reported the significant effects of the pH for the interaction between catalysts and contaminants due to their different surface charge below pH_{PZC} of different materials (*46*, *61*, *62*). For the photocatalytic decomposition of benzoic acid using $TiO_2/SiO_2/Fe_3O_4$, acidic condition was more favorable than neutral or alkaline condition (*46*).

Moreover, the $Fe_3O_4/SiO_2/TiO_2$ was employed to remove the herbicide glyphosate in soil under solar light (*18*). In this study, the effects of loading of Fe_3O_4, moisture content in the soil, loading of catalyst, soil thickness, and light intensity on the degradation of glyphosate were investigated. Among the samples (0-10%), 0.5% $Fe_3O_4/SiO_2/TiO_2$ showed the highest photocatalytic activity. Even though Fe_3O_4 enhanced the light absorbance in the visible light region, the electron-hole recombination increased at higher loadings of Fe_3O_4. For the moisture content, 30-50% was optimal for the glyphosate degradation because glyphosate and O_2 cannot diffuse to the photocatalysts at higher or lower moisture content. The optimal loading of catalyst was 4 mg TiO_2/g soil. At higher loading of catalyst than 0.2 g, the amount of catalyst was not a critical parameter compared to moisture content, soil thickness, and light intensity. The degradation of glyphosate was inversely proportional to the soil thickness due to the light penetration but the glyphosate degradation increased at higher light intensity.

Conclusions

The research studies discussed in this book chapter have shown the superior performance of iron-based nanomaterials in removing emerging contaminants. Fe(VI) reactions with emerging contaminants have demonstrated that Fe(VI) is selective to react with electron-rich organic moieties. The reaction rate of Fe(VI) depends on the pH of the solution due to the change in electron density on the target reactive groups upon protonation/deprotonation. Kinetic and pathway studies are needed to advance our understanding of the relationship between iron-based nanomaterials and contaminant species. Ferrites can be used as sorbents as well as reagents for a Fenton-like reactions to decompose water contaminants. Some of the ferrite materials can have photocatalytic effect. SHS is an efficient technique to prepare ferrite-based catalyst. Iron-based magnetic TiO_2 photocatalysts possess high photoactivity and can be easily separated from the treated solution and reused. Coating between TiO_2 and magnetic core decreases the photodissolution of iron and enhances the photocatalytic activity and reusability of TiO_2-composited magnetic iron oxide. Most of the techniques/methods mentioned in the chapter are still at an experimental or pilot stage. In the future research, they should be improved to achieve both technical efficiency and cost effectiveness for full-scale water treatment.

Acknowledgments

All authors have contributed equally in this chapter.

D. D. Dionysiou would like to acknowledge the Cyprus Research Promotion Foundation through Desmi 2009-2010 which was co-funded by the Republic of Cyprus and the European Regional Development Fund of the EU under contract number NEA IPODOMI/STRATH/0308/09. D. D. Dionysiou would also like to acknowledge funding from the US National Science Foundation (CBET 1236331) for support on his work on iron-based photocatalytic materials. C. Han acknowledges the Graduate School Dean's Fellowship, which supports University of Cincinnati doctoral students in their final year of degree work. B. Ren acknowledges the scholarship from China Scholarship Council for PhD student (No. 201206260144).

Disclaimer

The research described herein was developed by the author, an employee of the U.S. Environmental Protection Agency (EPA), on his own time. It was conducted independent of EPA employment and has not been subjected to the Agency's peer and administrative review. Therefore, the conclusions and opinions drawn are solely those of the author and are not necessarily the views of the Agency.

References

1. Cundy, A. B.; Hopkinson, L.; Whitby, R. L. D. *Sci. Total Environ.* **2008**, *400*, 42–51.
2. Sharma, V. K.; Kazama, F.; Jiangyong, H.; Ray, A. K. *J. Water Health* **2005**, *3*, 45–58.
3. Li, X.-Q.; Elliott, D. W.; Zhang, W.-X. *Crit. Rev. Solid State* **2006**, *31*, 111–122.
4. Reynolds, G. W.; Hoff, J. T.; Gillham, R. W. *Environ. Sci. Technol.* **1990**, *24*, 135–142.
5. Gillhan, R. W.; O'Hannesin, S. F. *Ground Water* **1991**, *29*, 752–761.
6. Gillhan, R. W.; O'Hannesin, S. F. *Ground Water* **1994**, *32*, 958–967.
7. Zhang, W.-X. *J. Nanopart. Res.* **2003**, *5*, 323–332.

8. Mench, M.; Vangronsveld, J.; Beckx, C.; Ruttens, A. *Environ. Pollut.* **2006**, *144*, 51–61.
9. Kumpiene, J.; Ore, S.; Renella, G.; Mench, M.; Lagerkvist, A.; Maurice, C. *Environ. Pollut.* **2006**, *144*, 62–69.
10. Boparai, H. K.; Joseph, M.; O'Carroll, D. M. *J. Hazard. Mater.* **2011**, *186*, 458–465.
11. Sharma, V. K. *Adv. Environ. Res.* **2002**, *6*, 143–156.
12. Ghernaout, D.; Ghernaout, B.; Naceur, M. W. *Desalination* **2001**, *271*, 1–10.
13. Jiang, J.; Lolyd, B. *Water Res.* **2002**, *36*, 1397–1408.
14. Lu, H.-C.; Chang, C.-E.; Vong, W.-W.; Chen, H.-T.; Chen, Y.-L. *Int. J. Phys. Sci.* **2011**, *6*, 855–865.
15. Mathew, T.; Malwadkar, S.; Pai, S.; Sharanappa, N.; Sebastian, C. P.; Satyanarayana, C. V. V.; Bokada, V. V. *Catal. Lett.* **2003**, *91*, 217–224.
16. Parson, J. G.; Lopez, M. L.; Peralta-Videa, J. R.; Gardea-Torresday, J. L. *Microchem. J.* **2009**, *91*, 100–106.
17. Álvarez, P. M.; Jaramillo, J.; López-Piñero, F.; Plucinski, P. K. *Appl. Catal., B* **2010**, *100*, 338–345.
18. Xu, X.; Ji, F.; Fan, Z.; He, L. *Int. J. Environ. Res. Public Health* **2011**, *8*, 1258–1270.
19. Chalasani, R.; Vasudevan, S. *ACS Nano* **2013**, *7*, 4093–4104.
20. Noubactep, C.; Caré, S.; Crane, R. *Water, Air, Soil Pollut.* **2012**, *223*, 1363–1382.
21. Sharma, V. K. *Coord. Chem. Rev.* **2013**, *257*, 495–510.
22. Sharma, V. K.; Li, X. Z.; Graham, N.; Doong, R.-A. *J. Water Supply: Res. Technol.* **2008**, *57*, 419–426.
23. Jiang, J.-Q.; Zhou, Z.; Sharma, V. K. *Microchem. J.* **2013**, *110*, 292–300.
24. Jiang, J.-Q.; Stanford, C.; Alsheyab, M. *Sep. Purif. Technol.* **2009**, *68*, 227–231.
25. Li, C.; Li, X. Z.; Graham, N.; Gao, N. Y. *Water Res.* **2008**, *42*, 109–120.
26. Sharma, V. K.; Anquandah, G. A. K.; Nesnas, N. *Environ. Chem. Lett.* **2009**, *7*, 115–119.
27. Lee, Y.; Yoon, J.; Gunten, A. V. *Environ. Sci. Technol.* **2005**, *39*, 8978–8984.
28. Anquandah, G. A. K.; Sharma, V. K.; Knight, D. A.; Batchu, S. R.; Gardinali, P. R. *Environ. Sci. Technol.* **2011**, *45*, 10575–10581.
29. Yang, B.; Ying, G.-G.; Zhao, J.-L.; Zhang, L.-J.; Fang, Y.-X.; Nghiem, L. D. *J. Hazard. Mater.* **2011**, *186*, 227–235.
30. Sharma, V. K.; Mishra, S. K.; Nesnas, N. *Environ. Sci. Technol.* **2006**, *40*, 7222–7227.
31. Sharma, V. K.; Mishra, S. K.; Ray, A. K. *Chemosphere* **2006**, *62*, 128–134.
32. Anquandah, G. A. K.; Sharma, V. K.; Panditi, V. R.; Gardinali, P. R.; Kim, H.; Oturan, M. A. *Chemosphere* **2013**, *91*, 105–109.
33. Sharma, V. K.; Sohn, M.; Anquandah, G. A. K.; Nesnas, N. *Chemosphere* **2012**, *87*, 644–648.
34. Zhang, Y. Y.; Deng, J. H.; He, C.; Huang, S. S.; Tian, S. H.; Xiong, Y. *Environ. Technol.* **2010**, *31*, 145–154.
35. Mendez-Arriaga, F.; Esplugas, S.; Gimenez, J. *Water Res.* **2010**, *44*, 589–595.

36. Cabrera, L. I.; Somoza, Á.; Marco, J. F.; Serna, C. J.; Puerto Morales, M. *J. Nanopart. Res.* **2012**, *14*, 1–14.
37. Prousek, J. *Pure Appl. Chem.* **2007**, *79*, 2325–2338.
38. Pereira, M. C.; Oliveira, L. C. A.; Murad, E. *Clay Miner.* **2012**, *47*, 285–302.
39. Choi, Y.; Lee, J. J. *Phys. Rev. B* **2009**, *404*, 692–694.
40. Fu, Y.; Wang, X. *Ind. Eng. Chem. Res.* **2011**, *50*, 7210–7218.
41. Han, C.; Pelaez, M.; Likodimos, V.; Kontos, A.; Falaras, P.; O'Shea, K.; Dionysiou, D. *Appl. Catal., B* **2011**, *107*, 77–87.
42. Pelaez, M.; Nolan, N. T.; Pillai, S. C.; Seery, M. K.; Falaras, P.; Kontos, A. G.; Dunlop, P. S. M.; Hamilton, J. W. J.; Byrne, J. A.; O'Shea, K.; Entezari, M. H.; Dionysiou, D. D. *Appl. Catal., B* **2012**, *125*, 331–349.
43. Wang, T. C.; Lu, N.; Li, J.; Wu, Y. *Environ. Sci. Technol.* **2011**, *45*, 9301–9307.
44. Chang, C.-F.; Man, C.-Y. *Ind. Eng. Chem. Res.* **2011**, *50*, 11620–11627.
45. Choi, K.-H.; Oh, S.-L.; Jung, J.-H.; Jung, J.-S. *J. Appl. Phys.* **2012**, *111*, 07B504.
46. Su, T.-L.; Chiou, C.-S.; Chen, H.-W. *Int. J. Photoenergy* **2012**, *2012*, 909678 (8 pages).
47. Cheng, T. C.; Yao, K. S.; Yeh, N.; Chang, C. I.; Hsu, H. C.; Gonzalez, F.; Chang, C. Y. *Thin Solid Films* **2011**, *519*, 5002–5006.
48. Srinivasan, S. S.; Wade, J.; Stefanakos, E. K. *J. Nanomater.* **2006**, *2006*, 45712 (1–4 pages).
49. Ida, S.; Yamada, K.; Matsunaga, T.; Hagiwara, H.; Matsumoto, Y.; Ishihara, T. *J. Am. Chem. Soc.* **2010**, *132*, 17374–17345.
50. Dom, R.; Subasri, R.; Radha, K.; Borse, P. H. *Solid State Commun.* **2011**, *151*, 470–473.
51. Boumaza, S.; Boudjemaa, A.; Bouguelia, A.; Bouarab, R.; Trari, M. *Appl. Energy* **2010**, *87*, 2230–2236.
52. Xu, S.; Shangguan, W.; Yuan, J.; Shi, J.; Chen, M. *Sci. Technol. Adv. Mater.* **2007**, *8*, 40–46.
53. Kitano, M.; Hara, M. *J. Mater. Chem.* **2010**, *20*, 627–641.
54. Chen, S.; Zhao, W.; Liu, W.; Zhang, H.; Yu, X.; Chen, Y. *J. Hazard. Mater.* **2009**, *172*, 1415–1423.
55. Casbeer, E.; Sharma, V. K.; Li, X.-Z. *Sep. Purif. Technol.* **2012**, *87*, 1–14.
56. Yuan, J.; Lü, Y.-K.; Li, Y.; Li, J.-P. *Chem. Res. Chinese U.* **2010**, *26*, 278–282.
57. Chen, F.; Xie, Y. D.; Zhao, J. C.; Lu, G. X. *Chemosphere* **2001**, *44*, 1159–1168.
58. Beydoun, D.; Amal, R.; Low, G. K. C.; McEvoy, S. *J. Phys. Chem. B* **2002**, *104*, 4387–4396.
59. Abramson, S.; Srithammavanh, L.; Siaugue, J. M.; Horner, O.; Xu, X.; Cabuil, V. *J. Nanoparticle* **2009**, *11*, 459–465.
60. Yu, X.; Liu, S.; Yu, J. *Appl. Catal., B* **2011**, *104*, 12–20.
61. Velegraki, T.; Mantzavinos, D. *Chem. Eng. J.* **2008**, *140*, 15–21.
62. Chen, H. W.; Ku, Y.; Kuo, Y. L. *Water Res.* **2007**, *41*, 2069–2078.

Chapter 9

Adsorption and Reduction of Some Aromatic Contaminants with Nanoscale Zerovalent Irons and Further Treatments

Yuh-fan Su, Chih-ping Tso, Yu-huei Peng, and Yang-hsin Shih*

Department of Agricultural Chemistry, National Taiwan University, Taipei 10617, Taiwan
***E-mail: yhs@ntu.edu.tw**

The nanoscale zerovalent iron (NZVI) has been used as a useful tool for the treatment of contaminated sites. NZVI and Pd/Fe nanoparticles were synthesized to investigate the removal of recalcitrant aromatic contaminants such as decabrominated diphenyl ether (DBDE), hexachlorobenzene (HCB), pentachlorophenol (PCP), and Congo red (CR) and some geochemical effects were also evaluated. These recalcitrant aromatic compounds can be fast removed with NZVIs. The degradation with NZVIs is favorable in an acid condition. Increasing the dose of NZVI particles enhanced the dechlorination rates of HCB. With an increase in temperature, the degradation rate increases. The degradation kinetics of HCB was not influenced by the HCO_3^- ion but was enhanced in the presence of the Cl^- ion due to its corrosion promotion. The NO_3^- competes with the reactants so it inhibits the degradation reaction. The enhancement effect of Cu^{2+} and Ni^{2+} ions on the degradation of PCP by Pd/Fe nanoparticles results from the deposition of reduced forms of copper and nickel on Pd/Fe surfaces. The debromination of PBDEs from para positions is more difficult than that from meta or ortho positions. Furthermore, adsorption process plays an important role on the removal mechanism on these halogenated aromatic compounds with NZVIs. The effect of the combination of microorganisms or sequential Fenton treatments on the mineralization of some contaminants was also evaluated.

Anaerobic microbes co-incubated with MZVI could lead to the enrichment of heterotrophic microbial populations and increase DBDE removal. Rapid decolorization of CR by NZVI and sequential Fe(II)(produced from NZVI)/H_2O_2 process can effectively decompose CR. To combine other treatment processes, the application of NZVI can be more attractive in pollutant removal. These better understandings serve as a useful reference for remediation design and prediction of treatment efficiency of aromatic contaminants with NZVIs.

Introduction

Polybrominated diphenyl ethers (PBDEs), one of emerging environmental contaminants, have been widely used as additive flame-retardants in various industrial products to reduce their fire hazards by interfering with the combustion of the polymeric materials. The concern about the risks of PBDEs has increased because the widely measured PBDEs in the environment, some PBDEs identified as endocrine disruptors (*1*), and PBDEs detected in human blood, milk, and tissue (*2*). The major industrial PBDE product is decabromodiphenyl ether (DBDE). Due to its extreme hydrophobicity and the largest number of bromine atoms, DBDE will accumulate in organic media (*3*). On the other hand, hexachlorobenzene (HCB) is one of the twelve persistent organic pollutants (POPs), which required the inventory and the action plan by the Stockholm Convention. Moreover, it is recalcitrant to degradation and has harmful effects on human beings and the environment (*4*). Pentachlorophenol (PCP), a carcinogenic compound, is a generally ionized chlorinated organic compound in neutral conditions that has been widely used as general pesticide, wood preservative, and industrial chemical. Although the use of PCP has been banned since 1984, it is still found in the environment. In view of their toxicity to the ecosystem and potential risk to human health, a new technology for efficient and effective removal of DBDE, HCB, and PCP is urgently needed.

Zerovalent iron (ZVI), generally in microscale (called MZVI), has been intensively studied for the remediation of a wide range of halogenated organic pollutants (*5–9*). Matheson and Tratnyek (*6*) presented three different reductive pathways for halogenated compounds in anoxic conditions. The main and general dehalogenation reaction can be represented by

$$Fe^0 + RX + H^+ \rightarrow Fe^{2+} + RH + X^- \quad (1)$$

ZVI has been used to remediate halogenated organic compound contaminated groundwater through the use of permeable barriers; however, the particle size and reactivity of conventional ZVI may limit its application. In the last decade, bimetallic particles have been exploited to dehalogenate organic compounds due to their high efficiencies compared with the primary metal alone. It has been suggested that supplying a second catalytic metal such as Pd, Pt, Ag, Ni, or Cu on primary zerovalent metal could prevent toxic byproduct formation by dehalogenating organic pollutants via hydrogen reduction rather than via electron transfer (*10*, *11*).

Since Wang and Zhang (*8*) presented that nanoscale zerovalent iron (NZVI), which has high specific surface area, showed much more reactivity for the transformation of halogenated organic compounds than commercial MZVI (*8*, *12–14*), NZVI has been used in the field remediation. Nanoscale bimetallic technology also offers new potential in treatment of organic and inorganic pollutants. Nanoscale Pd/Fe has the catalytic properties of palladium for hydrodechlorination (*15*) and can be used in groundwater remediation of chlorinated solvents. The site geochemistry can also affect the chemical reduction kinetics and efficacy of NZVI (*16*) for trichloroethylene (TCE).

For example, common ions in water, soil and groundwater will affect the degradation of halogenated compounds by NZVIs. Furthermore, in many remediation sites, some heavy metal ions were mixed contamination with hazardous halogenated compounds. Some cations such as Cu^{2+} can enhance the dechlorination of carbon tetrachloride by iron nanoparticles and led to the production of more benign products (*17*). Previous studies have showed that major anions can affect Fe^0 reactivity toward groundwater contaminants (*18*, *19*). Liu et al. (*16*) indicated that anions decreased NZVI reactivity with TCE in an increasing order: $Cl^- < SO_4^{2-} < HCO_3^- < HPO_4^{2-}$ at pH 8.9. Fan et al. (*20*) reported that SO_4^{2-} had a negative effect on the decolorization of methyl orange with NZVI. However, Lim and Zhu (*21*) concluded that SO_4^{2-} anions have no influence on dechlorination of trichlorobenzene by Pd/Fe. The anion effect depends on different experimental conditions including pH range (*22*, *23*), composition of precipitations (*21*, *24*, *25*), and anion concentrations (*16*). However, studies on the effect of cations and anions on degradation of persistent polyhalogenated organic pollutants by NZVI and Pd/Fe nanoparticles are scarce.

In the natural system, biodegradation is a very environmental friendly treatment. Previous studies have successfully combined biodegradation and metal degradation mechanisms in treating halogenated compounds (*26–30*). We also tried to evaluate the combined effects of MZVI and anaerobic sludge in PBDE degradation (*31*, *32*). On the other hand, the possible more toxic degradation byproducts after ZVI treated were produced so the further treatment such as the complete mineralization should be proposed. After the reduction process of with NZVI, zerovalent irons oxidized and formed ferrous ion and iron (hydro)oxides. Thus, we tried to add H_2O_2, which can react with ferrous ions in the reaction solution and then generate the hydroxyl free radicals, and then to conduct a Fenton oxidation (*33*).

The objectives of this chapter were to present our synthesized NZVI particles, the removal kinetics and mechanisms of recalcitrant aromatic contaminants such as DBDE, HCB, and Congo red (CR) (*33–37*) and the effect of environmental factors on them. We also synthesized Pd/Fe bimetallic nanoparticles and studied the effect of material properties and environmental factors on the removal of HCB and PCP (*38*, *39*). Furthermore, the effect of the combination of microorganisms or sequential Fenton treatments on the complete mineralization of some contaminants or degradation byproducts was also evaluated (*31*, *32*).

Effect of Nanoscale Material Properties

The NZVI and Pd/Fe nanoparticles synthesized had a grain size range from 20 nm to 100 nm and a specific surface area of approximately 46 m^2 g^{-1} (*34–36*, *38*, *39*). These as-prepared particles form chains of beads, which is attributed to magnetic interactions between the primary iron particles (*8*, *40*, *41*).

For DBDE, the chemical reduction kinetics by NZVI is much faster than that of commercial microscale zerovalent iron (MZVI) (*34*). In 90 min, around 40% of DBDE was transformed by MZVI; while more than 90% of DBDE was degraded by NZVI. The first-order rate constant of NZVI is about 7-fold higher than that of MZVI. The surface area normalized rate constants were defined as rate constants normalized by iron surface area per aqueous solution volume. The surface area normalized rate constant of NZVI is similar to that of MZVI indicating that the increased reactivity of the NZVI is mainly attributed to the higher surface area of the NZVI.

For HCB, the nanoscale Pd/Fe bimetallic particles reduced HCB more rapidly than NZVI (*38*), which resulted from the enhancement of electron release on Fe surface by the potential difference between Pd and Fe and the catalysis role of Pd (*12*, *42*).

In general, increasing the dose of NZVI particles enhanced the degradation rates and efficiencies of these aromatic contaminants (*35*, *36*), which was attributed to the increase in surface area and reaction sites on NZVI particles to accelerate the degradation reaction (*20*, *43*).

Effect of Environmental Factors

pH

Comparing the removal efficiencies of HCB by NZVI for the cases of pH 13 and 3, the latter had a more than 60% removal of HCB after a treatment time of around 48 h. In the former (pH 13), however, only about 10% HCB were degraded even when its treatment time was doubled (*36*). The degradation of HCB with NZVI is favorable in an acid condition. Similar phenomenon was also observed for the reduction of many contaminants by ZVIs such as DBDE (*34*), nitrate (*44*), and TCE (*45*). At high pH, a decrease in reactivity is expected with the build-up of iron (hydro)oxides on the surface of NZVI particles (*6*). Consequently, the dehalogenation process was inhibited due to the decrease of the electron transfer from ZVI surface. The enhancement effect of the acidic condition might be that a lower pH gives rise to a greater extent of acid washing of ion (hydro)oxides on metallic iron surfaces (*46*). That is, lowering pH would dissolve away passive layers at the surface of NZVI yielding fresh reactive sites for degradation reaction. Furthermore, iron corrosion in acid conditions could be accelerated, producing enough hydrogen (or hydrogen atoms), which favors the hydrogenation reaction.

Temperature

A progressively better removal of DBDE and HCB by NZVI was observed as the temperature increased. The rate constants for HCB increased from 0.050 h^{-1} at 5 °C to 0.12 h^{-1} at 45 °C (*36*), which clearly suggests that temperature can facilitate the degradation of polychlorinated compounds in aqueous solution. Compared to literature data, the relatively small activation energy of degradation of HCB by NZVI, 16.6 kJ mol^{-1}, indicating the less temperature dependency of the application of NZVI in the remediation. Furthermore, due to the rapid degradation ability of NZVIs and a limited temperature range in the general environment, the effect of temperature on the removal of polyhalogenated aromatic compounds in the subsurface could be quiet small.

Anions

NO_3^- inhibited the HCB degradation by NZVI particles due to the competitive process between of HCB with NO_3^- on NZVI surface in aqueous solutions (*37*). After the measurement of aqueous NO_3^- concentrations, NO_3^- ions were reduced in this study by that the NO_3^- concentration decreased to non-detectable level at the end of the dechlorination of HCB with NZVI. Cl^- and SO_4^{2-} ions enhanced HCB degradation by NZVI particles due to their corrosion promotion (*37*). But SO_4^{2-} ion inhibited reactivity of Pd/Fe nanoparticles toward PCP (*39*). The inhibitory effect could be attributed to SO_4^{2-} may form inner-sphere complexes on iron oxides which may adsorb on the iron surface and block reactive sites (*16*, *47*). Furthermore, SO_4^{2-} was reduced by ZVI and produced HS^- (*48*) and consequently resulted in the poisoning of Pd catalysts. HCO_3^- ion can enhance or inhibit NZVI reactivity toward contaminants due to accelerated corrosion of iron (*49*), pH buffer action of the H_2CO_3/HCO_3^- system (*50*), passivation of reactive sites from precipitation of $FeCO_3$ and Fe-anion complexes (*16*). However, HCO_3^- ion had no significant impact on the degradation kinetics of HCB by NZVI particles (*37*).

Cations

A schematic diagram illustrating the effect of cations in solutions on degradation of aromatic contaminants by NZVI and Pd/Fe nanoparticles is shown in Figure 1. Aqueous Cu^{2+} and Ni^{2+} ions could facilitate the degradation kinetics and efficiencies of PCP by Pd/Fe nanoparticles due to the deposition of reduced forms of copper and nickel on Pd/Fe surfaces (*39*). The positive standard potentials of the reduction of Cu^{2+} and Ni^{2+} ions coupled with the oxidation of Fe suggested that these redox reactions are spontaneous processes. Some portion of Cu^{2+} and Ni^{2+} ions were reduced to zerovalent copper and nickel and coated on the Pd/Fe surface. The presence of copper and nickel on Pd/Fe nanoparticles were confirmed by ICP–MS analysis. XANES absorption spectra were also confirmed their reduced forms of copper and nickel on Pd/Fe. Cu coatings facilitate electron transfer and promote corrosion of the Fe^0 (*51*). Like palladium, nickel is known as hydrodehalogenation and hydrogenation catalyst (*52*, *53*). The precipitation of Ni on Pd/Fe could act as new reactive sites and then enhance the degradation of

PCP due to their catalytic effect. The presence of Fe^{3+} ions also could enhance degradation of PCP by Pd/Fe nanoparticles (*39*). Hydrolysis of the added Fe^{3+} resulted in the production of H^+ ions and decreased the pH (*54, 55*) which likely accounted for the enhancement in degradation of PCP. In contrast, the Fe^{2+} ions would inhibit the reduction of HCB by NZVI might be attributed to Fe^{2+} ions enhance the formation of passive precipitates of Fe_3O_4 that coated on the surface of NZVI (*37*). Mg^{2+} and Na^+ ions showed no influence on degradation of HCB by NZVI and PCP by Pd/Fe nanoparticles (*37, 39*). The standard reduction potential of Mg^{2+} (-2.372 V) and Na^+ (-2.71 V) ions is less than that of Fe^{2+} ions (-0.44 V), it is believed that Mg^{2+} and Na^+ ions cannot affect reactivity of iron metal by reduction of ions.

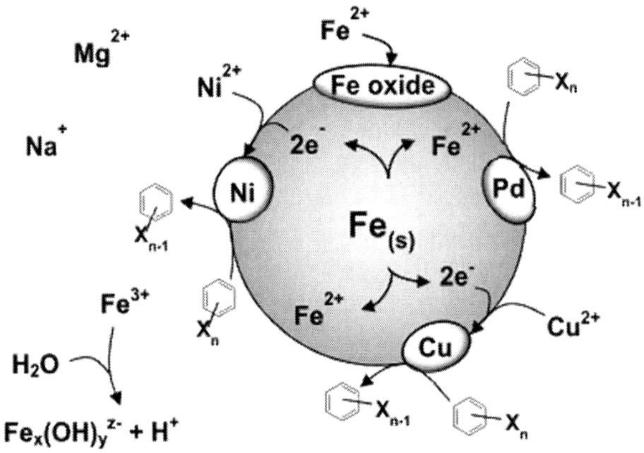

Figure 1. Schematic diagram illustrating the effect of cations on degradation of halogenated aromatic contaminants by NZVI and Pd/Fe nanoparticles.

Removal Mechanism and Degradation Pathway

The reductive dehalogenation by successive loss of halogen atoms on halogenated aromatic compounds is the main mechanism of the degradation pathway for these polyhalogenated aromatics with NZVIs, e.g. DBDE with NZVI (*34*), HCB with NZVI (*36*) and Pd/Fe nanoparticles (*38*). For example, for HCB, the first step in the reductive transformation is loss of one chlorine atom, resulting in penta-chlorobenzene (PCB). PCB continued dechlorination to form less chlorinated benzenes.

The reductive debromination process and by-products of DBDE with NZVI were also identified (*34*). The first step is also a loss of one bromine atom on DBDE, resulting in nona-BDEs. Each of the three nona-BDEs continued debromination to form lower brominated BDEs. According to these byproduct identifications, the bromines at ortho or meta positions are more susceptible to debromination than those at the para position (*34*). The PBDEs with bromines at the para positions were mostly preserved in all these identified intermediates. The preferential accumulation of para-brominated products (2,3′,4,4′,6-BDE119,

2,2′,4,4′,5,6′-BDE154, 2,2′,3,4,4′,5′,6-BDE183, 2,2′,3,4,4′,6′,6-BDE184), indicating that the debromination from the para position was the most unfavorable pathway. BDE119 sequentially debrominated to several BDEs with two para-bromines (4,4′-BDE15, 2,4,4′-BDE28, 2,2′,4,4′-BDE47, 2,4,4′,6-BDE75) and with one para-bromine (4-BDE3, 2,4-BDE7, 2,2′,4-BDE17), suggesting that para-debromination is not easy with NZVI (*34*).

At an acid condition, the degradation products follow the same trend of reductive debromination to form mono-BDE (*34*). However, at pH 10, the sequential debromination process was also observed (only nona- to penta-BDEs) but the continued reductive debromination of pentabromo congeners did not occur. The passive layers could form on the iron surface in an alkaline condition to decrease the electron transfer from the ZVI surface. Consequently, the subsequent debromination process was inhibited.

The reductive debromination process of DBDE with NZVI in different pH conditions showed different kinetics and efficacy (*34*). With the increase of pH, the residual DBDE fraction and identified debrominated byproduct fraction in aqueous phase increased (*34*). Only 18% of DBDE were adsorbed on NZVI surface at pH 5. However, around 43% of DBDE were adsorbed on NZVI when aqueous pH higher than 7. Overall, adsorption on NZVI plays a role on the removal of DBDE due to its high hydrophobicity and the (hydro)oxides formed on NZVI after reactions .

Combination of Biodegradation and Metal Degradation Treatments

We evaluated that the combined effects of MZVI and anaerobic sludge in DBDE and 4-monobrominated diphenyl ether (BDE-3) degradation (*31*, *32*). In this co-incubation, 63 % and 29 % enhancement of removal ability when compared to the single component conditions were found. Although microbes could hinder the accessibility of MZVI to DBDE and reduce the removal ability in the initial stage, the existence of reducing-bacteria and hydrogen-bacteria enhanced the degradation ability of MZVI, indicating microorganisms contribute to enhance the degradation ability of MZVI.

The anaerobic microbes could alter of oxidation state of metals and get support from metal oxides with the debromination ability when co-incubated with MZVI (*56*, *57*). Such synergistic effects due to the contribution from the coexisted microscale iron particle and microorganisms are illustrated in Figure 2. In conjunction with the oxidation of ZVI, the halogenated compounds were reduced. Preliminary dehalogenation by ZVI of a highly halogenated compound such as DBDE generated less halogenated byproducts that, in turn, could be susceptible to increase the chemicals' degradablity for microbes and then enhance microbial degradation. Therefore, pretreatment by ZVI is able to elevate the biodegradation capability, from which the degradation of halogenated compounds would be performed through co-metabolic and halorespiration processes (*27*, *28*). Besides, H_2 produced from the oxidation of iron with water could act as the electron donor in anaerobic processes (reaction 3 in Figure 2). Furthermore, the oxidation of Fe(II) could also supply energy for denitrification and the co-metabolic dehalogenation reaction (reaction 1 in Figure 2). The reduction of iron could prolong the half-life of the provided ZVI. The promotion of halorespiration is helpful for biodegradation. Iron-reducing bacteria enriched could maintain the iron cycle and enhance the reactivity of ZVI (reaction 2 in Figure 2) (*30*, *57–59*). The synergistic effect of microbes and MZVI for PBDE degradation was found (*32*) and could be applied in the remediation.

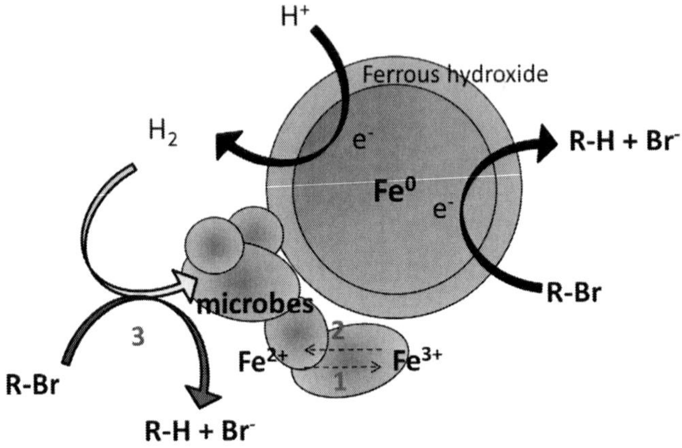

Figure 2. Conceptual model for the debromination of brominated organic compound by microscale iron and microorganisms under anaerobic conditions. Reaction 1: the oxidation of Fe(II) could supply energy for denitrification and the co-metabolic dehalogenation reaction; reaction 2: iron-reducing bacteria enriched could maintain the iron cycle and enhance the reactivity of ZVI; reaction 3: H_2 produced from the oxidation of iron with water could act as the electron donor in anaerobic processes.

Complete Mineralization by the Sequential Oxidation Treatment

After the reduction of recalcitrant organic contaminants with NZVIs, NZVIs were oxidized and formed ferrous ions and iron (hydro)oxides. The reduced contaminants could still keep their structure and might have toxic concern. A further treatment could be needed. Opportunely, after adding hydrogen peroxide, the ferrous ions produced from NZVIs in the reaction solution can generate the hydroxyl free radicals, showing a Fenton oxidation:

$$Fe^{2+} + H_2O_2 \rightarrow Fe^{3+} + OH\bullet + OH^- \quad (4)$$

$$RH + OH\bullet \rightarrow R\bullet + H_2O \quad (5)$$

$$R\bullet + Fe^{3+} \rightarrow R^+ + Fe^{2+} \quad (6)$$

$$Fe^{2+} + OH\bullet \rightarrow Fe^{3+} + OH^- \quad (7)$$

A sequential NZVI/H_2O_2 system is performed to enhance the mineralization of organic contaminants such as azo-dyes (*33*). For example, many azo-dyes such as Congo red (CR), could be fast removed by a suitable amount of NZVI (*33, 35*). However, the intermediate analysis suggests the presence of broken azo-bonds in CR with NZVI surface and thus generated a concern pollutant of its byproduct, benzidine. The sequential NZVI/H_2O_2 process can decompose CR effectively due to the Fenton reaction. The rapid removal kinetics and efficiency suggested that the sequential NZVI/H_2O_2 process can decompose azo-dyes and its byproducts effectively. For the comparison to the pre-treatment of CR with NZVI, $FeCl_2$ was used in the $FeCl_2$/H_2O_2 system as the dissolved iron source, indicating the removal efficiency was less than 20 %. Fenton reaction cannot effectively decompose the original structure of CR due to the recalcitrant nature of CR. After the reactions with NZVI, the degraded byproducts can be more easily mineralized than CR, indicating that the combined treatment of initial reduction with NZVI followed by oxidation with H_2O_2 can effectively mineralize azo-dyes from aqueous solutions. This further treatment process can be applied into the other recalcitrant organic contaminants after the reduction process with ZVIs.

Conclusions

The synthesized NZVI and Pd/Fe nanoparticles can rapidly reduce these recalcitrant organic contaminants such as PBDEs, HCB, PCP, and azo-dyes from water. The increase of temperature slightly enhances the degradation so only a relatively small activation energy is needed for the chemical reduction process by NZVIs. On the other hand, the temperature may not contribute a lot on the chemical reduction of these contaminants due to the limited temperature range in the general environment. The degradation with NZVIs is generally favorable in an acid condition. The reductive dehalogenation mechanism is the main pathway for the degradation of these polyhalogenated aromatic compounds with NZVIs. For PBDEs, the debromination of PBDEs from para positions is more difficult than that from meta or ortho positions. With the increase of aqueous pH, the chemical reduction of PBDEs diminished. Adsorption on NZVIs also plays a role on the removal of PBDEs.

For the dissolved ions in water, the degradation kinetics of HCB with NZVI was not influenced by the HCO_3^-, Mg^{2+}, and Na^+ ions but was enhanced in the presence of the Cl^- and SO_4^{2-} ions due to their corrosion promotion. The NO_3^- competes with the reactants so it inhibits the degradation reaction. In the presence of Cu^{2+} ions resulted from the reduced form of copper on NZVI surfaces so increase the degradation. For PCP, the similar enhancement effect of Cu^{2+} and Ni^{2+} ions on Pd/Fe nanoparticles also resulted from the deposit of copper and nickel on Pd/Fe surfaces.

The co-incubation of anaerobic sludge with MZVI enhanced 63% and 29% removal ability toward DBDE compared to each only condition because MZVI leads to the enrichment of heterotrophic microbial populations bearing nitrate- or iron-reducing activities. The interaction between MZVI and microbes contributed to the synergistic effect.

Rapid decolorization behaviors of CR with NZVI were observed but the mineralization of CR with NZVI is very low. Sequential NZVI/H_2O_2 processes can effectively decompose CR due to the Fenton reaction. Combination of the reduction power of NZVI and its oxidized ions with H_2O_2 can be considered an alternative for treating many other recalcitrant organic compounds.

In summary, fast removal of these recalcitrant aromatic compounds with NZVIs was demonstrated. Furthermore, adsorption process plays an important role on the removal mechanism on these halogenated aromatic compounds with NZVIs. On the other hand, to combine other treatment processes, the application of NZVI can be more attractive in pollutant removal. Our results contribute to a better understanding of aromatic contaminant degradation and serve as a useful reference for remediation design and prediction of treatment efficiency of aromatic contaminants with NZVIs.

Acknowledgments

The authors gratefully acknowledge the financial support of the National Science Council of Taiwan.

References

1. Darnerud, P. O.; Eriksen, G. S.; Johannesson, T.; Larsen, P. B.; Viluksela, M. *Environ. Health Perspect.* **2001**, *109*, 49–68.
2. Hites, R. A. *Environ. Sci. Technol.* **2004**, *38*, 945–956.
3. Schecter, A.; Papke, O.; Tung, K. C.; Staskal, D.; Birnbaum, L. *Environ. Sci. Technol.* **2004**, *38*, 5306–5311.
4. Murakami, T.; Shimura, M.; Kasai, E. *Environ. Sci. Technol.* **2008**, *42*, 7459–7463.
5. Keum, Y. S.; Li, Q. X. *Environ. Sci. Technol.* **2005**, *39*, 2280–2286.
6. Matheson, L. J.; Tratnyek, P. G. *Environ. Sci. Technol.* **1994**, *28*, 2045–2053.
7. Zhang, W. X. *J. Nanopart. Res.* **2003**, *5*, 323–332.
8. Wang, C. B.; Zhang, W. X. *Environ. Sci. Technol.* **1997**, *31*, 2154–2156.
9. Yan, W. L.; Lien, H. L.; Koel, B. E.; Zhang, W. X. *Environ. Sci.-Process Impacts* **2013**, *15*, 63–77.
10. Li, T.; Farrell, J. *Environ. Sci. Technol.* **2000**, *34*, 173–179.
11. Xu, X. H.; Zhou, H. Y.; He, P.; Wang, D. H. *Chemosphere* **2005**, *58*, 1135–1140.
12. Lien, H. L.; Zhang, W. X. *Colloids Surf., A* **2001**, *191*, 97–105.
13. Lowry, G. V.; Johnson, K. M. *Environ. Sci. Technol.* **2004**, *38*, 5208–5216.
14. Song, H.; Carraway, E. R. *Environ. Sci. Technol.* **2005**, *39*, 6237–6245.
15. Lien, H. L.; Zhang, W. X. *Appl. Catal., B* **2007**, *77*, 110–116.
16. Liu, Y.; Phenrat, T.; Lowry, G. V. *Environ. Sci. Technol.* **2007**, *41*, 7881–7887.
17. Lien, H. L.; Jhuo, Y. S.; Chen, L. H. *Environ. Eng. Sci.* **2007**, *24*, 21–30.
18. Devlin, J. F.; Allin, K. O. *Environ. Sci. Technol.* **2005**, *39*, 1868–1874.

19. Scherer, M. M.; Richter, S.; Valentine, R. L.; Alvarez, P. J. J. *Crit. Rev. Environ. Sci. Technol.* **2000**, *30*, 363–411.
20. Fan, J.; Guo, Y. H.; Wang, J. J.; Fan, M. H. *J. Hazard. Mater.* **2009**, *166*, 904–910.
21. Lim, T. T.; Zhu, B. W. *Chemosphere* **2008**, *73*, 1471–1477.
22. Fan, X. M.; Guan, X. H.; Ma, J.; Ai, H. Y. *J. Environ. Sci.* **2009**, *21*, 1028–1035.
23. Huang, Y. H.; Zhang, T. C. *Water Res.* **2005**, *39*, 1751–1760.
24. Bi, E. P.; Bowen, I.; Devlin, J. F. *Environ. Sci. Technol.* **2009**, *43*, 5975–5981.
25. Kohn, T.; Livi, K. J. T.; Roberts, A. L.; Vikesland, P. J. *Environ. Sci. Technol.* **2005**, *39*, 2867–2879.
26. Murugesan, K.; Bokare, V.; Jeon, J. R.; Kim, E. J.; Kim, J. H.; Chang, Y. S. *Bioresour. Technol.* **2011**, *102*, 6019–6025.
27. Oh, S. Y.; Chiu, P. C.; Kim, B. J.; Cha, D. K. *Water Res.* **2005**, *39*, 5027–5032.
28. Saxe, J. P.; Lubenow, B. L.; Chiu, P. C.; Huang, C. P.; Cha, D. K. *Water Environ. Res.* **2006**, *78*, 19–25.
29. Tseng, L. H.; Li, M. H.; Tsai, S. S.; Lee, C. W.; Pan, M. H.; Yao, W. J.; Hsu, P. C. *Chemosphere* **2008**, *70*, 640–647.
30. Van Nooten, T.; Springael, D.; Bastiaens, L. *Environ. Sci. Technol.* **2008**, *42*, 1680–1686.
31. Shih, Y. H.; Chou, H. L.; Peng, Y. H. *J. Hazard. Mater.* **2012**, *213*, 341–346.
32. Shih, Y. H.; Chou, H. L.; Peng, Y. H.; Chang, C. Y. *Bioresour. Technol.* **2012**, *108*, 14–20.
33. Shih, Y. H.; Tso, C. P. *Environ. Eng. Sci.* **2012**, *29*, 929–933.
34. Shih, Y. H.; Tai, Y. T. *Chemosphere* **2010**, *78*, 1200–1206.
35. Shih, Y. H.; Tso, C. P.; Tung, L. Y. *J. Environ. Eng. Manage.* **2010**, *20*, 137–143.
36. Shih, Y. H.; Hsu, C. Y.; Su, Y. F. *Sep. Purif. Technol.* **2011**, *76*, 268–274.
37. Su, Y. F.; Hsu, C. Y.; Shih, Y. H. *Chemosphere* **2012**, *88*, 1346–1352.
38. Shih, Y. H.; Chen, Y. C.; Chen, M. Y.; Tai, Y. T.; Tso, C. P. *Colloids Surf., A* **2009**, *332*, 84–89.
39. Shih, Y. H.; Chen, M. Y.; Su, Y. F. *Appl. Catal., B* **2011**, *105*, 24–29.
40. Zhang, L.; Manthiram, A. *Appl. Phys. Lett.* **1997**, *70*, 2469–2471.
41. Phenrat, T.; Saleh, N.; Sirk, K.; Tilton, R. D.; Lowry, G. V. *Environ. Sci. Technol.* **2007**, *41*, 284–290.
42. Kim, Y. H.; Carraway, E. R. *Environ. Technol.* **2003**, *24*, 69–75.
43. Zhang, X.; Lin, Y. M.; Chen, Z. L. *J. Hazard. Mater.* **2009**, *165*, 923–927.
44. Huang, Y. H.; Zhang, T. C. *Water Res.* **2004**, *38*, 2631–2642.
45. Liu, Y. Q.; Lowry, G. V. *Environ. Sci. Technol.* **2006**, *40*, 6085–6090.
46. Yang, G. C. C.; Lee, H. L. *Water Res.* **2005**, *39*, 884–894.
47. Su, C. M.; Puls, R. W. *Environ. Sci. Technol.* **2004**, *38*, 2715–2720.
48. Blowes, D. W.; Ptacek, C. J.; Benner, S. G.; McRae, C. W. T.; Bennett, T. A.; Puls, R. W. *J. Contam. Hydrol.* **2000**, *45*, 123–137.
49. Reardon, E. J. *Environ. Sci. Technol.* **1995**, *29*, 2936–2945.
50. Wieckowski, A.; Ghali, E.; Szklarczyk, M.; Sobkowski, J. *Electrochim. Acta* **1983**, *28*, 1619–1626.

51. Hou, M. F.; Wan, H. F.; Liu, T. L.; Fan, Y. N.; Liu, X. M.; Wang, X. G. *Appl. Catal., B* **2008**, *84*, 170–175.
52. Roy, H. M.; Wai, C. M.; Yuan, T.; Kim, J. K.; Marshall, W. D. *Appl. Catal., A* **2004**, *271*, 137–143.
53. Semmelhack, M. F.; Helquist, P.; Jones, L. D.; Keller, L.; Mendelson, L.; Ryono, L. S.; Smith, J. G.; Stauffer, R. D. *J. Am. Chem. Soc.* **1981**, *103*, 6460–6471.
54. Satapanajaru, T.; Anurakpongsatorn, P.; Pengthamkeerati, P.; Boparai, H. *Water, Air, Soil Pollut.* **2008**, *192*, 349–359.
55. Calza, P.; Maurino, V.; Minero, C.; Pelizzetti, E.; Sega, M.; Vincenti, A. *J. Photochem. Photobiol., A* **2005**, *170*, 61–67.
56. Hennebel, T.; De Gusseme, B.; Boon, N.; Verstraete, W. *Trends Biotechnol.* **2009**, *27*, 90–98.
57. Weber, K. A.; Achenbach, L. A.; Coates, J. D. *Nat. Rev. Microbiol.* **2006**, *4*, 752–764.
58. El Fantroussi, S.; Naveau, H.; Agathos, S. N. *Biotechnol. Prog.* **1998**, *14*, 167–188.
59. McCormick, M. L.; Bouwer, E. J.; Adriaens, P. *Environ. Sci. Technol.* **2002**, *36*, 403–410.

Chapter 10

Pollutant Degradation by Nanoscale Zero Valent Iron (nZVI): Role of Polyelectrolyte Stabilization and Catalytic Modification on nZVI Performance

A. W. McPherson,[1] M. N. Goltz,[1] and A. Agrawal[*,2]

[1]Department of Systems and Engineering Management, Air Force Institute of Technology, Wright Patt Air Force Base, Ohio 45433
[2]Department of Earth & Environmental Sciences, Wright State University, Dayton, Ohio 45435
[*]E-mail: abinash.agrawal@wright.edu

Nanoscale zero-valent iron (nZVI) is an attractive new groundwater remediation tool, which potentially offers increased reactivity toward and improved access to subsurface contaminants. Polyelectrolyte stabilization and bimetal catalyzation are two of the most promising methods for increasing the performance of nZVI. Addition of polyelectrolyte stabilizers to nZVI decreases particle agglomeration and reduces particle size, increasing net reactivity and enhancing transport in column studies. Amendment of nZVI with metallic catalysts improves reactivity toward almost all groundwater contaminants and, in the case of chlorinated organic compounds, reduces the formation of toxic byproducts. Here, these two methods are reviewed in depth. Physical processes involved are explored, the state-of-the-art is presented, and opportunities for future research are discussed. A general lack of comparability between past studies and the need for greater standardization of nZVI testing procedures is identified. For polyelectrolyte stabilization, prospects for future research include comparative studies that evaluate the performance of many polyelectrolytes side-by-side, use of engineered co-polymers to improve non-aqueous phase liquid (NAPL) targeting by nZVI, and resolution of poor nZVI transport observed at flowrates, ionic strengths, and hydraulic conductivity heterogeneity conditions representative of the subsurface. For bimetal catalyzation, opportunities for future research include comparative studies that evaluate the performance of many catalysts side-by-side, assessment of untested or largely overlooked catalysts such as Ag, and resolution of economic and health issues associated with the use of costly or toxic catalysts such as Pd and Ni, respectively.

Introduction

Nanoscale zero-valent iron (nZVI) is an exciting and rapidly developing new tool for the treatment of contaminated groundwater. Its potential applications include improved access to subsurface zones, rapid source zone remediation, economy over traditional remediation techniques, and site-specific tailoring of the nanoparticles to enhance performance. Synthesizing such small particles, only 10^{-9} m in diameter, vastly increases their reactive surface area per gram of iron. The specific surface area of nZVI is typically tens of square meters per gram, one to two orders of magnitude higher than micro-scale iron particles (*1*). As a result, nZVI particles that are much more reactive, in comparison to micro-scale iron particles, are well-suited for *in situ* remediation of a variety of contaminants. nZVI has been shown to effectively degrade halogenated organics, pesticides, antibiotics, organophosphates, organic dyes, explosive compounds, heavy metals, perchlorate, and nitrate (*2*, *3*). The mechanisms of contaminant treatment/removal by nZVI, which include reduction, sorption, complexation, and co-precipitation, may be contaminant-dependent (*2*, *3*). Further, it is hypothesized that smaller, nanoscale iron particles will be mobile in the subsurface, thereby facilitating their transport to target contaminated zones following their injection in the subsurface. The majority of interest in nZVI is focused on its application via direct injection or recirculating wells to treat/degrade the contaminants located in subsurface source zones, particularly those that are inaccessible to conventional treatment methods due to depth or site-use (*2*, *4*–*6*). nZVI may also be injected to form permeable reactive barriers (*2*, *7*, *8*).

A wide variety of approaches are used to synthesize nZVI particles, but they can generally be separated into two categories: top-down, in which coarser iron is refined into nano-scale particles, and bottom-up, in which nano-scale particles are grown from Fe ion precursors. Examples of top-down synthesis include laser ablation, spark discharge generation, and milling. Examples of bottom-up synthesis include vacuum sputtering, electrodeposition of iron ions, and the reduction of oxide compounds (*2*, *3*). For laboratory-scale and smaller field-scale use, the ‘borohydride reduction’ method originally proposed by Wang and Zhang in 1997 is most-widely used (*1*, *5*, *6*). In contrast, two of the leaders in commercial-scale synthesis of nZVI, Golder Associates Inc. and Toda Kogyo

Ltd., use different methods. Golder uses a top-down ball mill system while Toda Kogyo uses hydrogen to reduce pre-synthesized iron oxide nanoparticles (*2*, *9*). A large number of bench-scale studies employed nZVI provided by Toda Kogyo nanoparticles.

Numerous in-depth reviews of various aspects of contaminant removal processes by nZVI have been published in recent years, which include discussions of surface chemistry, particle characterization, potential toxicity, target contaminants for remediation, *in situ* implementability, application in permeable reactive barriers (PRBs), and small-scale field studies (*2–4*, *7*, *9–19*). Given the rapid pace of nZVI research in the last decade, however, some of these reviews are now unavoidably dated or limited in scope. Two recent developments that seem significant and may hold great promise, but have not yet been fully examined, are: (i) the role of stabilizing nZVI particles by polyelectrolytes in terms of improving their reactivity and transport characteristics, and (ii) the effect of modifying nZVI particles with a secondary catalyst metal for improvement in contaminant destruction. This work summarizes the published literature relevant to these aspects of nZVI synthesis and identifying the prospects for future research in this subject area.

Role of Polyelectrolyte Stabilization on nZVI Performance

Perhaps the most significant challenge to implementing nZVI injection for subsurface remediation is the tendency of individual nZVI particles to agglomerate quickly into discrete micro-scale aggregates or larger chain aggregates (*20*), where agglomeration of nanoparticles occurs due to attractive interparticle van der Waals and magnetic forces (*19*, *20*). The effect of nZVI agglomeration is two-fold; (i) the available reactive surface area of nZVI particles is significantly reduced, thus negatively impacting their reactivity, and (ii) the transport of larger, microscale iron aggregates in porous media becomes severely restricted (*19*). Modifying nZVI particles with a polyelectrolyte coating during their synthesis has been conclusively shown to overcome interparticle attractive forces, thereby inhibiting their aggregation and improving stability, transport, and reactivity (*21–24*). Use of polyelectolytes to stabilize nZVI was pioneered by Mallouk and co-workers beginning in 2000, where carbon nanoparticles and polyacrylic acid (PAA) resin were used as supports for nZVI (*25*, *26*). Since 2004 (*21*), there has been rapid growth in the use of polyelectrolyte coated, or stabilized, nZVI; to date, no fewer than 17 different stabilizing agents have been used (See Table 1). The benefits of using polyelectrolytes to stabilize nZVI include their low cost, wide availability, ease of implementation, non-toxic nature, and the possible promotion of long-term, complementary reductive *in situ* biodegradation of contaminants (*6*). Several polyelectrolytes, such as carboxymethyl cellulose (CMC) and xanthan gum, are food-grade additives (*22*, *27*, *28*). Other polyelectrolytes, such as chitosan and soy protein, are derived from natural sources (*28*, *29*).

Table 1. List of polyelectrolytes stabilizers for nZVI

Polyelectrolyte	*Reference(s)*
CMC (Carboxymethyl Cellulose)	(*6, 22–24, 27, 31–34, 39, 41, 44, 56–58, 60, 73, 84, 87*)
PAA (Polyacrylic Acid)	(*21, 23, 28, 32, 39, 40, 47, 103*)
PAP (Polyaspartate)	(*20, 31, 34, 43, 49, 51, 55, 103*)
PSS (Polystyrene Sulfonate)	(*20, 23, 31, 43, 48, 50*)
PAM (Polyacrylamide)	(*23*)
PVP (Polyvinylpyrrolidone)	(*24*)
PV3A (Polyvinyl alcohol-co-vinyl acetate-co-itaconic acid)	(*28, 53*)
OMA (Olefin Maleic Acid)	(*41, 42, 105*)
Triblock ($PMAA_{42}$-$PMMA_{26}$-PSS_{462})	(*51, 52*)
Triblock ($PMAA_{48}$-$PMMA_{17}$-PSS_{650})	(*49, 51, 52*)
Triblock ($PMAA_{15}$-$PBMA_{43}$-PSS_{811})	(*50*)
Guar Gum	(*24, 54, 104*)
Xanthan Gum	(*36, 37*)
Cellulose Acetate	(*92*)
Starch	(*22, 26*)
Soy Proteins	(*28*)
Chitosan	(*29, 80*)

As their name suggests, polyelectrolytes are chains of individual polymers, with each repeating polymer unit bearing an electrolyte group. The long-chained polymers typically have a very high molecular weight, on the order of thousands of grams per mole. When introduced to water, the electrolyte groups disassociate, leaving the polymer units either positively or negatively charged, and releasing their complementary ions to solution (*30*). Polymer adhesion to nZVI occurs through either physisorption, driven by dispersive or electrostatic forces, or chemisorption, by covalent bonding to the surface (*31*). The charged polymer units aid the polyelectrolyte's adsorption to the nZVI surface via physisorption. In contrast, nZVI stabilization by CMC and PAA occurs by chemisorption, specifically, mono- and bidentate chelating, bidentate bridging, and hydrogen bonding (*27, 32*). Because polyelectrolyte chains are flexible, they may adsorb to nZVI particles at several points per chain, resulting in high adsorption energies and slow desorption. Kim et al. (2009) showed that for polyaspartate (PAP), CMC, and polystyrene sulfonate (PSS), less than 30% (w/w) desorbed over four months (*31*). This rate is insignificant in the context of both laboratory and field-scale use of nZVI, where the reactive particles aggregate within hours to days. While many polyelectrolytes are generally identified by type, others are further identified by their molecular weights. The distinction is an important one; the variability in performance between different polyelectrolytes of different molecular weights of the same type may be as much or greater than between type (*20*, *24*, *31*, *33*, *34*).

Polyelectrolyte coatings on nZVI can successfully prevent agglomeration and stabilize the nZVI particles by increasing the repulsive forces between individual nanoparticles, and in some cases, by also forming gel-like networks between the particles (*35–37*). The colloidal stability of disaggregated nZVI nanoparticles in aqueous solutions is commonly understood through *Derjaguin, Landau, Verwey and Overbeek (DLVO) theory*, under which repulsive steric and electrostatic double layer (EDL) forces oppose attractive van der Waals and magnetic forces. Because these forces act over different distances, their sum, which predicts whether aggregation will occur or not, is dependent on the distance between particles. EDL forces arise when the charged surface of a nanoparticle attracts oppositely charged aqueous ions from the surrounding fluid. This "cloud" or double layer of charged ions around a particle develops a repulsive electrical force towards similarly charged particles. Polyelectrolytes, with their ionic polymer units, increase the charge surrounding the nanoparticle, and also, the repulsive force. In addition, steric repulsion may arise from the interaction of the polyelectrolyte coating with that of another particle, or with a surface, and the associated thermodynamic penalty (*20*, *32*). Often, the combined effect of EDL and steric forces is referred to as electrosteric repulsion (*20*, *31*).

In addition to the stabilization effect of polyelectrolytes on nZVI, the presence of polyelectrolytes during the precipitation-synthesis of nanoparticles may also affect particle nucleation and, thus, nanoparticle size. When nanoparticles are precipitated in aqueous solution, many small crystallites initially form which act as nuclei for further growth. Polyelectrolytes can mediate faster and more numerous nucleation and slow particle growth, which yields more numerous, smaller diameter particles (*38*). For nZVI, such effects were reported twice (*27*, *33*). In contrast, Cirtiu et al. (2011) reported that the presence of certain polyelectrolytes (CMC, PAA, PSS, and polyacrylamide or PAM), while inhibiting agglomeration, increased individual particle size compared to unstabilized nZVI (*23*); it was proposed that such polymers had chelated aqueous Fe ions, resulting in an artificial increase in Fe concentration locally at discrete nucleation points prior to precipitation.

CMC polyelectrolytes of various molecular weights have been shown to produce nZVI particles that are smaller than those facilitated by several other polyelectrolytes, e.g., PAP, PSS, PAA, polyvinylpyrrolidone (PVP), PAM, and guar gum. Cirtiu et al. (2011) investigated the differences between pre- and post-synthesis stabilization of nZVI by CMC, PAM, PSS, and PAA (*23*); they reported that while post-synthesis stabilization generally resulted in smaller particle diameters, more stable colloidal suspensions were produced when nZVI was pre-stabilized, i.e., polyelectrolytes were added prior to nZVI synthesis. This difference between pre- versus post-nZVI synthesis amendment of polyelectrolytes has made the comparison of performance of different polyelectrolytes difficult, in terms of nZVI particle size and colloidal stability, since both synthesis methods were used in recent studies (*20*, *23*, *24*, *31*, *39*). In post-synthesis stabilization studies, CMC performs poorest in terms of colloidal stability, with PSS and PAP proving best (*20*, *31*). However, pre-synthesis stabilization studies suggest that CMC outperforms PAA, PAM, and PSS, though not guar gum, in terms of nZVI stability (*23*, *24*, *39*).

Transport of Polyelectrolyte-Stabilized nZVI

The size and stability of the iron nanoparticles are used as predictors of their transport potential through porous media. The transport of polyelectrolyte-stabilized nZVI through saturated porous media has been studied extensively in the last 5 years. In 1-D column and 2-D bench-scale studies, the transport of stabilized nZVI is conclusively superior to untreated nZVI (*19*, *28*, *31*, *32*, *40–42*). The transport behavior of nZVI stabilized by different polyelectrolyte types is largely similar (*28*, *31*, *32*, *39*). Yet, small differences between polyelectrolyte types may cause differences in the adsorbed polyelectrolyte layer thickness and allow nZVI agglomeration (*41*, *43*). Further, extrapolating bench-scale column results of nZVI breakthrough curves for predicting nZVI field-scale transport may be questionable. For example, He et al. (2009) concluded that although CMC-stabilized nZVI showed excellent transport in a variety of media-filled columns, groundwater flow modeling at field scale suggested that at a realistic groundwater flow-rate of 0.1 m/day, 99% of the nZVI would be removed within 16 cm (*44*). Field studies, to be discussed later, support this conclusion of limited *in situ* nZVI transport (*6*, *45*).

In column studies, the following factors influenced transport of stabilized nZVI:

Aggregation Tendencies

nZVI is inherently magnetic, whereas its oxides are less so (*46*). By comparing the stability and transport of less-magnetic hematite particles to nZVI, it was shown that magnetic attractive forces between nZVI particles increase agglomeration, significantly reducing their transport (*41*, *47*, *48*). This is explained by the fact that magnetic forces are stronger and act over longer distances than van der Waals forces, thus playing a dominant role in causing aggregation (*47*).

Nanoparticle Concentration

Phenrat et al. (2009) reported that at a low concentration of 0.3 g/L all particles were mobile, but at higher nZVI concentrations (1-6 g/L), agglomeration during transport occurred (*48*). Raychoudhury et al. (2010) report similar results, with increased retardation and, in the case of PAA stabilized particles, decreased effluent concentrations as the nZVI concentration was increased from 0.1 to 3.0 g/L (*39*). Particle agglomeration, followed by straining is the most likely mechanism for capture of nZVI at high concentrations within porous media.

Ionic Strength

In a study with PAP and triblock-copolymer stabilized nZVI (*49*), nZVI transport was reduced between 7 and 38 fold for cation concentrations from 10 to 100 mM. In the presence of cations, triblock-copolymer clearly outperformed PAP, with 20 to 59 fold better transport. Lin et al. (2010) reported that at 40 mM cation concentration, the transport of PAA-stabilized nZVI decreased by about 40%, but this had no effect on CMC-stabilized nZVI (*32*). In summary, an increase in ionic strength may cause a decreased in the interparticle repulsive force, leading to greater nZVI agglomeration thus affecting its transport.

Transport Velocity

A decrease in pumping rate generally inhibited the transport of stabilized nZVI in bench scale studies. However, the majority of column studies were conducted at flow velocities that are artificially high (>1.0 m/day) compared to typical groundwater flow (*28*, *31*, *48*). Lin et al. (2010) showed that even at these high velocities, decreasing the flow-rate from 38.2 to 6.28 m/day decreased transport by ~5% (*32*). Raychoudhury et al. (2010) also demonstrated that decreasing the flow from 14.4 m/day to a realistic value of 0.29 m/day decreased transport 20% to 30% (*39*). This effect is explained by an increase in stagnant zones in the pore spaces and decreased drag torques which, when high, result in captured particle detachment from collector surfaces (*32*, *39*, *48*).

In addition, numerous fate and transport studies of stabilized nZVI have also been conducted under 2-D flow conditions that may be more representative of the subsurface. An investigation with 4 g/L PAA-stabilized nZVI suspension in homogeneous porous media demonstrated its transport without retardation, yet greater density of nZVI slurry caused the nZVI to travel downward and laterally (*40*). Phenrat et al. (2010) found that in heterogeneous porous media, stabilized nZVI follows preferential flowpaths and collects in pore spaces that have insufficient fluid shear (*41*). In 2011, the same research group reported that stabilized nZVI is capable of targeting NAPL entrapped in porous media, likely due to partitioning to the NAPL-water interface (*42*). In a series of studies from Carnegie Mellon University, the use of a combination of polymers, so-called triblock copolymers that work in concert to stabilize nZVI and preferentially deliver the particles to water and NAPL interfaces, was examined (*49–52*). The first polymer, poly(methacrylic acid) (PMAA), strongly adsorbs to iron oxide surfaces, anchoring the copolymer in place. The second, poly(methyl methacrylate or butyl methacrylate) (PMMA or PMBA, respectively), has low polarity, imparting hydrophobicity and an affinity for NAPL phases. Finally, PSS, an anion, ensures strong electrostatic repulsion from other particles and the predominantly negatively charged surfaces of porous media. The results clearly showed improved nZVI stability and transport in sand columns, but also limited NAPL targeting and a 2 to 10 fold decrease in reactivity toward trichloroethylene (TCE) (*51*, *52*). Although the utilization of specially designed copolymer coatings represents a novel approach to nZVI delivery, the issues of decreased reactivity and poor targeting have not been resolved.

Xanthan gum is yet another polyelectrolyte that has shown noteworthy performance in improving particle stability and nZVI transport. At 6 g/L, it could stabilize nZVI at concentrations as high as 15 g/L for up to 10 days, preventing both aggregation and sedimentation (*36*). In contrast to other stabilizers, which depend on establishing steric and electrostatic repulsion between particles, xanthan gum works by forming a polymer network between particles. In packed sand column studies, xanthan-gum-stabilized nZVI reached a steady-state elution of 88% of the original 20 g/L inlet concentration, with a 92% overall mass recovery (*37*). The transport of such high concentrations of nZVI with little loss to the porous media suggests exciting possibilities for *in situ* mobility.

Reactivity of Polyelectrolyte-Stabilized nZVI

The reactivity of stabilized nZVI toward contaminants may become inhibited due to polyelectrolyte coatings on the particle surface. Phenrat et al. (2009) examined this effect for nZVI stabilized with PSS, CMC, and PAP, and found the reactivity toward TCE decreased between 4 and 25 fold (*34*), which was attributed to (a) blocking of reactive sites on nZVI surface, and (b) TCE partitioning to the polyelectrolyte coatings, effectively decreasing the aqueous TCE concentration. However, such loss in reactivity is expressed on a nZVI surface-area-normalized basis. Because polyelectrolyte stabilization of nZVI also decreased mean particle size and prevented aggregation, both of which dramatically increased the overall reactive surface area per unit nZVI mass (*22*, *24*, *53–55*), the net effect of polyelectrolyte stabilization was an impressive gain in nZVI reactivity on a mass-normalized basis, making polyelectrolyte-stabilized nZVI a "new class" of particles altogether (*22*, *24*, *27*, *56–60*).

Most studies investigating the effect of polyelectrolyte coatings on nZVI reactivity were done for only one polyelectrolyte class, making comparison across different classes difficult. In addition to Phenrat et al. (2009), two other studies compared nZVI reactivity for several polyelectrolyte classes. Sakulchaicharoen et al. (2010) found CMC, PVP, and guar gum coatings resulted in smaller nanoparticles relative to uncoated nZVI, yet their effects on nZVI reactivity were mixed. For example, application of guar gum at 0.05% (w/w) improved nZVI reactivity ~2 fold, but an increase to 0.10% decreased reactivity ~3 fold. Similarly, PVP360K (PVP, MW = 360,000 g/mol) improved reactivity 10 fold, but lower molecular weight PVP40K reduced reactivity ~4 fold. Notably, CMC90K, CMC250K, and CMC700K all outperformed uncoated nZVI and all other polyelectrolyte treatments both in terms of nanoparticle size and reactivity, with CMC250K proving best. In a study of perchlorate degradation by Xiong et al. (2007), starch and CMC stabilizers improved degradation by 1.8 and 5.5 fold over uncoated nZVI, respectively (*56*).

Role of Secondary Metal on nZVI Performance

As of 2011, approximately 40% of nZVI remediation projects in the U.S. employed catalyzed nZVI (*2*). Numerous bench-scale investigations have demonstrated that modifying nZVI with another catalytic metal (e.g., palladium, etc.) can enhance nanoparticle reactivity toward degradation of a variety of drinking water contaminants by chemical reduction, when compared to nZVI without such modification (*2*, *61–65*). The nZVI modification, referred to as 'catalyzation' henceforth in this chapter, can also enable reduction of particularly recalcitrant contaminants that otherwise would not degrade with bare or 'uncatalyzed' nZVI. In contrast, 'catalyzed' nZVI can accelerate and/or alter the dominant degradation pathway, and, in the case of chlorinated hydrocarbons, yield saturated hydrocarbons as end products, with fewer harmful chlorinated intermediates (*2*, *66*). Modifying nZVI particles with a secondary catalyst metal, such as Pd, can make the nanoparticles much more reactive, thereby becoming rather short-lived. As a result, catalyzed nZVI may be effective only for active treatments (i.e., source zone remediation, for rapid destruction of contaminants) and not for long-term, passive remediation in permeable reactive barriers (*2*, *8*). The potential applications of catalyzed nZVI have been discussed in part elsewhere (*2*, *13*, *67*); however, no article has yet reviewed in full the breadth of studies focused on nZVI enhancement by catalyzation. Prior to the development of nZVI, a great number of studies investigated the modification of micro-scale zero-valent iron (ZVI) with a catalyst metal to prepare a bimetallic reductant for enhancing its performance. Much of this earlier work guided initial efforts at catalyzing nZVI. However, a high specific surface area combined with high reactivity, short life-span and the tendency of nZVI to agglomerate may limit the relevance of these earlier studies. Thus, this chapter focuses on the effect of secondary catalyst metals on nZVI performance potential at field-scale.

To catalyze nZVI, typically a lesser amount of a second metal, anywhere from 0.1% to 25% (w/w), is precipitated/deposited onto the surface of the iron nanoparticles. A galvanic cell is established between the two metals, with Fe acting as a sacrificial anode, transferring electrons to and thereby protecting the catalyst cathode from oxidation (*2*, *68*). This relationship can be predicted by the standard reduction-oxidation (redox) potential of each metal. As shown in Table 2, with the exception of Zn, Fe has a higher redox potential than the secondary metals used to catalyze nZVI, indicating that Fe^0 is thermodynamically favored to be oxidized. As will be shown, Pt, Pd, and Ag, Ni, are highly effective catalysts, which may owe in part to their ability to not only accept two electrons from Fe^0, but to also capture an additional electron from the already oxidized Fe^{2+} as well. In the case of heavy metals, which cannot be degraded *per se*, redox potential may be used to predict how effective nZVI might be at sequestering them. For metals with a standard redox potential for oxidation close to Fe, such as Cd, removal is by sorption or surface complexation only. However, for metals with a potential much lower the Fe, such as Pb, Ni and Cu, sorption followed by reduction is likely (*2*).

Figure 1 depicts the reductive degradation of a chlorinated organic compound by catalyzed nZVI, where Fe^0 reacts with water to produce molecular hydrogen (H_2) and hydroxide anions under anaerobic condition. The hydroxide anions may react with Fe^{2+} to form iron oxide species or, if released to the bulk water, can raise the aqueous pH (*45*). The H_2 produced may adsorb to the catalyst surface (*66*, *69*) and it may also interact with the nZVI surface (*66*, *69*, *70*). Contaminant reduction can occur at the catalyst surface (fast reaction) and at the nZVI surface (slower reaction) with electrons donated either by Fe^0 that is in direct contact with the catalyst or by H_2 produced from anaerobic corrosion of Fe; in case of hydrodechlorination, hydrogen replaces the liberated chloride ion (*67*, *69*, *71*). Following factors may cause nZVI reactivity to decline over time: (a) flow of electrons from Fe^0 core to catalyst is diminished as Fe^0 is spent, and (b) growth of the iron oxide layer inhibits electron transfer from the zero-valent core and quenches the catalyst (*72*, *73*).

As in most redox reactions, pH affects the reactivity of catalyzed nZVI in degrading chlorinated hydrocarbons. In three studies, an acidic to neutral condition (pH 4 to 7) was found favorable (*60*, *73*, *74*). Under weakly acidic conditions, reduction of water by nZVI can produce copious amounts of H_2 quickly (*60*, *73*) and simultaneously oxidize Fe^0 to Fe^{2+}. After Fe^0 is consumed, available H_2 may serve as an electron donor to continue catalytic contaminant degradation (*60*, *73*). However, if gaseous H_2 rapidly accumulates on the particle surface, it may inhibit contaminant degradation by blocking access to the reactive sites (*73*). On the other hand, under slightly basic conditions, greater $[OH^-]$ may contribute to development of an iron oxide precipitate layer on the nZVI surface, and inhibit electron transfer for reduction (*60*, *73*, *74*). Furthermore, basic conditions may not generate sufficient H_2 necessary for effective catalytic degradation of contaminants (*73*).

Catalyst Metals for nZVI Enhancement

Several studies have examined the reactivity of various catalysts under similar conditions. These studies are especially useful in elucidating the relative reactivity of different secondary catalyst metals in nZVI-catalyst combinations and of contaminant-dependent changes in catalyst reactivity (*56*, *57*, *61–65*, *75–77*). The first such study with nanoscale particles, by Zhang et al. (1998), investigated the reduction of chlorinated ethenes by 20 g/L of Pd catalyzed nZVI (Pd-nZVI) (*61*), and reported complete degradation of tetrachloroethene (PCE), trichloroethene (TCE), *trans*-dichloroethene (*trans*-DCE), and vinyl chloride (VC) within 3 hours with no chlorinated byproducts present. In contrast, degradation of TCE by Pt-nZVI took 4 times longer than Pd-nZVI. Furthermore, 100 g/L of Ni-nZVI took more than twice as long as Pd-nZVI to degrade *trans*-DCE. In another study, Kim and Carraway (2003) compared TCE degradation by Pd-nZVI, Ni-nZVI, and Cu-nZVI (*62*); TCE at a concentration of 100 mg/L was degraded by 25 g/L of Pd- and Ni-nZVI within 1 hour and 13 days, respectively. No chlorinated byproducts were observed. As seen in Table 3 (See Ref. (*62*)), surface-area-normalized reaction rates for all three catalysts indicate the order of catalyst reactivity is Pd > Ni > Cu. Taken together, these findings suggest that for chlorinated ethenes the order of catalyst reactivity is Pd > Pt > Ni > Cu.

Table 2. Standard reduction potentials of metallic reductants

Reduction Half-Reaction	*Standard Redox-Potential, mV*
$Co^{3+} + e^- \rightarrow Co^{2+}$	1820
$Pt^{2+} + 2e^- \rightarrow Pt^0$	1200
$Pd^{2+} + 2e^- \rightarrow Pd^0$	990
$Ag^+ + e^- \rightarrow Ag^0$	797
$Fe^{3+} + e^- \rightarrow Fe^{2+}$	769
$Cu^+ + e^- \rightarrow Cu^0$	520
$Cu^{2+} + 2e^- \rightarrow Cu^0$	340
$Pb^{2+} + 2e^- \rightarrow Pb^0$	-130
$Sn^{2+} + 2e^- \rightarrow Sn^0$	-140
$Ni^{2+} + 2e^- \rightarrow Ni^0$	-250
$Co^{2+} + 2e^- \rightarrow Co^0$	-280
$Cd^{2+} + 2e^- \rightarrow Cd^0$	-400
$Fe^{2+} + 2e^- \rightarrow Fe^0$	-441
$Cr^{3+} + 3e^- \rightarrow Cr^0$	-740
$Zn^{2+} + 2e^- \rightarrow Zn^0$	-760
$Mn^{2+} + 2e^- \rightarrow Mn^0$	-1180
$Al^{3+} + 3e^- \rightarrow Al^0$	-1680
$Mg^{2+} + 2e^- \rightarrow Mg^0$	-2370

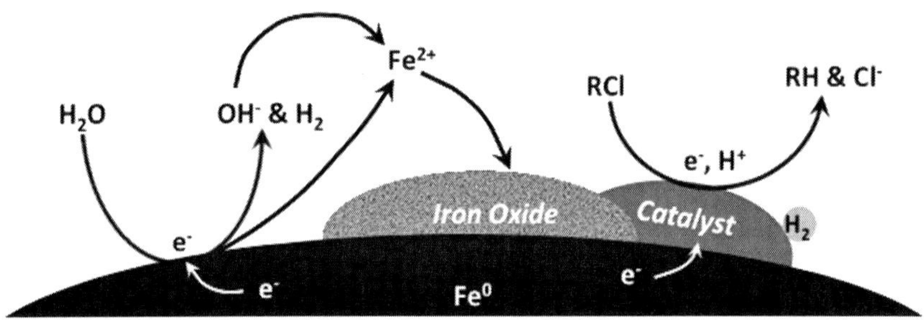

Figure 1. Idealized schematic of reductive dechlorination by catalyzed nZVI.

Table 3. Summary of catalyzed nZVI studies with chlorinated and brominated aliphatic hydrocarbons as target contaminants

Contaminant	*Catalyst Metal*	*Catalyst Loading, % (w/w)*	*nZVI Loading, g/L*	*Initial Conc. mg/L*	*Removal Efficiency*		*k_{SA}, L/hr/m^2 or (k_m), (L/hr/g)*	*Chlorinated End Products*	*Ref.*
Tetrachloroethylene (PCE)	Pd	0.10	5.0	20.0	94%	1.5 hr	1.2E-02	None	*(68)*
	Pd	1.0	0.25	30.0	–		2.3E-01	None	*(71)*
	Pd	0.10	0.10	15.0	80%	2 hr	2.4E+00	TCE	*(58)*
	Pd	–	20.0	20.0	100%	0.25 hr	–	None	*(61)*
Trichloroethene (TCE)	Pd	0.10	5.0	20.0	100%	1.5 hr	1.8E-02	None	*(68)*
	Pd	0.10	0.10	25.0	100%	1.5 hr	6.7E-01	DCE, Trace	*(22)*
	Pd	–	50.0	20.0	100%	0.25 hr	1.0E-01	None	*(1)*
	Pd	0.10	0.10	50.0	100%	1 hr	1.6E+00	None	*(27)*
	Pd	5.0	5.0	120.0	–		5.4E-03	–	*(71)*
	Pd	0.10	0.10	20.0	70%	1 hr	6.4E-03	–	*(57)*
	Pd	0.10	0.10	15.0	80%	2 hr	2.4E+00	–	*(58)*
	Pd	0.10	0.10	50.0	100%	0.35 hr	2.5E-01	–	*(24)*
	Pd	0.05	2.00	17.0	100%	2.5 hr	–	–	*(35)*
	Pd	–	20.0	20.0	100%	0.25 hr	–	None	*(61)*
	Pd	0.05	–	100.0	–		1.3E-01	None	*(62)*
	Ni	20.0	2.5	24.0	100%	3 hr	9.8E-02	None	*(69)*

Contaminant	Catalyst Metal	Catalyst Loading, % (w/w)	nZVI Loading, g/L	Initial Conc. mg/L	Removal Efficiency		k_{SA}, L/hr/m² or (k_m), (L/hr/g)	Chlorinated End Products	Ref.
	Ni	25.0	2.5	10.0	100%	2 hr	5.7E-02	None	*(91)*
	Ni	21.4	0.60	80.0	–		(3.1E+00)	–	*(92)*
	Ni	6.6	1.0	350.0	100%	44 hr	–	None	*(97)*
	Ni	0.05	–	100.0	–		1.2E-02	None	*(62)*
	Pt	–	20.0	20.0	100%	1 hr	–	None	*(61)*
	Cu	0.03	–	100.0	–		2.0E-03	None	*(62)*
1,1-Dichloroethene (1,1-DCE)	Pd	0.10	5.0	20.0	92%	1.5 hr	1.2E-02	None	*(68)*
cis-1,2-Dichloroethene (*c*-DCE)	Pd	0.10	5.0	20.0	100%	1.25 hr	1.8E-02	None	*(68)*
	Cu	0.03	–	100.0	–		2.0E-03	None	*(62)*
1,1-Dichloroethene (1,1-DCE)	Pd	0.10	5.0	20.0	92%	1.5 hr	1.2E-02	None	*(68)*
cis-1,2-Dichloroethene (*c*-DCE)	Pd	0.10	5.0	20.0	100%	1.25 hr	1.8E-02	None	*(68)*
	Pd	–	20.0	20.0	100%	1 hr	–	None	*(61)*
trans-1,2-Dichlroethene (*t*-DCE)	Pd	0.10	5.0	20.0	100%	1.5 hr	1.5E-02	None	*(68)*
	Pd	–	100.0	9.7	100%	0.25 hr	–	None	*(61)*

Continued on next page.

Table 3. (Continued). Summary of catalyzed nZVI studies with chlorinated and brominated aliphatic hydrocarbons as target contaminants

Contaminant	*Catalyst Metal*	*Catalyst Loading, % (w/w)*	*nZVI Loading, g/L*	*Initial Conc. mg/L*	*Removal Efficiency*		k_{SA}, *L/hr/m²* *or* (k_m), *(L/hr/g)*	*Chlorinated End Products*	*Ref.*
	Ni	–	100.0	9.7	98%	2 hr	–	None	*(61)*
Vinyl Chloride (VC)	Pd	–	20.0	20.0	100%	3 hr	–	None	*(61)*
Hexachloroethane (HCA)	Pd	0.50	5.0	30.0	100%	1.75 hr	2.0E-02	None	*(79)*
Pentachloroethane (PCA)	Pd	0.50	5.0	30.0	~100%	1.5 hr	2.6E-02	None	*(79)*
1,1,2,2-Tetrachloroethane (1,1,2,2-TeCA)	Pd	0.50	5.0	~25.0	~100%	1.25 hr	8.8E-03	None	*(79)*
1,1,1,2-Tetrachloroethane (1,1,1,2-TeCA)	Pd	0.50	5.0	~25.0	~100%	1.5 hr	2.1E-02	None	*(79)*
1,1,1-Trichloroethane (1,1,1-TCA)	Pd	0.50	5.0	~25.0	100%	7 hr	5.4E-03	None	*(79)*
	Pd	0.10	0.10	15.0	50%	2 hr	4.3E-01	–	*(58)*
1,2-Dichloroethane (1,2-DCA)	Pd	0.50	5.0	~25.0	<5%	24 hr	–	–	*(79)*
Carbon Tetrachloride (CT)	Pd	0.10	12.5	15.4	100%	1 hr	9.0E-03	CF, DCM	*(78)*
	Pd	0.20	10.0	100.0	100%	3 hr	3.6E-03	–	*(73)*
	Pd	3.8	1.2	15.4	–		(5.3E+00)	CF	*(64)*
	Ni	25.0	2.5	20.0	100%	1 hr	9.2E-02	CF, DCM	*(90)*

Contaminant	*Catalyst Metal*	*Catalyst Loading, % (w/w)*	*nZVI Loading, g/L*	*Initial Conc. mg/L*	*Removal Efficiency*		*k_{SA}, L/hr/m^2 or (k_m), (L/hr/g)*	*Chlorinated End Products*	*Ref.*
	Ni	2.4	1.2	15.4	–		(2.1E+01)	CF	(*64*)
	Cu	2.1	1.2	30.8	100%	0.1 hr	(4.4E+01)	CF, DCM	(*64*)
Chloroform (CF)	Pd	0.10	12.5	14.3	92%	1 hr	6.5E-03	DCM	(*78*)
	Pd	0.20	10.0	100.0	100%	3 hr	2.7E-03	–	(*73*)
Dichlromethane (DCM)	Pd	0.10	12.5	6.5	30%	2 hr	–	–	(*78*)
	Pd	0.20	10.0	100.0	~15%	8 hr	3.5E-05	–	(*73*)
Carbon Tetrabromide (CTB)	Ni	20.0	2.5	21.2	100%	0.5 hr	2.0E-01	BM	(*95*)
Bromoform (BF)	Ni	20.0	2.5	21.6	100%	1 hr	4.9E-02	BM	(*95*)
Dibromomethane (DBM)	Ni	20.0	2.5	21.9	100%	3.5 hr	1.1E-02	BM	(*95*)

In a comparative study of carbon tetrachloride (CT) degradation by Pd-, Ni-, and Cu-nZVI (*64*), three different synthesis methods were employed for preparing bimetallic nZVI, including the hydrogen reduction process (HRP) and the solution deposition process (SDP). The authors reported the order of catalyst reactivity to depend on the synthesis process: for HRP-nZVI, the order of catalyst reactivity was Cu > Ni > Pd (Table 3; See Ref. (*64*)). However, for SDP-nZVI (not included in Table 3), this order was reversed, agreeing with previous studies. This discrepancy with HRP-synthesized nZVI was attributed to segregation of the catalyst metal on the particle surface, leading the catalyst to act as separate, individual particles, which was shown to be less reactive (*61*). Both chloroform (CF) and dichloromethane (DCM) were observed as byproducts with HRP-synthesized nZVI.

Other studies have investigated the degradation of recalcitrant cyclic compounds by catalyzed nZVI. The study by Wang et al. (2010) used Pd-, Ni-, and Cu-nZVI to degrade 1,2,3,4-tetrachlorodibenzo-*p*-dioxin (1,2,3,4-TCDD), one of a group of chlorinated dioxins and furans (*65*). Interestingly, Pd-nZVI, fully degraded 1,2,3,4-TCDD within 10 days from an initial concentration of 0.35 mg/L, but nZVI, Ni-nZVI and Cu-nZVI were unreactive toward 1,2,3,4-TCDD. With Pd-nZVI, 7 different chlorinated byproducts were identified, yet >95% of the parent compound degraded to a non-chlorinated byproduct. Zhou et al. (2010) reported the ability of Pd-, Pt-, Ni-, Cu-, and Co- nZVI to degrade chlorophenols, a widespread industrial chemical and contaminant (*77*). The order of catalyst reactivity for degradation of 2,4,6-trichlorophenol (TCP) is reported to be Pd > Ni > Pt > Co > Cu (Table 4; See Ref. (*77*)). Phenol was the major byproduct (80% to 100%) of degradation for the three chlorophenols investigated.

Rivero-Huguet and Marshall (2009) compared the performance of a wide variety of bimetallic reductants, including Pd-nZVI and Cu-nZVI, for the remediation of hexavalent chromium (Cr(VI)), a heavy metal toxin widely found at industrial sites (*63*); for Pd-nZVI and Cu-nZVI the surface-area-normalized reaction kinetics of Cr(VI) degradation were comparable, with reactivity about twice that of nZVI (Table 4; See Ref. (*63*)). Effectiveness of nZVI remediation of Cr(VI) was highly dependent on pH, showing a 50-fold decrease in reaction kinetics with pH increase from 2 to 6.

Pollutant Degradation with Pd-nZVI

Among the various catalyst metals investigated for bimetallic nZVI, Pd and Ni are most common, and Pd is clearly the most reactive known catalyst for nZVI. As reported in about 30 published investigations, Pd is shown to enhance nZVI reduction of chlorinated ethenes, ethanes, and methanes; brominated methanes; chlorinated phenols; polychlorinated biphenyls; polybrominated diphenyl ethers; dioxins; chlorinated benzene, chlorinated cyclohexane, and atrazine; hexavalent chromium and selenate; and nitrate (*1*, *5*, *6*, *22*, *24*, *27*, *29*, *35*, *57*, *58*, *60–66*, *68*, *71–73*, *77–87*). However, Pd-nZVI did not significantly improve perchlorate degradation (*88*).

Table 4. Summary of catalyzed nZVI studies with chlorinated and brominated aromatics.

Contaminant	*Catalyst Type*	*Catalyst Loadin- g% (w/w)*	*nZVI Loading, mg/L*	*Initial Conc., mg/L*	*Removal Efficiency*		*k_{SA}, hr/m² or (k_m), (L/hr/g)*	*Chlorinated End Products*	*Ref.*
Pentachlorophenol (PCP)	Pd	0.05	12.5	5.0	97%	1.7 hr	(4.0E-01)	–	(*85*)
	Ni	20.0	2.5	50.6	100%	0.33 hr	–	None	(*74*)
	Ni	25.0	1.3	50.0	80%	20 hr	–	CP	(*98*)
2,4,6-Trichlorophenol (246TCP)	Pd	0.50	5.0	20.0	94%	1 hr	2.4E-02	DCP, CP	(*77*)
	Ni	0.50	5.0	20.0	56%	1 hr	–	–	(*77*)
	Pt	0.50	5.0	20.0	~18%	1 hr	–	–	(*77*)
	Co	0.50	5.0	20.0	~15%	1 hr	–	–	(*77*)
	Cu	0.50	5.0	20.0	~9%	1 hr	–	–	(*77*)
2,4-Dichlorophenol (24DCP)	Pd	0.50	5.0	20.0	100%	1 hr	3.4E-02	CP, <6%	(*77*)
4-Chlorophenol (4CP)	Pd	0.50	5.0	20.0	100%	1 hr	1.1E-01	None	(*77*)
	Ni	10.0	0.40	80.4	100%	1 hr	(2.5E+01)	None	(*93*)
Polychlorinated Biphenyls (PCBs)	Pd	–	50.0	5.0	100%	17 hr	–	–	(*1*)
	Pd	0.10	1.0	2.5	75%	100 hr	3.1E-04	Present	(*22*)
	Pd	0.25	10.0	0.25	100%	56 d	(7.5E-04)	2,3-DCB	(*86*)

Continued on next page.

Table 4. (Continued). Summary of catalyzed nZVI studies with chlorinated and brominated aromatics.

Contaminant	*Catalyst Type*	*Catalyst Loadin-g% (w/w)*	*nZVI Loading, mg/L*	*Initial Conc., mg/L*	*Removal Efficiency*		k_{SA}, *hr/m²* *or* (k_m), *(L/hr/g)*	*Chlorinated End Products*	*Ref.*
Polybrominated Diphenyl Etheres (PBDEs)	Pd	0.25	10.0	0.25	100%	2.5-15 d	(~5.5E-02)	Lower PBDEs	*(86)*
1,2,3,4-Tetrachlorodibenzo-*p*-dioxin (1,2,3,4-TeCDD)	Pd	0.50	5.0	0.50	~75%	48 hr	3.0E-04	Lower CDDs	*(66)*
	Pd	0.01	20.8	0.35	100	10 d	1.5E-02	Lower CDDs	*(65)*
	Pd	0.02	20.8	0.35	–		1.3E-02	–	*(65)*
	Pd	0.04	20.8	0.35	–		1.2E-02	–	*(65)*
	Ni	–	20.8	0.35	Non-Reactive		–	–	*(65)*
	Cu	–	20.8	0.35	Non-Reactive		–	–	*(65)*
1,2,3-Trichlorodibenzo-*p*-dioxin (1,2,3-TriCDD)	Pd	0.50	5.0	0.50	~93%	48 hr	1.5E-03	Lower CDDs	*(66)*
1,2-Dichlorodibenzo-*p*-dioxin (1,2-DiCDD)	Pd	0.50	5.0	–	–		2.7E-03	Mono-CDDs	*(66)*
1,2,4-Trichlorobenzene (1,2,4-TCB)	Pd	1.0	1.7	30.8	100%	1 hr	1.5E-01	None	*(80)*
	Pd	0.10	0.71	15.0	100%	0.75 hr	3.9E-01	12DCB, <5%	*(81)*
	Pd	0.10	0.71	20.0	–		2.7E-01	DCBs, MCB	*(82)*
1,2-Dichlorobenzene (12DCB)	Pd	0.10	0.71	15.0	100%	0.5 hr	5.0E-01	None	*(81)*

Contaminant	*Catalyst Type*	*Catalyst Loading% (w/w)*	*nZVI Loading, mg/L*	*Initial Conc., mg/L*	*Removal Efficiency*		*k_{SA}, hr/m² or (k_m), (L/hr/g)*	*Chlorinated End Products*	*Ref.*
1,3-Dichlorobenzene (13DCB)	Pd	0.10	0.71	15.0	–		5.6E-01	–	(*81*)
1,4-Dichlorobenzene (14DCB)	Pd	0.10	0.71	15.0	–		8.9E-01	–	(*81*)
Monochlorobenzene (MCB)	Pd	0.10	0.71	Chart 15.0	100%	0.25 hr	1.1E+00	None	(*81*)
p-Nitrochlorobenzene (*p*-NCB)	Pd	0.20	0.20	50.0	100%	0.5 hr	(7.8E+00)	Chloroani-line	(*60*)
γ-Hexachlorocyclohexane (Lindane)	Pd	0.20	0.50	5.0	100%	5 hr	2.0E+00	None	(*83*)
	Pd	0.80	0.50	1.0	100%	0.85 hr	–	–	(*87*)
Atrazine	Pd	0.80	0.50	1.0	100%	0.15 hr	–	–	(*87*)

The modification of nZVI with even small amounts of Pd showed a dramatic effect on the nanoparticle reactivity, with the majority of studies having used a Pd loading of <1.0% on a mass basis (See Tables 3 and 4). Pd is a precious metal, with a price over $20,000 per kg. Numerous studies reported the optimal loading of Pd on nZVI to range between 0.2 and 1.0% (*29*, *57*, *58*, *60*, *73*, *80*, *81*, *87*). The destruction of 1,2,4-trichlorobenzene (1,2,4-TCB) with 1% Pd on nZVI was superior to 0.1% and 0.5% Pd, with a 10-fold increase in reactivity when the Pd loading increased from 0.1% to 1.0% (*80*). In a subsequent study, a linear increase in 1,2,4-TCB degradation kinetics was observed with increasing Pd loading on nZVI from 0.01% to 0.5% (*81*). He and Zhao (2008) reported the degradation kinetics of TCE to increase 10-fold with 0.05% Pd on nZVI compared to non-catalyzed nZVI (*57*); however, increasing Pd loading to 0.1% only improved the degradation rate constant by another 30%, with no additional gains observed at 0.2% Pd loading. Cho and Choi (2010) reported similar findings with TCE removal efficiency peaking at 0.1% Pd and not improving further at 0.2% or 0.4% Pd (*58*). For the degradation of lindane, a persistent pesticide, a Pd loading of 0.8% was optimal (*87*). Wang et al. (2009) conducted a thorough investigation of Pd loading on nZVI for degradation of CT, CF, and DCM (*73*); for all three contaminants, the removal efficiency dramatically peaked at 0.2% Pd, followed by a steady decrease as loading was incrementally increased to 3.0%. Dong et al. (2011) reported a step-wise increase in Pd loading from 0.05% to 0.2% resulted in consistent improvement in nZVI reactivity towards nitrochlorobenzene degradation (*60*). The degradation of PCE with Pd nZVI at 1.0% Pd loading was considered optimal (*29*), which is in agreement with a similar study by Zhu et al. (2006) (*80*).

Pd nZVI has been demonstrated to undergo profound structural changes upon reaction with water, contaminants, and other electron acceptors. The question of catalyst stability in water and its effects on Pd nZVI reactivity was raised by Zhu and Lim (2007) after they observed release of Fe into solution over the course of degradation experiments (*81*). Structural changes of Pd nZVI in an aqueous environment were further investigated by Yan et al. (2010), who observed Pd nZVI with 1.5% Pd as having numerous 2 to 5 nm islets resting on an amorphous oxide shell above a metallic core (*72*). Upon reaction with water, Pd migrates to the metallic iron core interface, accompanied by oxidation and outward diffusion of iron species. Within 24 hrs, the Pd catalyst becomes completely covered by an outer iron-oxide matrix. This view of the inactivation of Pd nZVI is supported by another study, which found a 4-fold decrease in reactivity by the third cycle of 1,2,4-TCB degradation (*81*). An attempt to reactivate spent Pd nZVI by HCl and $NaBH_4$ treatments was not fully successful.

Several factors other than pH might impact degradation of contaminants by Pd nZVI. Typically, Pd nZVI is synthesized by depositing Pd on the nZVI surface. However, a simple mixture of distinct, monometallic Fe and Pd nanoparticles showed twice the reactivity toward TCE as Pd nZVI (*73*). It was hypothesized that TCE is catalytically reduced at the Pd nanoparticle surface by H_2 produced from reduction of water by nZVI. Shih et al. (2011) reported that addition of Cu and Ni cations to Pd nZVI increased its reactivity toward pentachlorophenol (*85*); this was attributed to reduction of Cu^{2+} or Ni^{2+} ions and formation of trimetallic nZVI particles, where Ni cations had a greater effect on reactivity than Cu cations. Shih et al. (2011) also showed that sulfate amendment can diminish Pd nZVI reactivity by 42%, possibly due to blocking of reactive sites. Natural organic matter (NOM) and humic acid (HA) were also shown to reduce Pd nZVI reactivity. Dong et al. (2011) showed that the addition of 10 mg/L of HA reduced reactivity seven-fold due to adsorbed HA blocking the reactive sites (*60*). The amendment of 50 mg/L of NOM also reduced Pd nZVI reactivity 9-fold; however, this was attributed to competition by NOM for diatomic H_2 (*82*).

Pollutant Degradation with Ni-nZVI

After Pd, Ni is the most-well studied catalyst metal for nZVI enhancement. As previously discussed, Ni-nZVI has shown comparable, though not equal, reactivity to Pd-nZVI for several classes of contaminants. Ni has the advantage of being a non-precious, less expensive metal than Pd for chemical reagents. As reported in about 20 published investigations, bimetallic Ni-nZVI is shown to enhance degradation or reduction of chlorinated ethenes, chlorinated methanes, brominated methanes, chlorinated phenols, dioxins, monoazo dye, nitrate, and heavy metals (*21, 62, 64, 65, 69, 74, 76, 77, 89–99*). In the case of perchlorate, however, addition of Ni to nZVI inhibited degradation (*56*). This was explained by its competing reaction with water, which preferentially consumed nZVI and is consistent with the study of perchlorate degradation by Pd-nZVI and Ag-nZVI (*88*).

Amendment of nZVI with Ni during synthesis has a dramatic effect on its ability to catalyze degradation. As seen in Tables 3 and 4, the majority of investigations examined Ni loading on nZVI near 20%; in cases with variable loading of Ni on nZVI, the most reactive loading is reported. Tee et al. (2005) investigated the effect of varying Ni loading on nZVI from 0% to 100% (*91*), and observed a 120-fold increase in nZVI reactivity toward TCE as Ni loading increased from 0% to 25%, followed by a 60-fold decrease as Ni loading increased further to 80%. Wu and Ritchie (2006) reported a similar rise (~30 fold) in nZVI reactivity toward TCE as Ni loading increased to 14.3%, but it declined ~3-fold as Ni loading increased further to ~35% (*92*), suggesting an optimal Ni loading between 14.3 and 21.4%. Zhang et al. (2006) reported pentachlorophenol (PCP) removal by Ni-nZVI increased from 32 to 98% as Ni loading increased from 0 to 10%, with marginal improvement in reactivity as Ni loading increased further to 20% (*1*). A 430-fold increase in Ni-nZVI reactivity toward *p*-chlorophenol was noted as Ni loading increased from 0 to 10%; higher loadings were not investigated (*93*). Finally, in contrast to other studies, Barnes et al. (2010) found 8.5% Ni loading optimal for TCE degradation, although 15% Ni performed comparably (*97*).

The majority of nZVI studies synthesize nanoparticles via the aqueous reduction of metallic ions by $NaBH_4$. In the case of Ni-nZVI synthesis, however, a co-precipitation method of Fe and Ni is reported; that is, nanoparticles are formed in a solution containing both metallic ion precursors. This method incorporates Ni into the core of the particles, where it likely cannot serve as a catalyst, as effectively as at the surface. In contrast, for post-synthesis Ni deposition on nZVI particles the catalyst may be entirely on the surface. One study investigating the outcomes of these two Ni-nZVI synthesis methods found modest differences in terms of particle size and reactivity (*92*). The size of Ni-nZVI particles with post-synthesis Ni deposition ranged from 7-11 nm, while Ni-nZVI particles synthesized by co-precipitation were 11-24 nm. The larger co-precipitated particles expressed a smaller specific surface area and would likely not transport as well *in situ*. Furthermore, the mass-normalized reactivity of Ni-nZVI synthesized by Ni deposition was 5-fold higher than Ni-nZVI with co-precipitated Ni and Fe.

The foremost concern with amending nZVI with Ni is the possible health issues associated with disassociation of Ni from the particles and into water. Chronic exposure to Ni in drinking water may cause decreased body weight and heart and liver damage (*100*). The U.S. EPA set the maximum contaminant level (MCL) for Ni at 0.1 mg/L, although this was remanded in 1995 (*100*). In 2011, the World Health Organization set a drinking water guideline of 0.07 mg/L for Ni (*101*). Full-scale implementation of nZVI involves injection of many kilograms of particles into the subsurface, of which Ni would be a significant mass-fraction (perhaps ~5-10%). Wu and Ritchie (2006) studied the dissolution of Ni in batch reactors used for Ni nZVI degradation of TCE (*92*), and observed a gradual increase in aqueous Ni concentration over time. Barnes et al. (2010) also observed a rise in aqueous Ni concentration to between 1.8 and 6.8 mg/L, depending on the Ni loading, concurrent with TCE degradation by Ni-nZVI (*97*). As will be discussed in the next section, nZVI concentrations injected in field-scale operations commonly exceed 1.0 g/L. Paired with the fact that optimal Ni loading in Ni nZVI catalyst for subsurface injection may be higher than the Ni levels examined by Barnes et al. (2010), even higher levels of dissolved Ni, well in excess of drinking water guidelines, could be expected near the injection site during field-scale remediation.

Field-Scale Use of Stabilized and Catalyzed nZVI

Three field studies using Pd-nZVI, two of which also employed a polyelectrolyte stabilizer, have been reported in the literature (*5*, *6*, *16*, *45*, *102*). The earliest of these, by Elliott and Zhang (2001), sought to remediate chlorinated hydrocarbon contamination, predominantly TCE, at an active manufacturing facility in Trenton, NJ (*5*). The size of the test site was small, measuring 4.5 m by 3.0 m and 9 m to confining clay and bedrock. TCE concentrations at the site consistently tested between 0.45 and 0.80 mg/L. Pd-nZVI with a Pd loading of 0.33% was used. Initial laboratory tests indicated that amendment of nZVI with Pd decreased the half-life of TCE by a factor of 60. In the field, nZVI was injected in two phases, using a set of recirculating wells and gravity-fed into an injection well, respectively. nZVI injection concentrations ranged from 0.75 to 1.5 g/L, with a total of 1.7 kg injected over two days.

CAH concentrations and site conditions were monitored at the injection well and at three surrounding monitoring wells. Oxidation-reduction potential (ORP) was measured *in situ* indicating a reducing environment conducive to reductive dechlorination. Post-injection, the ORP shifted from historically positive values between +150 and +250 mV to strongly reducing conditions (-360 mV) at the injection well and slightly reducing conditions (-20 to -75 mV) at the monitoring wells. While the pH increased from a range of 4.6–5.2 to 5.1–7.7, dissolved oxygen (DO) was completely eliminated, almost certainly by reaction with nZVI. A pumping test 6 weeks after injection confirmed no significant change in hydraulic conductivity due to nZVI injection. TCE removal efficiency varied both spatially and temporally, ranging from 1.5% to 96.5% over the three weeks following injection. Rebound of TCE concentrations to pre-injection levels was observed over several months, suggesting limited longevity of the nZVI treatment.

The second field study, by Henn and Waddill (2008), treated a source area on the Naval Air Station in Jacksonville, FL that was contaminated with a combination of chlorinated ethenes and ethanes (*102*). The test site was small, with a 98 m^2 area and 6 m depth to confining clay. Remediation was implemented by Tetra Tech NUS, Inc.. and a proprietary formulation of polyelectrolyte-coated Pd-nZVI was used. The maximum total concentration of all contaminants was 550 mg/kg in soil and 80 mg/L in groundwater. As in Elliott and Zhang (2001), both gravity-fed direct-injection wells and recirculating wells were used to deliver nZVI to the source area. Recirculating wells produced favorable iron delivery and enhanced mass transfer of sorbed and NAPL phase contaminants to the aqueous phase, where they could be degraded. A total of 150 kg of nZVI was injected over two weeks at concentrations ranging from 2.0 to 4.5 g/L.

Following Pd-nZVI injection, ORP dropped from initial levels between +100 and -100 mV to strongly reducing conditions between -200 and -550 mV. While DO levels declined significantly, the pH remained stable between 6 and 7. Despite an initial pumping test indicating no difference in soil permeability at the time of injection, a subsequent test one year later indicated a 45% reduction in soil permeability. Aqueous removal efficiencies of individual contaminants ranged from 65% to 99%, with some degrading to non-chlorinated end products. Total mass reduction, including sorbed phase, NAPL, and dissolved phases, was estimated between 16% and 62%, with 23% reduction as the most likely estimate. Furthermore, mass flux of contaminants from the source area was reduced. Rebound of aqueous contaminant concentrations was observed within a year, suggesting nZVI longevity of between 6 and 9 months.

In a field study by He et al. (2010), groundwater treatment was sought at a former manufacturing facility contaminated with chlorinated ethenes and polychlorinated biphenyls (PCBs) (*6*). For the pilot-scale test, a CMC-stabilized Pd-nZVI was chosen; the selection was based on a series of laboratory studies for optimizing the reactivity and transport of nZVI (*22*, *27*, *33*, *44*, *57*). The test-site was small, measuring 5 m by 3 m, with injection and monitoring wells reaching to a depth of 15 m. Monitoring wells were placed 1.5 m (MW-1) and 3.0 m (MW-2) downgradient of the injection well. The concentrations of the primary contaminants, PCE, TCE, *cis*-DCE, VC, and Aroclor 1242 (PCBs), were 1.2-12.0 mg/L, 1.6-23.8 mg/L, 8.5-20.0 mg/L, 1.1-2.2 mg/L, and 6.9-97.4 μg/L, respectively. Direct-injection, first gravity-fed and later under pressure, was completed to accomplish two separate injections of Pd-nZVI; delivery under pressure resulted in better transport of Pd-nZVI within the aquifer. Nanoparticles were synthesized on-site for maximum reactivity. For the first injection, approximately 570 liters of 0.2 g/L nZVI was delivered. A month later, another 570 liters of 1.0 g/L nZVI was injected. Pd loading for both was 0.1%.

After injection, ORP in the two monitoring wells declined drastically, to highly negative values (-350 and -180 mV, respectively), but recovered to only slightly negative pre-injection values (-50 to -20 mV) within 10 days. Decrease in ORP was matched by a complete, though transient, consumption of DO. nZVI successfully transported to MW-1, with >10% of the original concentration reaching the well after the first injection and >80% after the second, but <5% reached MW-2, only 3.0 m away, in both cases. For chlorinated ethenes, nZVI injection resulted in drastic reductions in aqueous concentrations, between 70% and ~100%, within the first 10 days. However, by day 13 concentrations rebounded significantly. Over the next 20 months, concentrations continued to slowly decline and remained below pre-injection levels. In the case of Aroclor 1242, nZVI injection saw an ~80%, long-term decline in its concentration in MW-2, but no degradation at MW-1. While not effective for long-term control of a plume, the short life-span and high reactivity of stabilized Pd-nZVI appears to make it effective for source-zone remediation.

A common observation in the three case studies summarized above was that nZVI injection also facilitated conditions conducive to long-term biotic destruction of contaminants. Elliot and Zhang (2001) noted that by establishing moderately-reducing ORP, pH increase to near neutral, and reducing DO facilitated by CMC-stabilized Pd-nZVI provided favorable conditions for the growth of anaerobic, iron-reducing bacteria that may degrade chlorinated hydrocarbons. Henn and Waddill (2008) also noted that favorable ORP conditions persisted at their study site and this may have promoted biological degradation at the site. Ultimately, the site remedy transitioned to monitored natural attenuation, a self-sustaining, long-term remedy. He et al. (2010) also reported favorable site conditions and appreciable, long-term biotic degradation of PCE and TCE throughout a 20-month monitoring period; the authors hypothesized that CMC and molecular H_2 produced by nZVI's reaction with water served as electron donors for microbial degradation of chlorinated hydrocarbons.

Ongoing Research Needs

For nZVI to become a widely accepted remediation tool, the cost of implementation must be lowered or its efficiency must be improved, likely both. nZVI is synthesized by energy and reagent intensive processes, which make it relatively expensive compared to micro-scale particles and industrial byproducts such as iron filings. In 2011, the current price of nZVI was estimated to be between \$50 and \$200 per kg, which may not be cost-effective for widespread implementation; a price range between \$10 and \$25 per kg is needed to be competitive with other remediation techniques (*2*, *3*). Prospective improvements in nZVI efficiency may include increased reactivity, transportability, NAPL targeting, and the successful implementation at higher concentrations of stabilized nZVI *in situ*. Given the rapid development of nZVI in the past 15 years, these advancements may become feasible within the next decade.

In order to overcome these challenges, a greater degree of coordination and experimental planning is needed. Crane and Scott (2011) suggested the need in both academia and industry for a standardized testing and reporting framework for different nZVI preparations. Tables 3 and 4, included here, further advance this point. For example, surface-area normalized pseudo first-order degradation rate constants (k_{SA}), a commonly used parameter for comparing reactivity of various nZVI formulations, have been shown to vary by more than three orders-of-magnitude for a given target contaminant. In a minority of cases, this indicates genuine differences in nZVI formulations. This variability may also be attributed to differences in specific surface area, although these differences are typically less than one order-of-magnitude for stabilized nZVI. Most likely, the observed variability is due to differences in testing parameters: the nZVI synthesis method, post-synthesis handling and storage, nZVI loading, catalyst loading, and initial contaminant concentration. Lack of control for these parameters across the literature makes comparison of results difficult, if not impossible. Furthermore, many studies did not report one or more of these fundamental parameters. Of the studies reviewed here, a quarter did not report the degradation rate constant (i.e. k_{SA}, k_{m}, or k_{obs}), while others did not report chlorinated byproduct generation, leaving the question of nZVI's role in reducing overall toxicity unanswered. As an alternative to large-scale standardization of testing and reporting parameters, a greater number of comparative studies (i.e., those which include a wide range of catalysts, stabilizers, etc.) could further elucidate the relative performance of different nZVI formulations.

Although polyelectrolyte stabilization clearly improved nZVI transport in laboratory settings, the relative performance of different types and molecular weights of polyelectrolytes is unclear and often varies between studies. This is also true in terms of polyelectrolyte stabilized nZVI reactivity; few studies include multiple polyelectrolyte types and molecular weights. Also, the use of copolymers, specially engineered polymer combinations, remains largely unexplored. However, the most pressing question for polyelectrolyte stabilized nZVI is whether or not concentrations relevant to effective remediation can be successfully transported *in situ* at flow rates, ionic strengths, and hydraulic conductivity heterogeneities that are representative of subsurface conditions.

As discussed earlier, comparative studies of catalyst metals are limited. No study compared all potential catalysts together under similar conditions. Relative catalyst performance may also be contaminant-dependent, confounding comparison between the studies that do exist. Furthermore, as evidenced by the Pd and Ni catalyst studies, the optimal catalyst loading varies between metals. Comparisons of many catalysts at only one loading may underestimate the true potential of an individual catalyst. Although Pd and Ni have rightly received the majority of research attention, other effective catalysts may have been overlooked. Specifically, Ag was shown to outperform Ni and provide comparable reactivity to Pd, with the benefit of being much less expensive than Pd (*63*). Finally, it remains to be answered whether injecting nZVI catalyzed by metals that are either precious or raise health concerns is a wise practice from an economic or environmental viewpoint. The question of catalyst leaching, specifically Ni leaching, into groundwater should be conclusively answered before widespread implementation. Furthermore, trade-offs may exist between the increased reactivity provided by bimetallic nZVI and longevity of nZVI *in situ*.

Field-scale studies of nZVI are few, and many possibilities for future research exist. Ultimately, field studies are needed to discover how nZVI can be successfully transported to reach and mix with target contaminants *in situ*. Many laboratory studies report impressive contaminant degradation results, but only at nZVI loadings well in excess of those successfully used in field scale tests. In contrast, recent studies reporting impressive gains in nZVI reactivity by catalyzing or stabilizing the particles have yet to be scaled to nZVI loadings above 1.0 g/L. This disparity between laboratory findings and practical field-scale use of nZVI should be bridged with research focused on synthesizing nZVI that takes advantage of stabilization and catalyzation at higher nZVI loadings on the order of 5 to 20 g/L. Finally, given the important role natural attenuation can play in management of contaminated sites, the role of nZVI injection in long-term microbial degradation of contaminants should be further investigated.

References

1. Wang, C. B.; Zhang, W. X. Synthesizing nanoscale iron particles for rapid and complete dechlorination of TCE and PCBs. *Environ. Sci. Technol.* **1997**, *31*, 2154–2156.
2. Crane, R. A.; Scott, T. B. Nanoscale zero-valent iron: Future prospects for an emerging water treatment technology. *J. Hazard. Mater.* **2012**, *211-212*, 112–125.
3. Li, X.; Elliott, D. W. Zero-valent iron nanoparticles for abatement of environmental pollutants: Materials and engineering aspects. *Crit. Rev. Solid State Mater. Sci.* **2006**, *31*, 111–122.
4. Nyer, E. K.; Vance, D. B. Nano-scale iron for dehalogenation. *Ground Water Monit. Rem.* **2001**, *21*, 41–46.
5. Elliott, D. W.; Zhang, W. X. Field assessment of nanoscale bimetallic particles for groundwater treatment. *Environ. Sci. Technol.* **2001**, *35*, 4922–4926.
6. He, F.; Zhao, D.; Paul, C. Field assessment of carboxymethyl cellulose stabilized iron nanoparticles for in situ destruction of chlorinated solvents in source zones. *Water Res.* **2010**, *44*, 2360–2370.
7. Tratnyek, P. G.; Johson, R. L. Nanotechnologies for environmental cleanup. *Nano Today* **2006**, *1*, 44–48.
8. Henderson, A. D.; Demond, A. H. Long-term performance of zero-valent iron permeable reactive barriers: A critical review. *Environ. Eng. Sci.* **2007**, *24*, 401–423.
9. Geiger, C. L.; Carvalho-Knighton, K. M; Novaes-Card, S.; Maloney, P.; DeVor, R. A Review of Environmental Applications of Nanoscale and Microscale Reactive Metal Particles. In *Environmental Applications of Nanoscale and Microscale Reactive Metal Particles*; Geiger, C. L.,

Carvalho-Knighton, K. M., Eds.; American Chemical Society: Washington, D.C., 2009; pp 1–20.

10. Zhang, W. Nanoscale iron particles for environmental remediation: An overview. *J. Nanopart. Res.* **2003**, *5*, 323–332.
11. Cundy, A. B.; Hopkinson, L.; Whitby, R. L. D. Use of iron-based technologies in contaminated land and groundwater remediation: A review. *Sci. Total Environ.* **2008**, *400*, 42–51.
12. Cook, S. M. *Assessing the use and application of zero-valent iron nanoparticle technology for remediation at contaminated sites*; U.S. EPA, Office of Superfund and Technology Innovation: Washington, DC, 2009.
13. Xinhong, Q.; Zhanqiang, F. Degradation of Halogenated Organic Compounds by Modified Nano Zero-Valent Iron. *Prog. Chem.* **2010**, *22*, 291–297.
14. Li, L.; Fan, M.; Brown, R. C.; Van Leeuwen, J. H.; Wang, J.; Wang, W.; Song, Y.; Zhang, P. Synthesis, properties, and environmental applications of nanoscale iron-based materials: A review. *Crit. Rev. Environ. Sci. Technol.* **2006**, *36*, 405–431.
15. Glazier, R.; Venkatakrishnan, R.; Gheorghiu, F.; Walata, L.; Nash, R.; Zhang, W. X. Nanotechnology takes roots. *Civil Eng.* **2003**, *73*, 64–69.
16. Gavaskar, A.; Tatar, L.; Condit, W. *Cost and performance report, nanoscale zero-valent iron technologies for source remediation*; Contract Rep. CR-05-007-ENV; Naval Facilities Engineering Command: Port Hueneme, CA, 2005.
17. Zhang, W. X.; Elliott, D. W. Applications of iron nanoparticles for groundwater remediation. *Remediation* **2006**, *16*, 7–21.
18. Comba, S.; Molfetta, A. D.; Sethi, R. A comparison between field applications of nano-, micro-, and millimetric zero-valent iron for the remediation of contaminated aquifers. *Water, Air, Soil Pollut.* **2011**, *215*, 595–607.
19. Petosa, A. R.; Jaisi, D. P.; Quevedo, I. R.; Elimelech, M.; Tufenkji, N. Aggregation and deposition of engineered nanomaterials in aquatic environments: Role of physiochemical interactions. *Environ. Sci. Technol.* **2010**, *44*, 6532–6549.
20. Phenrat, T.; Saleh, N.; Sirk, K.; Kim, H. J.; Tilton, R. D.; Lowry, G. V. Stabilization of aqueous nanoscale zerovalent iron dispersions by anionic polyelectrolytes: Adsorbed anionic polyelectrolyte layer properties and their effect on aggregation and sedimentation. *J. Nanopart. Res.* **2008**, *10*, 795–814.
21. Schrick, B.; Hydutsky, B. W.; Blough, J. L.; Mallouk, T. E. Delivery Vehicles for Zerovalent Metal Nanoparticles in Soil and Groundwater. *Chem. Mater.* **2004**, *16*, 2187–2193.
22. He, F.; Zhao, D. Preparation and characterization of a new class of starch-stabilized bimetallic nanoparticles for degradation of chlorinated hydrocarbons in water. *Environ. Sci. Technol.* **2005**, *39*, 3314–3320.
23. Cirtiu, C. M.; Raychoudhury, T.; Ghoshal, S.; Moores, A. Systematic comparison of the size, surface characteristics and colloidal stability of zero

valent iron nanoparticles pre- and post-grafted with common polymers. *Colloids Surf., A.* **2011**, *390*, 95–104.

24. Sakulchaicharoen, N.; O'Carroll, D. M.; Herrera, J. E. Enhanced stability and dechlorination activity of pre-synthesis stabilized nanoscale Fe-Pd particles. *J. Contam. Hydrol.* **2010**, *118*, 117–127.
25. Ponder, S. M.; Darab, J. G.; Mallouk, T. E. Remediation of Cr(VI) and Pb(II) aqueous solutions using supported, nanoscale zero-valent iron. *Environ. Sci. Technol.* **2000**, *34*, 2564–2569.
26. Ponder, S. M.; Darab, J. G.; Bucher, J.; Caulder, D.; Craig, I.; Davis, L.; Edelstein, N.; Lukens, W.; Nitsche, H.; Rao, L.; Shuh, D. K.; Mallouk, T. E. Surface chemistry and electrochemistry of supported zerovalent iron nanoparticles in the remediation of aqueous metal contaminants. *Chem. Mater.* **2001**, *13*, 479–486.
27. He, F.; Zhao, D.; Liu, J.; Roberts, C. B. Stabilization of Fe-Pd nanoparticles with sodium carboxymethyl cellulose for enhanced transport and dechlorination of trichloroethylene in soil and groundwater. *Ind. Eng. Chem. Res.* **2007**, *46*, 29–34.
28. Jiemvarangkul, P.; Zhang, W. X.; Lien, H. L. Enhanced transport of polyelectrolyte stabilized nanoscale zero-valent iron (nZVI) in porous media. *Chem. Eng. J.* **2011**, *170*, 482–491.
29. Kustov, L. M.; Finashina, E. D.; Shuvalova, E. V.; Tkachenko, O. P.; Kirichenko, O. A. Pd-Fe nanoparticles stabilized by chitosan derivatives for perchloroethene dechlorination. *Environ. Int.* **2011**, *37*, 1044–1052.
30. Dobrynin, A. V.; Rubinstein, M. Theory of polyelectrolytes in solutions and at surfaces. *Prog. Polym. Sci.* **2005**, *30*, 1049–1118.
31. Kim, H. J.; Phenrat, T.; Tilton, R. D.; Lowry, G. V. Fe^0 nanoparticles remain mobile in porous media after aging due to slow desorption of polymeric surface modifiers. *Environ. Sci. Technol.* **2009**, *43*, 3824–3830.
32. Lin, Y. H.; Tseng, H. H.; Wey, M. Y.; Lin, M. D. Characteristics of two types of stabilized nano zero-valent iron and transport in porous media. *Sci. Total Environ.* **2010**, *408*, 2260–2267.
33. He, F.; Zhao, D. Manipulating the size and dispersability of zerovalent iron nanoparticles by use of carboxymethyl cellulose stabilizers. *Environ. Sci. Technol.* **2007**, *41*, 6216–6221.
34. Phenrat, T.; Liu, Y.; Tilton, R. D.; Lowry, G. V. Adsorbed polyelectrolyte coatings decrease Fe^0 nanoparticle reactivity with TCE in water: Conceptual model and mechanisms. *Environ. Sci. Technol.* **2009**, *43*, 1507–1514.
35. Lin, Y. H.; Tseng, H. H.; Wey, M. Y.; Lin, M. D. Characteristics, morphology, and stabilization mechanism of PAA250K-stabilized bimetal nanoparticles. *Colloids Surf., A.* **2009**, *349*, 137–144.
36. Comba, S.; Sethi, R. Stabilization of highly concentrated suspensions of iron nanoparticles using shear-thinning gels of xanthan gum. *Water Res.* **2009**, *43*, 3717–3726.
37. Vecchia, E. D.; Luna, M.; Sethi, R. Transport in porous media of highly concentrated iron micro- and nanoparticles in the presence of xanthan gum. *Environ. Sci. Technol.* **2009**, *43*, 8942–8947.

38. Shimmin, R. G.; Schoch, A. B.; Braun, P. V. Polymer size and concentration effects on the size of gold nanoparticles capped by polymeric thiols. *Langmuir* **2004**, *20*, 5613–5620.
39. Raychoudhury, T.; Naja, G.; Ghoshal, S. Assessment of transport of two polyelectrolyte-stabilized zero-valent iron nanoparticles in porous media. *J. Contam. Hydrol.* **2010**, *118*, 143–151.
40. Kanel, S. R.; Goswami, R. R.; Clement, T. P.; Barnett, M. O.; Zhao, D. Two dimensional transport of surface stabilized zero-valent iron nanoparticles in porous media. *Environ. Sci. Technol.* **2008**, *42*, 896–900.
41. Phenrat, T.; Cihan, A.; Kim, H. J.; Mital, M.; Illangasekare, T.; Lowry, G. V. Transport and deposition of polymer-modified Fe^0 nanoparticles in 2-D heterogeneous porous media: Effects of particle concentration, Fe^0 content, and coatings. *Environ. Sci. Technol.* **2010**, *44*, 9086–9093.
42. Phenrat, T.; Fagerlund, F.; Illangasekare, T.; Lowry, G. V.; Tilton, R. D. Polymer-modified Fe0 nanoparticles target entrapped NAPL in two dimensional porous media: Effect of particle concentration, NAPL saturation, and injection strategy. *Environ. Sci. Technol.* **2011**, *45*, 6102–6109.
43. Phenrat, T.; Kim, H. J.; Fagerlund, F.; Illangasekare, T.; Lowry, G. V. Empirical correlations to estimate agglomerate size and deposition during injection of a polyelectrolyte-modified Fe^0 nanoparticle at high particle concentration in saturated sand. *J. Contam. Hydrol.* **2010**, *118*, 152–164.
44. He, F.; Zhang, M.; Qian, T.; Zhao, D. Transport of carboxymethyl cellulose stabilized iron nanoparticles in porous media: Column experiments and modeling. *J. Colloid Interface Sci.* **2009**, *334*, 96–102.
45. Quinn, J.; Elliott, D.; O'Hara, S.; Billow, A. Use of nanoscale iron and bimetallic particles for environmental remediation: A review of field-scale applications. In *Environmental Applications of Nanoscale and Microscale Reactive Metal Particles*; Geiger, C. L., Carvalho-Knighton, K. M., Eds.; American Chemical Society: Washington, D.C., 2009; pp 263−283.
46. Phenrat, T.; Saleh, N.; Sirk, K.; Tilton, R. D.; Lowry, G. V. Aggregation and sedimentation of aqueous nanoscale zerovalent iron dispersions. *Environ. Sci. Technol.* **2007**, *41*, 284–290.
47. Jones, E. H.; Reynolds, D. A.; Wood, A. L.; Thomas, D. G. Use of electrophoresis for transporting nano-iron in porous media. *Groundwater* **2011**, *49*, 172–183.
48. Phenrat, T.; Kim, H. J.; Fagerlund, F.; Illangasekare, T.; Tilton, R. D.; Lowry, G. V. Particle size distribution, concentration, and magnetic attraction affect transport of polymer-modified Fe^0 nanoparticles in sand columns. *Environ. Sci. Technol.* **2009**, *43*, 5079–5085.
49. Saleh, N.; Kim, H. J.; Phenrat, T.; Matyjaszewski, K.; Tilton, R. D.; Lowry, G. V. Ionic strength and composition affect the mobility of surface-modified Fe^0 nanoparticles in water-saturated sand columns. *Environ. Sci. Technol.* **2008**, *42*, 3349–3355.
50. Sirk, K. M.; Saleh, N. B.; Phenrat, T.; Kim, H. J.; Dufour, B.; Ok, J.; Golas, P. L.; Matyjaszewski, K.; Lowry, G. V.; Tilton, R. D. Effect of

adsorbed polyelectrolytes on nanoscale zero valent iron particle attachment to soil surface models. *Environ. Sci. Technol.* **2009**, *43*, 3803–3808.

51. Saleh, N.; Sirk, K.; Liu, Y.; Phenrat, T.; Dufour, B.; Matyjaszewski, K.; Tilton, R. D.; Lowry, G. V. Surface modifications enhance nanoiron transport and NAPL targeting in saturated porous media. *Environ. Eng. Sci.* **2007**, *24*, 45–57.
52. Saleh, N.; Phenrat, T.; Sirk, K.; Dufour, B.; Ok, J.; Sarbu, T.; Matyjaszewski, K.; Tilton, R. D.; Lowry, G. V. Adsorbed triblock copolymers deliver reactive iron nanoparticles to the oil/water interface. *Nano Lett.* **2005**, *5*, 2489–2494.
53. Sun, Y. P.; Li, X. Q.; Zhang, W. X.; Wang, H. P. A method for the preparation of stable dispersion of zero-valent iron nanoparticles. *Colloids Surf., A.* **2007**, *308*, 60–66.
54. Tiraferri, A.; Chen, K. L.; Sethi, R.; Elimelech, M. Reduced aggregation and sedimentation of zero-valent iron nanoparticles in the presence of guar gum. *J. Colloid Interface Sci.* **2008**, *324*, 71–79.
55. Freyria, F. S.; Bonelli, B.; Sethi, R.; Armandi, M.; Belluso, E.; Garrone, E. Reactions of acid orange 7 with iron nanoparticles in aqueous solutions. *J. Phys. Chem.* **2011**, *115*, 24143–24152.
56. Xiong, Z.; Zhao, D.; Pan, G. Rapid and complete destruction of perchlorate in water and ion-exchange brine using stabilized zero-valent iron nanoparticles. *Water Res.* **2007**, *41*, 3497–3505.
57. He, F.; Zhao, D. Hydrodechlorination of trichloroethene using stabilized Fe-Pd nanoparticles: Reaction mechanism and effects of stabilizers, catalysts and reaction conditions. *Appl. Catal., B* **2008**, *84*, 533–540.
58. Cho, Y.; Choi, S. I. Degradation of PCE, TCE, and 1,1,1-TCA by nanosized Fe-Pd bimetallic particles under various experimental conditions. *Chemosphere* **2010**, *81*, 940–945.
59. Wang, Q.; Qian, H.; Yang, Y.; Zhang, Z.; Naman, C.; Xu, X. Reduction of hexavalent chromium by carboxymethyl cellulose-stabilized zero-valent iron nanoparticles. *J. Contam. Hydrol.* **2010**, *114*, 35–42.
60. Dong, T.; Luo, H.; Wang, Y.; Hu, B.; Chen, H. Stabilization of Fe-Pd bimetallic nanoparticles with sodium carboxymethyl cellulose for catalytic reduction of para-nitrochlorobenzene in water. *Desalination* **2011**, *271*, 11–19.
61. Zhang, W. X.; Wang, C. B.; Lien, H. L. Treatment of chlorinated organic contaminants with nanoscale bimetallic particles. *Catal. Today* **1998**, *40*, 387–395.
62. Kim, Y. H.; Carraway, E. R. Reductive dechlorination of TCE by zero valent bimetals. *Environ. Technol.* **2003**, *24*, 69–75.
63. Rivero-Huguet, M.; Marshall, W. D. Reduction of hexavalent chromium mediated by micro- and nano-sized mixed metallic particles. *J. Hazard. Mater.* **2009**, *169*, 1081–1087.
64. Chun, C. L.; Baer, D. R.; Matson, D. W.; Amonette, J. E.; Penn, R. L. Characterization and reactivity of iron nanoparticles prepared with added Cu, Pd, and Ni. *Environ. Sci. Technol.* **2010**, *44*, 5079–5085.

65. Wang, Z.; Huang, W.; Peng, P.; Fennell, D. E. Rapid transformation of 1,2,3,4-TCDD by Pd/Fe catalysts. *Chemosphere* **2010**, *78*, 147–151.
66. Kim, J. H.; Tratnyek, P. G.; Chang, Y. S. Rapid dechlorination of polychlorinated dibenzo-p-dioxins by bimetallic and nanosized zerovalent iron. *Environ. Sci. Technol.* **2008**, *42*, 4106–4112.
67. Lim, T. T.; Zhu, B. W. Practical applications of bimetallic nanoiron particles for reductive deahlogenation of haloorganics: Prospects and challenges. In *Environmental Applications of Nanoscale and Microscale Reactive Metal Particles*; Geiger, C. L., Carvalho-Knighton, K. M., Eds.; American Chemical Society: Washington, D.C., 2009; pp 245–261.
68. Lien, H. L.; Zhang, W. X. Nanoscale iron particles for complete reduction of chlorinated ethenes. *Colloids Surf., A* **2001**, *191*, 97–105.
69. Schrick, B.; Blough, J. L.; Jones, A. D.; Mallouk, T. E. Hydrodechlorination of trichloroethylene to hydrocarbons using bimetallic nickel-iron nanoparticles. *Chem. Mater.* **2002**, *14*, 5140–5147.
70. Lowry, G. V.; Reinhard, M. Pd-catalyzed TCE dechlorination in water: Effect of $[H_2](aq)$ and H_2-utilizing competitive solutes on the TCE dechlorination rate and product distribution. *Environ. Sci. Technol.* **2001**, *35*, 696–702.
71. Lien, H. L.; Zhang, W. X. Nanoscale Pd/Fe bimetallic particles: Catalytic effects of palladium on hydrodechlorination. *Appl. Catal., B* **2007**, *77*, 110–116.
72. Yan, W.; Herzing, A. A.; Li, X. Q.; Kiely, C. J.; Zhang, W. X. Structural evolution of Pd-doped nanoscale zero-valent iron (nZVI) in aqueous media and implications for particle aging and reactivity. *Environ. Sci. Technol.* **2010**, *44*, 4288–4294.
73. Wang, X.; Chen, C.; Chang, Y.; Liu, H. Dechlorination of chlorinated methanes by Pd/Fe bimetallic nanoparticles. *J. Hazard. Mater.* **2009**, *161*, 815–823.
74. Zhang, W.; Quan, W.; Wang, J.; Zhang, Z.; Chen, S. Rapid and complete dechlorination of PCP in aqueous solution using Ni-Fe nanoparticles under assistance of ultrasound. *Chemosphere* **2006**, *65*, 58–64.
75. Lin, C. J.; Lo, S. L.; Liou, Y. H. Dechlorinatino of trichloroethylene in aqueous solution by noble metal-modified iron. *J. Hazard. Mater.* **2004**, *B116*, 219–228.
76. Tee, Y. H.; Bachas, L.; Bhattacharyya, D. Degradation of trichloroethylene by iron-based bimetallic nanoparticles. *J. Phys. Chem.* **2009**, *C113*, 9454–9464.
77. Zhou, T.; Li, Y.; Lim, T. T. Catalytic hydrodechlorination of chlorophenols by Pd/Fe nanoparticles: Comparisons with other bimetallic systems, kinetics, and mechanism. *Sep. Purif. Technol.* **2010**, *76*, 206–214.
78. Lien, H. L.; Zhang, W. X. Transformation of chlorinated methanes by nanoscale iron particles. *J. Environ. Eng.* **1999**, *125*, 1042–1047.
79. Lien, H. L.; Zhang, W. X. Hydrodechlorination of chlorinated ethanes by nanoscale Pd/Fe bimetallic particles. *J. Environ. Eng.* **2005**, *131*, 4–10.
80. Zhu, B. W.; Lim, T. T.; Feng, J. Reductive dechlorination of 1,2,4-trichlorobenzene with palladized nanoscale Fe^0 particles supported on chitosan and silica. *Chemosphere* **2006**, *65*, 1137–1145.

81. Zhu, B. W.; Lim, T. T. Catalytic reduction of chlorobenzenes with Pd/Fe nanoparticles: Reactive sites, catalyst stability, particle aging, and regeneration. *Environ. Sci. Technol.* **2007**, *41*, 7523–7529.
82. Zhu, B. W.; Lim, T. T.; Feng, J. Influences of ampiphiles on dechlorination of a trichlorobenzene by nanoscale Pd/Fe: Adsorption, reaction kinetics, and interfacial interactions. *Environ. Sci. Technol.* **2008**, *42*, 4513–4519.
83. Nagpal, V.; Bokare, A. D.; Chikate, R. C.; Rode, C. V.; Paknikar, K. M. Reductive dechlorination of γ-hexachlorocyclohexane using Fe-Pd bimetallic nanoparticles. *J. Hazard. Mater.* **2010**, *175*, 680–687.
84. Zhang, M.; He, F.; Zhao, D.; Hao, X. Degradation of soil-sorbed trichloroethylene by stabilized zero valent iron nanoparticles: Effects of sorption, surfactants, and natural organic matter. *Water Res.* **2011**, *45*, 2401–2414.
85. Shih, Y. H.; Chen, M. Y.; Su, Y. F. Pentachlorophenol reduction by Pd/Fe bimetallic nanoparticles: Effects of copper, nickel, and ferric cations. *Appl. Catal., B* **2011**, *105*, 24–29.
86. Zhuang, Y.; Ahn, S.; Seyfferth, A. L.; Masue-Slowey, Y.; Fendorf, S.; Luthy, R. G. Dehalogenation of polybrominated diphenyl ethers and polychlorinated biphenyl by bimetallic, impregnated, and nanoscale zerovalent iron. *Environ. Sci. Technol.* **2011**, *45*, 4896–4903.
87. Joo, S. H.; Zhao, D. Destruction of lindane and atrazine using stabilized iron nanoparticles under aerobic and anaerobic conditions: Effects of catalyst and stabilizer. *Chemosphere* **2008**, *70*, 418–425.
88. Cao, J.; Elliott, D.; Zhang, W. X. Perchlorate reduction by nanoscale iron nanoparticles. *J. Nanopart. Res.* **2005**, *7*, 499–506.
89. Mondal, K.; Jegadeesan, G.; Lalvani, S. B. Removal of selenate by Fe and Ni-Fe nanosized particles. *Ind. Eng. Chem. Res.* **2004**, *43*, 4922–4934.
90. Feng, J.; Lim, T. T. Pathways and kinetics of carbon tetrachloride and chloroform reductions by nano-scale Fe and Fe/Ni particles: Comparison with commercial micro-scale Fe and Zn. *Chemosphere* **2005**, *59*, 1267–1277.
91. Tee, Y. H.; Grulke, E.; Bhattacharyya, D. Role of Ni/Fe nanoparticle composition on the degradation of trichloroethylene from water. *Ind. Eng. Chem. Res.* **2005**, *44*, 7062–7070.
92. Wu, L.; Ritchie, S. M. C. Removal of trichloroethylene from water by cellulose acetate supported bimetallic Ni/Fe nanoparticles. *Chemosphere* **2006**, *63*, 285–292.
93. Zhang, W. H.; Quan, X.; Zhang, Z. Y. Catalytic reductive dechlorination of p-chlorophenol in water using Ni/Fe nanoscale particles. *J. Environ. Sci.* **2007**, *19*, 362–366.
94. Bokare, A. D.; Chikate, R. C.; Rode, C. V.; Paknikar, K. M. Effect of surface chemistry of Fe-Ni nanoparticles on mechanistic pathways of azo dye degradation. *Environ. Sci. Technol.* **2007**, *41*, 7437–7443.
95. Lim, T. T.; Feng, J.; Zhu, B. W. Kinetic and mechanistic examinations of reductive transformation pathways of brominated methanes with nano-scale Fe and Ni/Fe particles. *Water Res.* **2007**, *41*, 875–883.

96. Barnes, R. J.; Riba, O.; Gardner, M. N.; Singer, A. C.; Jackman, S. A.; Thompson, I. P. Inhibition of biological TCE and sulphate reduction in the presence of iron nanoparticles. *Chemosphere* **2010**, *80*, 554–562.
97. Barnes, R. J.; Riba, O.; Gardner, M. N.; Scott, T. B.; Jackman, S. A.; Thompson, I. P. Optimization of nano-scale nickel/iron particles for the reduction of high concentration chlorinated aliphatic hydrocarbons. *Chemosphere* **2010**, *79*, 448–454.
98. Cheng, R.; Zhou, W.; Wang, J. L.; Qi, D.; Gou, L.; Zhang, W. X.; Qian, Y. Dechlorination of pentachlorophenol using nanoscale Fe/Ni particles: Role of nano-Ni and its size effect. *J. Hazard. Mater.* **2010**, *180*, 79–85.
99. Ryu, A.; Jeong, S. W.; Jang, A.; Choi, H. Reduction of highly concentrated nitrate using nanoscale zero-valent iron: Effects of aggregation and catalyst on reactivity. *Appl. Catal., B* **2011**, *105*, 128–135.
100. U.S. Enivonrmental Protection Agency. Technical factsheet on: Nickel. http://www.epa.gov/ogwdw/pdfs/factsheets/ioc/tech/nickel.pdf (accessed February 15, 2012).
101. World Health Organization. *Guidelines for drinking-water quality*, 4th ed.; WHO: Geneva, Switzerland, 2011. http://whqlibdoc.who.int/publications/2011/9789241548151_eng.pdf (accessed February 15, 2012).
102. Henn, K. W.; Waddill, D. W. Utilization of nanoscale zero-valent iron for source remediation-A case study. *Remediation J.* **2006**, *16*, 57–77.
103. Hydutsky, B. W.; Mack, E. J.; Beckerman, B. B.; Skluzacek, J. M.; Mallouk, T. E. Optimization of nano- and microiron transport through sand columns using polyelectrolyte mixtures. *Environ. Sci. Technol.* **2007**, *41*, 6418–6424.
104. Tiraferri, A.; Sethi, R. Enhanced transport of zerovalent iron nanoparticles in saturated porous media by guar gum. *J. Nanopart. Res.* **2009**, *11*, 635–645.
105. Xiu, Z. M.; Gregory, K. B.; Lowry, G. V.; Alvarez, P. J. J. Effect of bare and coated nanoscale zerovalent iron on tceA and vcrA gene expression in Dehalococcoides spp. *Environ. Sci. Technol.* **2010**, *44*, 7647–7651.

Chapter 11

Magnetic Bimetallic Fe/Ag Nanoparticles: Decontamination and Antimicrobial Agents

Virender K. Sharma,*,[1] Karolina M. Siskova,[2] and Radek Zboril[2]

[1]Department of Chemistry and Center of Ferrate Excellence, Florida Institute of Technology, Melbourne, Florida 32901, USA
[2]Regional Centre of Advanced Technologies and Materials, Departments of Experimental Physics and Physical Chemistry, Faculty of Science, Palacky University, Slechtitelu 11, 78371 Olomouc, Czech Republic
***E-mail: vsharma@fit.edu**

Bimetallic iron and silver containing nanoparticles (Fe-Ag NPs) have numerous applications in optical, medical, and remediation fields. This chapter reports the synthesis and characterization of bimetallic Fe-Ag NPs with core-shell structures. These bimetallic NPs can be applied to remove phosphate and reduce nitrate in the aqueous solution. Bimetallic Fe-Ag NPs also showed their effectiveness in transforming the halogenated aromatic contaminants in the environment. The silver NPs in the bimetallic Fe-Ag NPs generally enhance the removal of pollutants in water. Examples of antimicrobial and antifungal properties of bimetallic Fe-Ag NPs are given. Bimetallic Fe-Ag NPs are magnetic and can be easily separable after the water treatment. Bimetallic Fe-Ag NPs are promising materials for decontamination of water.

Introduction

Water plays an important role in almost every aspect of life such as good health, food security, industrial development, and clean hydroelectric energy. With increase in the global population, sustainable development would require adequate water supply to meet the demand for drinking water. Globally, more than one billion people do not have access to safe and clean water. Significantly, ~ 50% of the world's population does not have adequate water purification systems. In the world, the consumption of water has been increasing twice every twenty years, and if the present trend continues, about one-third of the population will be living in areas where basic water requirement demands for drinking, cooking, and sanitation could not be met by 2025 (*1*, *2*). Polluted or contaminated water resources may be causing diseases. For example, the United Nations report states that 10,000 to 20,000 children die each day from preventable water-related illness (*3*).

One of the recent approaches is to develop and apply new materials made out of active metal nanoparticles (NPs) such as silver, palladium, nickel, and silver (*4–6*). Zero-valent iron nanoparticles (nZFe or NZVI) have been studied extensively as decontaminating agent (*4–6*). Literature showed that nZVI can perform reduction and sorption process to decontaminate pollutants and inactivate microorganisms from water (*7–11*). However, the efficiency of decontamination is reduced because nZVI tend to react with the surrounding media and spontaneously aggregate (*12*). More recently, researchers are applying nZVI containing hybrid NPs to overcome the limitations associated with nZVI (*13–15*). Hybrids NPs consisting of multiple components generally have shown multiple functionalities to result in unique properties and applications that are difficult to obtain from single-component NPs (*13*, *16*). In the environmental remediation field, studies have demonstrated that combinations of nZVI with Pd, Pt, Ni, Cu, and Ag increased removal efficiency of nZVI (*17–26*). Bimetallic nZVI-Ag NPs have advantageous properties because nZVI and Ag NPs would enhance removal efficiency either cumulatively or synergistically because silver NPs also possess antibacterial and antifungal activities (*27–31*). Moreover, nZVI have magnetization values (*32*), hence, bimetallic nZVI-Ag can easily be separated after the treatment of contaminated water.

In the present chapter, we first present briefly the synthesis and characterization of nZVI-Ag NP, followed by examples of applications of these bimetallic nanoparticles to remove nutrients and organics and to inactivate a wide range of bacteria and fungus.

Synthesis and Characterization

A few research groups have successfully synthesized iron-silver bimetallic nanoparticles with the objective of decontamination of contaminants such as nutrient and organics (*20*, *21*, *33–37*). In a recent work, bimetallic nZVI-Ag NPs were synthesized with an objective of achieving phosphate removal as well as inactivation of microbials (*38*). In the synthesis, four different Fe:Ag ratios bimetallic were prepared. Briefly, 12.5, 25, 50, or 100 mg of $AgNO_3$ was dissolved in 50 mL deionized water and transferred into 60 mL vials covered with septum. Vials had 100 mg nZVI under nitrogen environment. The reaction mixtures were immediately subjected to sonication, followed by shaking at 150 rpm for 15 min. Magnetic sepration was used to terminate the reaction between nZVI and $AgNO_3$ was terminated. Finally, washing using deionized water was repeated five times and samples were labeled as Fe-Ag1, Fe-Ag2, Fe-Ag3, and Fe-Ag4 based on the increasing concentration of silver in reaction mixtures. In this study, magnetic bimetallic NPs were characterized using transmission electron microscope (TEM), scanning electron microscope (SEM), X-ray diffraction (XRD), superconducting quantum interference device (SQUID), and ^{57}Mossbauer spectra (*38*). Four crystalline phases in the synthesis of bimetallic Fe-Ag were found, which were zero-valent iron (α-Fe), maghemite or magnetite (Fe_2O_3/Fe_3O_4), lepidocrocite (γ-FeOOH), and metallic silver (Ag^0). The evidences of iron phases were also obtained from the Mossbauer spectroscopy. The results of XRD analysis and TEM images are presented in Figure 1.

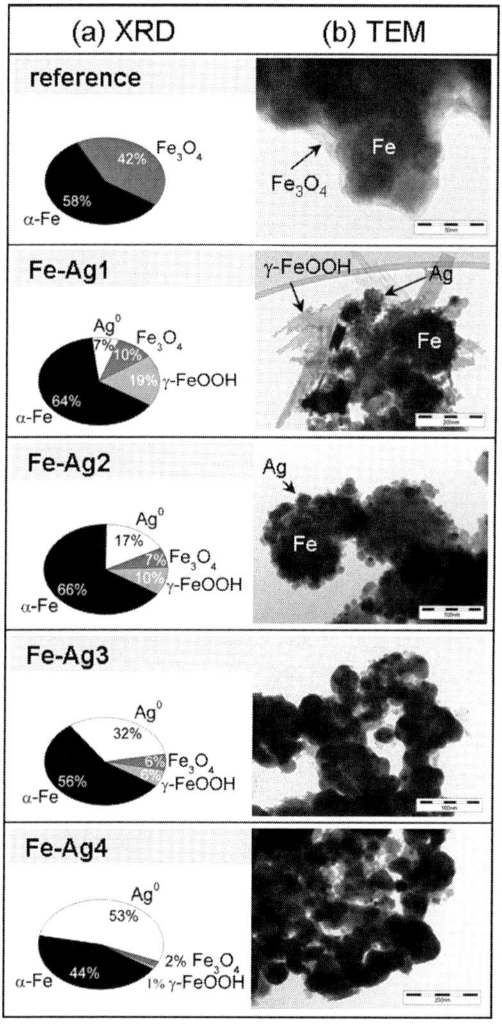

Figure 1. (a) Cake-graphs illustrating the phase composition based on XRD analysis: metallic silver (Ag0), zero-valent iron (α-Fe), lepidocrocite (γ-FeOOH), and magnetite (Fe3O4); and (b) TEM images of reference (partially oxidized nZVI) and Fe-Ag bimetallic NPs. Reproduced with permission from reference (38). Copyright (2013, American Chemical Society).

The percentage of Ag^0 increased with an increase in concentration of $AgNO_3$ (Fe-Ag1 to Ag-Ag4) (Figure 1a). Consequently, the levels of iron-bearing phases decreased. Increase in the concentration of $AgNO_3$ in the reaction mixture also increased the size of AgNPs (Figure 1b). In comparison with the reference, different morphologies and spatial arrangements of iron- and silver-containing phases, a thin iron oxide shell, formed by the oxidation of nZVI, on Fe^0 could be seen. Comparatively, Fe-Ag samples had a needle-shaped ferric oxy-hydroxides and nearly spherical metallic AgNPs on the surface of nZVI of Fe-Ag samples. The uniform distribution of AgNPs on Fe surface was confirmed by elemental analysis and SEM images (*38*). Importantly, saturation magnetization of nZVI in Fe-Ag samples was determined to be three times higher than that of magnetite of similar size. This would allow an easy separation of NPs from water by using a simple magnet. Other important feature of Fe-Ag NPs was that the dry powder of reference and the NPs did not change in their phase composition in air for 21 days. This suggests that Fe^0 core in the NPs was well protected by the Fe_3O_4/γ-FeOOH shell and AgNPs. In water, Fe-AgNPs showed changes in which contents of Fe^0 decreased and the levels Ag^0 slightly lowered. Therefore, bimetallic Fe-Ag NPs were found reactive in water (*38*).

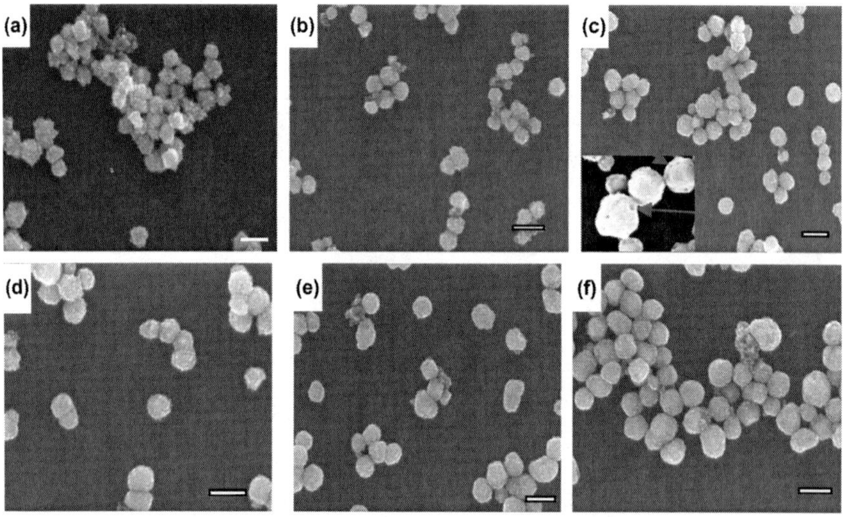

Figure 2. SEM images of Fe_2O_3/Au/Ag prepared by addition of $AgNO_3$ with different amount: 0.1 mL (a), 0.4 mL (b), 0.7 mL (c), 0.8 mL (d), 1.2 mL (e), and 1.6 mL (f). The inset from figure (c) showed the existence of pinholes on the surface of nanoparticles. Scale bars: 200 nm. Reproduced with permission from reference (39). Copyright (2012, Elsevier, Inc.).

Novel multifunctional Fe_2O_3-Ag-Au nanostructures were prepared (*39*). Using the varying amount of $AgNO_3$, Ag shell of different thickness was obtained. SEM images of nanostructures are shown in Figure 2, which show different thickness of Ag shell. In the initial coating of silver, the Ag shell attached to the Fe_2O_3/Au particles with several pinholes (Figure 2a). Subsequent increase in the concentration of $AgNO_3$, pinholes filled (Figures 2b, 2c). The shell thickness increased with the continuous growth of Ag shell (Figure 2d – f). It appears that each addition of 0.4 mL increased the diameter of ~ 10 nm of the Ag shell.

Nutrient Removal

The efficiency of bimetallic Fe-Ag NPs (see Figure 1) to remove phosphate was tested (Figure 3) (*38*). Before performing the tests, the stability of synthesized Fe-Ag NPs was confirmed by exposing NPs to sonication (Figure 3a). The Ag to Fe ratios remained constant within experimental error for up to 300 hours (Figure 3a). Significantly, the highest amount NPs (Fe-Ag-4) did not release any silver. Results in Figure 3a demonstrate a strong sorption capacity of nZVI for silver and the Fe-Ag NPs had high stability against sonication. This further suggests no release of silver would occur during applications of Fe-Ag NPs for treatment and subsequent separation of NPs can be carried out.

Figure 3b presents removal of phosphate by magnetic bimetallic Fe-Ag NPs. Both reference (i.e. nZVI) and bimetallic NPs were able to remove phosphate from water (Figure 3b). The general trend of removal of phosphate was Fe-Ag1 > Fe-Ag2 ~Fe-Ag3 > FeAg-4~ nZVI. This order was opposite to the order of silver content in bimetallic NPs indicating thus that the silver NPs did not play any significant role in removing phosphate. The higher content of iron in Fe-Ag1 compared to other bimetallic NPs may thus explain the observed trend of removal efficiency. Three possible processes were suggested on the removal of phosphate: (i) release of iron ions from the reaction of nZVI in water, form complex with phosphate to form iron phosphate precipitates, (ii) the partial oxidation of Fe^0 would form iron three ions during the aging of bimetallic Fe-Ag-NPs to result in increased removal of phosphate, and (iii) phosphate directly adsorb onto the surfaces of NPs (*38*). These three processes may be working together to cause the removal of phosphate from water. Furthermore, these processes depend on pH, redox, concentration of phosphate, and typr of phosphate (inorganic and organic) and future studies may include further exploring the mechnaism of phosphate removal.

The study on the reduction of NO_3^- by bimetallic nZVI-Ag nanoparticles (ZVIIBMNPs) has been studied (*20*, *21*). The effect of pH, temperature, the dose of Fe-Ag NPs, and the concentration of ethylenediaminetetraacaetate (EDTA) has been investigated. Results were modeled to obtain optimum conditions for maximum reduction of NO_3^- (*20*). The maximum reduction capacity was 33 mg g^{-1} under the optimum condition of 60 °C, pH 4, 1.0 gL^{-1} Fe-Ag NPs, and 2.0 mM EDTA. Importantly, EDTA enhanced the reduction of NO_3^- (Figure 4). The trend of enhancement was up to 2.0 mM EDTA. At this concentration of EDTA, an increase of 48.7 % reduction of NO_3^- was seen compared to without EDTA.

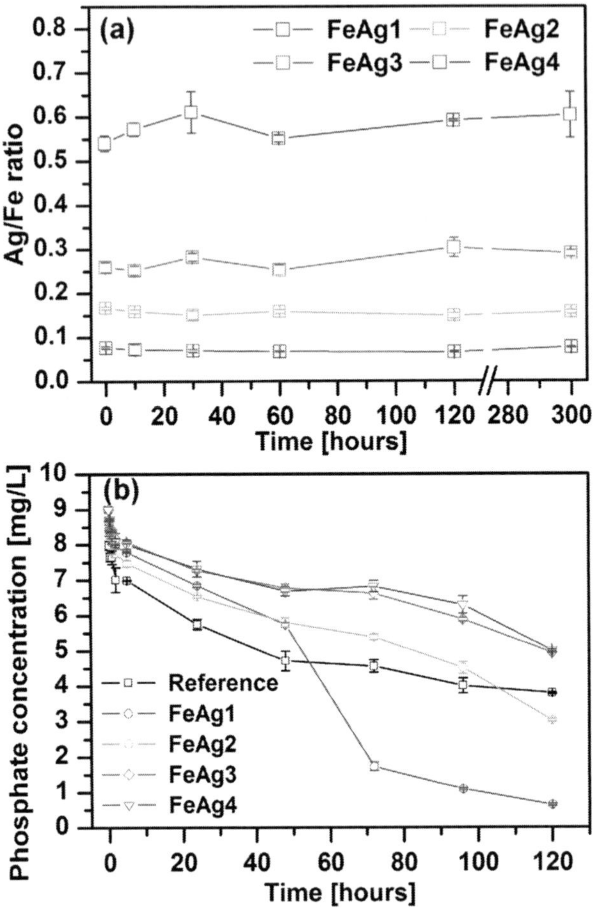

Figure 3. (a) Determination of Ag release from Fe-Ag NPs exposed to sonication expressed as dependence of Ag/Fe ratio on sonication time and (b) phosphate removing by bimetallic Fe-Ag NPs and/or reference sample (partially oxidized nZVI) from solutions with the initial phosphate concentration of 10 mg/L at pH 5.8, determined as residual PO_4^{3-} concentrations in time. Reproduced with permission from reference (38). Copyright (2013, American Chemical Society).

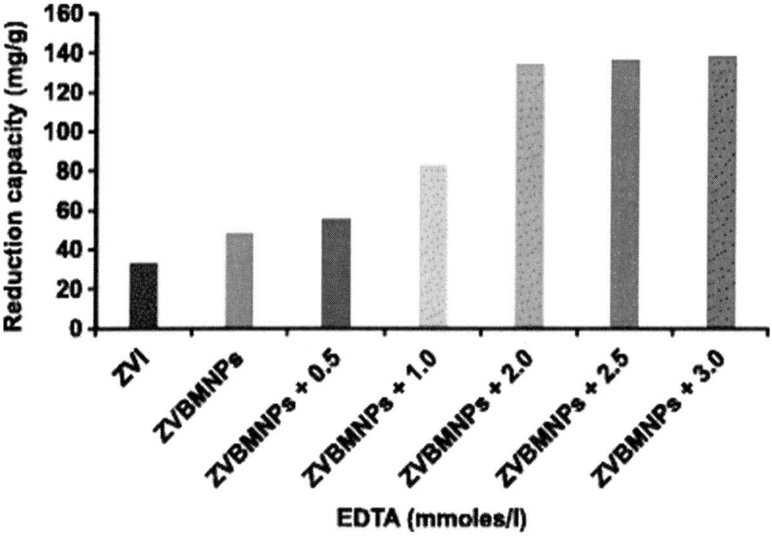

Figure 4. Effect of EDTA concentration on NO_3^- reduction by the ZVBMNPs (Fe-Ag NPs. Reproduced with permission from reference (20). Copyright (2012, Springer Inc.).

Figure 5 presents a scheme of the reduction of NO_3^- with and without EDTA. The reduction of NO_3^- to NH_4^+ took place through intermediate NO_2^- and N_2 species. No direct evidence of N_2 detection was provided. Other study could not measure N_2 species in the reduction of NO_3^- by ZVI (*40*). The presence of silver in the bimetallic Fe-Ag NPs increased the reduction to 27.2 % compared to reduction observed with only nZVI at similar surface area of both nanoparticles. It seems that silver acted as a catalyst to accelerate the reduction of NO_3^- ion. The oxidation of Fe^0 results in the formation of passive iron oxide layer, which could reduce the reduction efficiency of bimetallic Fe-Ag NPs. However, EDTA could interact with iron to form dissolved iron complexed species (Figure 5). Therefore, the negative effect of iron oxide(s) layer on the reduction of NO_3^- by Fe^0 was decreased in the presence of EDTA (Figure 4). The EDTA may also reduce the agglomeration of iron nanoparticles in order to have more active sites than that of without EDTA. This phenomenon would be beneficial for the reduction of NO_3^- efficiently.

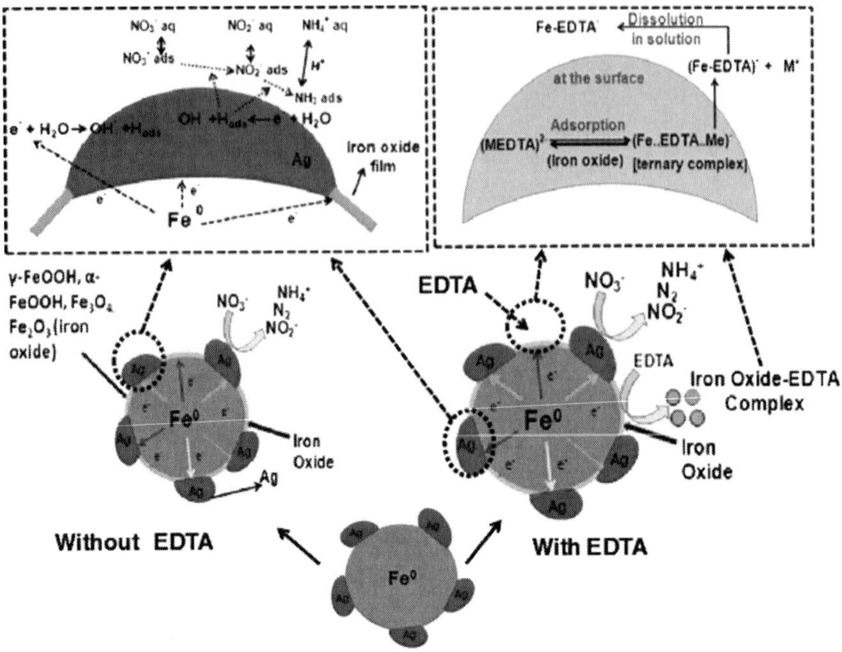

Figure 5. Schematic of EDTA catalyzed NO ₃⁻ reduction by Fe–Ag ZVBMNPs in aqueous medium. Reproduced with permission from reference (20). Copyright (2012, Springer Inc.).

Transformation of Organics

A few studies on the transformation of chlorinated organic compounds by bimetallic Fe-Ag NPs have been performed (*33*, *34*, *41*). Bimetallic Fe-Ag NPs (1 % Ag) were used to transform chlorinated benzenes, tetrachlorobenzenes (TeCB), pentachloro benzene (PCB), and hexachlorobenzene (HCB). Dechlorination of HCB is shown in Figure 6a (*41*). Control experiments using Fe^0 did not show any significant degradation of HCB. Bimetallic Fe-Ag NPs transformed HCB into lower molecular weight chlorinated benzenes than the parent molecule (Figure 6a). Among different products generated during dechlorination, 1,4 dichlorobenzene (DCB), 1, 2, 4-trichlorobenzene (TCB), and 1,2,4,5-TeCB were the major products (Figure 6a). The rate of dechlorination of HCB was positively related to the amount of silver in Fe-Ag NPs (*41*). Comparatively, conventional iron of micrometer size even at ten times more iron than bimetallic NPs had almost no transformation of HCB (Figure 6b). However, ~ 70% sorption of HCB onto Fe particles occurred in the first 4 hours. The transformation products appeared after a long period of time. After 24 h, less than 1 % of pentachlorobenzene (PeCB) was generated. Trichlorobenzenes (TCB) and tetrachlorobenzenes (TeCB) were produced after 100 h. No DCB was observed in experiments with using only iron particles. Similar results of dechlorination of HCB using nanoscale iron particles were observed (*42*, *43*).

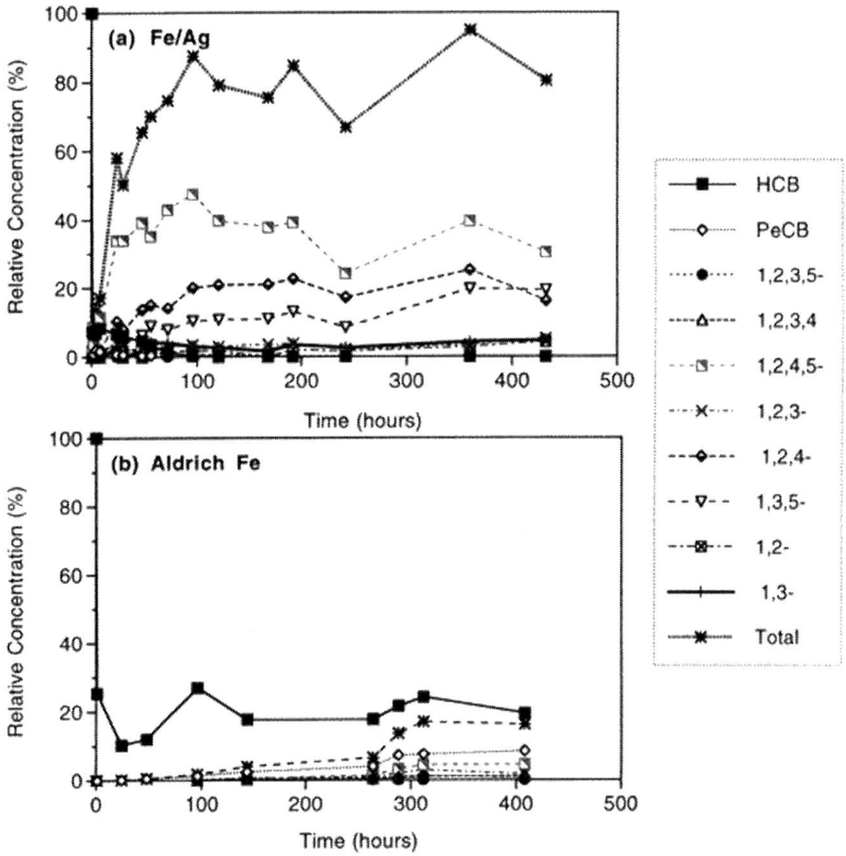

Figure 6. Transformation of HCB with (a) subcolloidal Fe/Ag particles and (b) Aldrich Fe particles (10 μm). Initial HCB concentration was 4 mg/L. Metal particle loading was 0.5 g/20 mL for the Fe/Ag and 5 g/20 mL for the Aldrich iron. Concentrations are normalized to the initial HCB concentration. The total in the figure refers to the sum of all daughter products in solution. Reproduced with permission from reference (41). (Copyright (2000, American Chemical Society).

A few studies also investigated these transformations under microwave energy (MW) and ultrasound (US) radiation (Figures 7 and 8). Figure 7 shows the debromination of polybrominated diphenyl ethers (PBDEs), BDE-209 and BDE-47. The removal of BDE-209 was complete within 10 min (Figure 7a). The Ag NPs in the bimetallic Fe-Ag NPs enhanced the debromination rate of BDE-209 under MW energy (Figure 7a). In the case of BDE-47, the removal efficiencies were 56 % and 78 % for Fe^0 and Fe-Ag NPs, respectively (Figure 7b). The results suggest that silver played a role in debromination of both BDE-209 and BDE-47 by bimetallic systems. A role of hydrogen atom has been postulated in debromination of PBDEs. The debromiated products were analyzed using gas chromatography-mass spectrometry (GC-MS) and liquid chromatography-mass spectrometry (LC-MS) techniques. The reduction of BDE-209 resulted in di- to nona-bromaited congeners while di- to tri-BDEs were seen in the degradation of BDE-47 (*33*). Reaction pathways included stepwise debromaination of PBDEs. Some studies using using iron nanoparticles of nanoscale without microwave also showed the effective treatment of HCB (*44–46*).

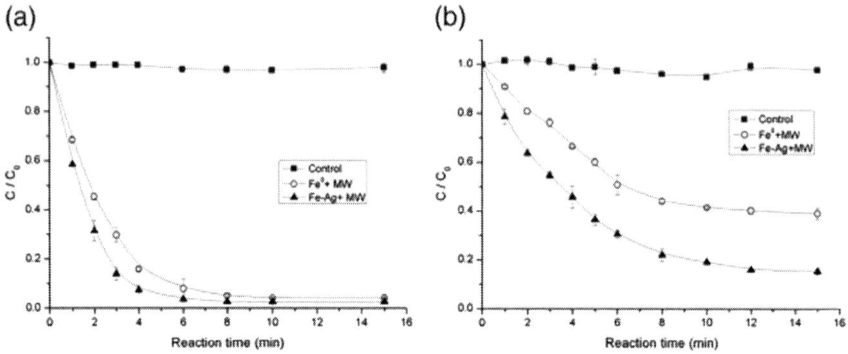

Figure 7. Degradation of PBDEs by MW, Fe0/MW or Fe–Ag/MW with decrease of (a) BDE-209 and (b) BDE-47 concentrations. [Fe0], 2 g L− 1; [Fe–Ag bimetallic], 2 g L− 1; Ag content, 1.0% (w/w); [PBDEs], 5 mg L− 1; pH, 6.5 ± 0.5; and MW energy, 800 W. Reproduced with permission from reference (33). Copyright (2012, Elsevier, Inc.).

The degradation of tetrabromobisphenol A (TBBPA) in Fe-Ag NPs suspension solution has also been studied under ultrasound conditions (Figure 8). In the absence of NPs, there was only 3 % degradation of TBBPA in 2 h. In the presence of bimetallic Fe-Ag NPs but without applying ultrasound, the initial decrease in TBBPA concentration in 10 min was followed by an increase as the reaction proceeded. This was possibly due to the initial adsorption of TBBPA onto NPs and a subsequent dissolution of parent compound from the surface of particles. Under US, the degradation of TBBPA occurred both in nZVI and Fe-Ag NPs, but the latter showed the complete removal in 20 min (Figure 8). The nZVI particles could degrade up to 60 % of TBBPA in 60 min. The results of Figure 8 suggest that both US and silver in bimetallic NPs were important in the rapid degradation of TBBPA. The major intermediates during the degradation of TBBPA were tri-BBPA, di-BBPA, mono-BBPA, and BPA (*37*).

Antimicrobial Activity

The magnetic bimetallic Fe-Ag NPs synthesized, shown in Figure 1, were tested for their antibacterial and antifungal activities (*38*). Tests were made under aerobic conditions and the standard dilution technique was employed to determine minimum inhibition concentrations (MIC) values. The values provided information on the toxicity of a broad spectrum of bacterial strains. The nZVI particles as reference were also tested by dispersing them into water. Silver ions were not present in the reference samples. Reference samples did not show any antibacterial and antifungal activity even at the 3.5 g L^{-1}, i.e. the highest concentration used in the study (*38*). The formation of passive FeOOH layer on Fe^0 shell (see TEM image in Figure 1b) may be responsible for no antimicrobial activity of the reference. Iron oxides usually did not show any antimicrobial effects (*11*).

Figure 9a shows clearly that all Fe-Ag NPs significantly inhibited growth of tested bacteria. The values of MIC values were in the range from 100 – 540 mg L^{-1} of the total mass of Fe-Ag NPs. The comparison of reference and bimetallic Fe-Ag having the same amounts of iron indicates that the silver NPs in bimetallic Fe-Ag NPs were solely responsible for the inhibition of growth of microorganisms. Figure 9a further showed that the values of MIC usually varied slightly on both the particular Fe-Ag bimetallic AgNPs and microbial strain. However, in some instances (e.g. *Pseudomonas aeruginosa* CCM3955 and *Staphylococcus epidermidis*), there was no significance difference in the values of MIC values for the different Fe-Ag NPs. A role of FeOOH layer on AgNPs was suggested to cause such behavior of bimetallic NPs against some bacterial strains. Figure 9b shows the microbial toxicity based on the Ag content in Fe-Ag NPs. The FeAg1 had the smallest size and had the most effective antibacterial activity (Figure 9b). This is consistent with literature that the smaller the AgNPs, the higher the antibacterial property (*27*, *47–49*). There was also the possibility that the larger amount needle-shaped γ-FeOOH nanostructures on the surface of Fe-Ag NPs improved the interaction of AgNPs with the cell of bacterial strains (*38*).

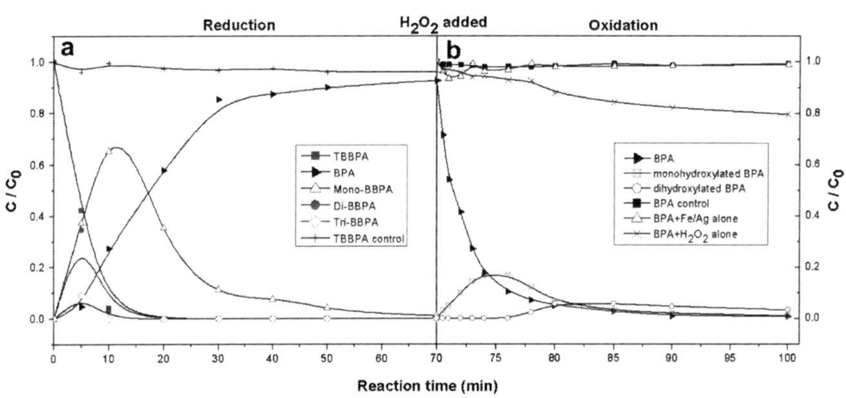

Figure 8. (a) Temporal disappearance of TBBPA and appearance of by-products in aqueous solution during Fe–Ag–US reduction treatment. [Fe-Ag bimetallic] = 0.8 g L−1, Ag content (wt %) = 1%, and [TBBPA]0 = 5 mg L−1. (b) Temporal change of BPA and its hydroxylated products concentration in the solution during Fe-Ag/H2O2/US oxidation treatment. [Fe-Ag bimetallic] = 0.8 g L−1; flow rate of H2O2 = 2 mg L−1 min−1, lasted 10 min; [BPA]0 = 4.64 mg L−1; US conditions: 40 kHz, 100 W. Reproduced with permission from reference (36). Copyright (2011, Elsevier, Inc.).

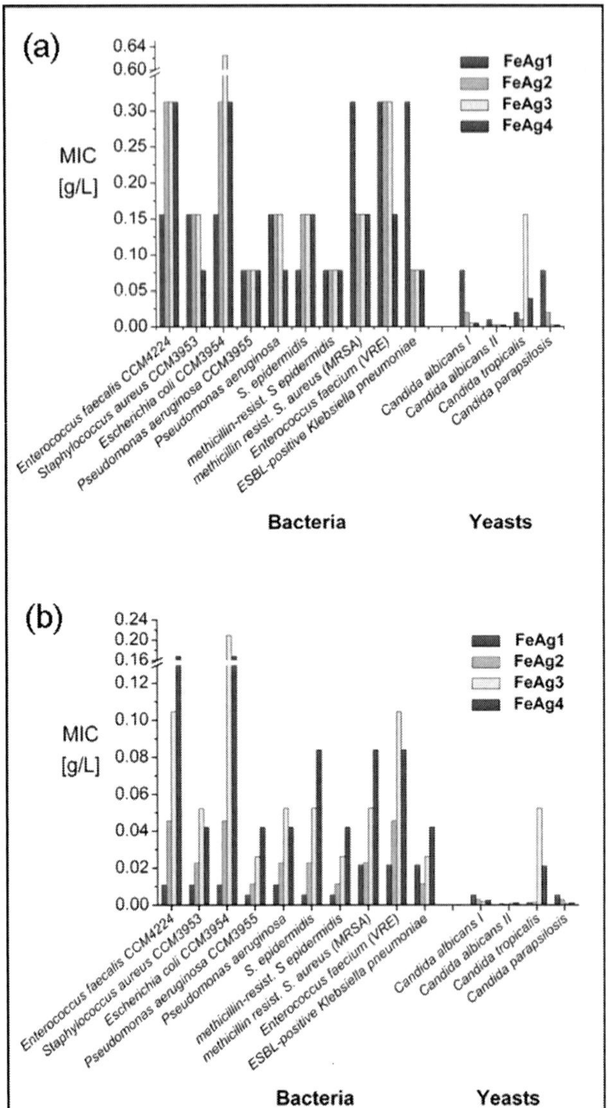

Figure 9. Antibacterial and antifungal activities of Fe-Ag NPs against tested bacteria and yeast: (c) MIC and (d) Ag-related MIC values. The reference MIC values were also tested, however, even at the highest concentration employed (i.e., 3.5 g/L), it did not show any antibacterial and antifungal activity against all microorganisms included in the study. Reproduced with permission from reference (38). Copyright (2013, American Chemical Society).

The values of MIC based on total mass were in the range from 3.3 – 270 mg L^{-1} for yeast (Figure 9a). These are lower than those obtained for bacterial strains. This is consistent with the results reported in the literature (*48*). The values of MIC were in the order as Fe-Ag1 > Fe-Ag2 > Fe-Ag3 > Fe-Ag4. This indicates the importance of the content of Ag in the studied bimetallic NPs. However, when the MIC values were recalculated using the Ag content in Fe-Ag NPs, both Fe-Ag1 and Fe-Ag2 had very high antifungal activity against all tested yeast (Figure 9b).

Conclusions

This review of currently available results appearing in the literature concerning magnetic Fe-Ag nanoparticles clearly demonstrates the usefulness and the effectiveness of these bimetallic nanoparticles in the removal of different types of contaminants, such as inorganic compounds (phosphates, nitrates) and halogenated organic species. Even microbial growth can be greatly inhibited by the application of Fe-Ag nanoparticles. It is demonstrated that Ag phase is (i) a powerful tool against microorganisms (bacteria, yeasts); (ii) able to catalyze debromination and/or dechlorination of halogenated organic compounds; and (iii) serving as a catalyst to accelerate the reduction of nitrate ion. On the contrary, Fe^0 phase is responsible for magnetic response of the whole bimetallic nanoparticle (possessing high values of magnetization) as well as for the reductive and sorption mode of action in pollutants removal.

Acknowledgments

The authors gratefully acknowledge the support by the Operational Program Research and Development for Innovations - European Regional Development Fund (project CZ.1.05/2.1.00/03.0058 of the Ministry of Education, Youth and Sports of the Czech Republic). The Authors also acknowledge the support by the Operational Program Education for Competitiveness - European Social Fund (project CZ.1.07/2.3.00/20.0056 of the Ministry of Education, Youth and Sports of the Czech Republic). K.M.S. thanks to the Grant Agency of the Czech Republic for financial support (project no. P108/11/P657). Authors would like to thank reviewers for their comments which improved the chapter greatly.

References

1. Elimelech, M.; Phillip, W. A. The future of seawater desalination: Energy, technology, and the environment. *Science* **2011**, *333*, 712–717.
2. Shannon, M. A.; Bohn, P. W.; Elimelech, M.; Georgiadis, J. G.; Marias, B. J.; Mayes, A. M. Science and technology for water purification in the coming decades. *Nature* **2008**, *452*, 301–310.
3. Schwarzenbach, R. P.; Egli, T.; Hofstetter, T. B.; von Gunten, U.; Wehrli, B. Global water pollution and human health. *Ann. Rev. Environ. Resources* **2010**, *35*, 109–136.

4. Karn, B.; Kuiken, T.; Otto, M. Nanotechnology and in situ remediation: A review of the benefits and potential risks. *Environ. Health Perspect.* **2009**, *117*, 1823–1831.
5. Li, X.; Brown, D. G.; Zhang, W. Stabilization of biosolids with nanoscale zero-valent iron (nZVI). *J. Nanopart. Res.* **2007**, *9*, 233–243.
6. Ju-Nam, Y.; Lead, J. R. Manufactured nanoparticles: An overview of their chemistry, interactions and potential environmental implications. *Sci. Total Environ.* **2008**, *400*, 396–414.
7. Noubactep, C.; Caré, S.; Crane, R. Nanoscale metallic iron for environmental remediation: Prospects and limitations. *Water, Air, Soil Pollut.* **2012**, *223*, 1363–1382.
8. Zhang, L.; Fang, M. Nanomaterials in pollution trace detection and environmental improvement. *Nano Today* **2010**, *5*, 128–142.
9. Chen, Q.; Gao, M.; Li, J.; Shen, F.; Wu, Y.; Xu, Z.; Yao, M. Inactivation and magnetic separation of bacteria from liquid suspensions using electrosprayed and nonelectrosprayed nZVI particles: Observations and mechanisms. *Environ. Sci. Technol.* **2012**, *46*, 2360–2367.
10. Diao, M.; Yao, M. Use of zero-valent iron nanoparticles in inactivating microbes. *Water Res.* **2009**, *43*, 5243–5251.
11. Marsalek, B.; Jancula, D.; Marsalkova, E.; Mashlan, M.; Safarova, K.; Tucek, J.; Zboril, R. Multimodal action and selective toxicity of zerovalent iron nanoparticles against cyanobacteria. *Environ. Sci. Technol.* **2012**, *46*, 2316–2323.
12. Phenrat, T.; Saleh, N.; Sirk, K.; Tilton, R. D.; Lowry, G. V. Aggregation and sedimentation of aqueous nanoscale zerovalent iron dispersions. *Environ. Sci. Technol.* **2007**, *41*, 284–290.
13. Peng, S.; Lei, C.; Ren, Y.; Cook, R. E.; Sun, Y. Plasmonic/magnetic bifunctional nanoparticles. *Angew. Chem., Int. Ed.* **2011**, *50*, 3158–3163.
14. Mattei, J. G.; Pelletier, F.; Ciuculescu, D.; Lecante, P.; Dupin, J. C.; Yaacoub, N.; Allouche, J.; Greneche, J. M.; Gonbeau, D.; Amiens, C.; Casanove, M. J. Formation of bimetallic FeBi nanostructured particles: Investigation of a complex growth mechanism. *J. Phys. Chem. C* **2013**, *117*, 1477–1484.
15. Mattei, J. G.; Pelletier, F.; Ciuculescu, D.; Lecante, P.; Amiens, C.; Casanove, M. J. Development of Bi-metallic Fe-Bi nanocomposites: Synthesis and characterization. *J. Nanosci. Nanotechnol.* **2012**, *12*, 8640–8646.
16. Sridharan, K.; Endo, T.; Cho, S.; Kim, J.; Park, T. J.; Philip, R. Single step synthesis and optical limiting properties of Ni-Ag and Fe-Ag bimetallic nanoparticles. *Opt. Mater.* **2013**, *35*, 860–867.
17. Ambashta, R. D.; Repo, E.; Sillanpää, M. Degradation of tributyl phosphate using nanopowders of iron and iron-nickel under the influence of a static magnetic field. *Ind. Eng. Chem. Res.* **2011**, *50*, 11771–11777.
18. Lee, C.; Sedlak, D. L. Enhanced formation of oxidants from bimetallic nickel-iron nanoparticles in the presence of oxygen. *Environ. Sci. Technol.* **2008**, *42*, 8528–8533.

19. Schrick, B.; Blough, J. L.; Jones, A. D.; Mallouk, T. E. Hydrodechlorination of trichloroethylene to hydrocarbons using bimetallic nickel-iron nanoparticles. *Chem. Mater.* **2002**, *14*, 5140–5147.
20. Singh, K. P.; Singh, A. K.; Gupta, S. Optimization of nitrate reduction by EDTA catalyzed zero-valent bimetallic nanoparticles in aqueous medium. *Environ. Sci. Pollut. Res.* **2012**, *19*, 3914–3924.
21. Singh, K. P.; Singh, A. K.; Gupta, S.; Rai, P. Modeling and optimization of reductive degradation of chloramphenicol in aqueous solution by zero-valent bimetallic nanoparticles. *Environ. Sci. Pollut. Res.* **2012**, *19*, 2063–2078.
22. Dien, N. T.; De Windt, W.; Buekens, A.; Chang, M. B. Application of bimetallic iron (BioCAT slurry) for pentachlorophenol removal from sandy soil. *J. Hazard. Mater.* **2013**, *252-253*, 83–90.
23. Douvalis, A. P.; Zboril, R.; Bourlinos, A. B.; Tucek, J.; Spyridi, S.; Bakas, T. A facile synthetic route toward air-stable magnetic nanoalloys with Fe-Ni/Fe-Co core and iron oxide shell. *J. Nanopart. Res.* **2012**, *14*, 1130–1146.
24. Kuang, Y.; Zhou, Y.; Chen, Z.; Megharaj, M.; Naidu, R. Impact of Fe and Ni/Fe nanoparticles on biodegradation of phenol by the strain Bacillus fusiformis (BFN) at various pH values. *Bioresour. Technol.* **2013**, *136*, 588–594.
25. Zhou, Y.; Kuang, Y.; Li, W.; Chen, Z.; Megharaj, M.; Naidu, R. A combination of bentonite-supported bimetallic Fe/Pd nanoparticles and biodegradation for the remediation of p-chlorophenol in wastewater. *Chem. Eng. J.* **2013**, *223*, 68–75.
26. Zhang, Z.; Hu, S.; Baig, S. A.; Tang, J.; Xu, X. Catalytic dechlorination of Aroclor 1242 by Ni/Fe bimetallic nanoparticles. *J. Colloid Interface Sci.* **2012**, *385*, 160–165.
27. Sharma, V. K.; Yngard, R. A.; Lin, Y. Silver nanoparticles: Green synthesis and their antimicrobial activities. *Adv. Colloid Interface Sci.* **2009**, *145*, 83–96.
28. Dallas, P.; Sharma, V. K.; Zboril, R. Silver polymeric nanocomposites as advanced antimicrobial agents: Classification, synthetic paths, applications, and perspectives. *Adv. Colloid Interface Sci.* **2011**, *166*, 119–135.
29. Panáček, A.; Kvítek, L.; Prucek, R.; Kolář, M.; Večeřová, R.; Pizúrová, N.; Sharma, V. K.; Nevěčná, T.; Zbořil, R. Silver colloid nanoparticles: Synthesis, characterization, and their antibacterial activity. *J. Phys. Chem. B* **2006**, *110*, 16248–16253.
30. Panacek, A.; Prucek, R.; Safarova, D.; Dittrich, M.; Richtrova, J.; Benickova, K.; Zboril, R.; Kvitek, L. Acute and Chronic Toxicity Effects of Silver Nanoparticles (NPs) on *Drosophila melanogaster*. *Environ. Sci. Technol.* **2012**, *45*, 4974–4979.
31. Kvitek, L.; Vanickova, M.; Panacek, A.; Soukupov, J.; Dittrich, M.; Valentova, E.; Prucek, R.; Bancirova, M.; Milde, D.; Zboril, R. Initial study on the toxicity of silver nanoparticles (NPs) against Paramecium caudatum. *J. Phys. Chem. C* **2009**, *113*, 4296–4300.
32. Siskova, K.; Tucek, J.; MacHala, L.; Otyepkova, E.; Filip, J.; Safarova, K.; Pechousek, J.; Zboril, R. Air-stable nZVI formation mediated by glutamic acid: Solid-state storable material exhibiting 2D chain morphology and high reactivity in aqueous environment. *J. Nanopart. Res.* **2012**, *14*, 805–817.

33. Luo, S.; Yang, S.; Sun, C.; Gu, J. Improved debromination of polybrominated diphenyl ethers by bimetallic iron-silver nanoparticles coupled with microwave energy. *Sci. Total Environ.* **2012**, *429*, 300–308.
34. Luo, S.; Yang, S.; Wang, X.; Sun, C. Reductive degradation of tetrabromobisphenol using iron-silver and iron-nickel bimetallic nanoparticles with microwave energy. *Environ. Eng. Sci.* **2012**, *29*, 453–460.
35. Luo, S.; Yang, S.; Xue, Y.; Liang, F.; Sun, C. Two-stage reduction/subsequent oxidation treatment of 2,2′,4,4′-tetrabromodiphenyl ether in aqueous solutions: Kinetic, pathway and toxicity. *J. Hazard. Mater.* **2011**, *192*, 1795–1803.
36. Luo, S.; Yang, S.; Sun, C.; Wang, X. Feasibility of a two-stage reduction/subsequent oxidation for treating Tetrabromobisphenol A in aqueous solutions. *Water Res.* **2011**, *45*, 1519–1528.
37. Luo, S.; Yang, S.; Wang, X.; Sun, C. Reductive degradation of tetrabromobisphenol A over iron-silver bimetallic nanoparticles under ultrasound radiation. *Chemosphere* **2010**, *79*, 672–678.
38. Markova, Z.; Siskova, K. M.; Filip, J.; Cuda, J.; Kolar, M.; Safarova, K.; Medrik, I.; Zboril, R. Air stable magnetic bimetallic Fe-Ag nanoparticles for advanced antimicrobial treatment and phosphorous removal. *Environ. Sci. Technol.* **2013**, *47*, 5285–5293.
39. Han, S.; Guo, Q.; Xu, M.; Yuan, Y.; Shen, L.; Yao, J.; Liu, W.; Gu, R. Tunable fabrication on iron oxide/Au/Ag nanostructures for surface enhanced Raman spectroscopy and magnetic enrichment. *J. Colloid Interface Sci.* **2012**, *378*, 51–57.
40. Yang, G. C.; Lee, H. Chemical reduction of nitrate by nanosized iron: Kinetics and pathways. *Water Res.* **2005**, *39*, 884–894.
41. Xu, Y.; Zhang, W. Subcolloidal Fe/Ag particles for reductive dehalogenation of chlorinated benzenes. *Ind. Eng. Chem. Res.* **2000**, *39*, 2238–2244.
42. Zhang, W.; Wang, C.; Lien, H. Treatment of chlorinated organic contaminants with nanoscale bimetallic particles. *Catal. Today* **1998**, *40*, 387–395.
43. Shih, Y.; Hsu, C.; Su, Y. Reduction of hexachlorobenzene by nanoscale zero-valent iron: Kinetics, pH effect, and degradation mechanism. *Sep. Purif. Technol.* **2011**, *76*, 268–274.
44. Su, Y.; Hsu, C.; Shih, Y. Effects of various ions on the dechlorination kinetics of hexachlorobenzene by nanoscale zero-valent iron. *Chemosphere* **2012**, *88*, 1346–1352.
45. Yan, W.; Herzing, A. A.; Li, X.-.; Kiely, C. J.; Zhang, W.-. Structural evolution of Pd-doped nanoscale zero-valent iron (nZVI) in aqueous media and implications for particle aging and reactivity. *Environ. Sci. Technol.* **2010**, *44*, 4288–4294.
46. Li, S.; Elliott, D. W.; Spear, S. T.; Ma, L.; Zhang, W.-. Hexachlorocyclohexanes in the environment: Mechanisms of dechlorination. *Crit. Rev. Environ. Sci. Technol.* **2011**, *41*, 1747–1792.

47. Morones, J. R.; Elechiguerra, J. L.; Camacho, A.; Holt, K.; Kouri, J. B.; Ramírez, J. T.; Yacaman, M. J. The bactericidal effect of silver nanoparticles. *Nanotechnology* **2005**, *16*, 2346–2353.
48. Prucek, R.; Tucek, J.; Kilianová, M.; Panácek, A.; Kvítek, L.; Filip, J.; Kolár, M.; Tománková, K.; Zboril, R. The targeted antibacterial and antifungal properties of magnetic nanocomposite of iron oxide and silver nanoparticles. *Biomaterials* **2011**, *32*, 4704–4713.
49. Baker, C.; Pradhan, A.; Pakstis, L.; Pochan, D. J.; Shah, S. I. Synthesis and antibacterial properties of silver nanoparticles. *J. Nanosci. Nanotechnol.* **2005**, *5*, 244–249.

Chapter 12

Application of Zerovalent Silicon for the Dechlorination of Chlorinated Hydrocarbons – A Review

Chun-Chi Lee and Ruey-an Doong*

Department of Biomedical Engineering and Environmental Sciences, National Tsing Hua University, Hsinchu, 30013, Taiwan
***E-mail: radoong@mx.nthu.edu.tw**

The dechlorination of chlorinated hydrocarbons by zerovalent metals has received much attention. In addition to iron, zerovalent silicon is a strong reductive material which can serve as an alternative for remediation of contaminated sites. However, little is known about the reaction kinetics and mechanism of chlorinated hydrocarbons by zerovalent silicon under anoxic conditions. In this review, parameters controlling the reactivity of silicon in dechlorination were discussed. In addition, the enhanced effects of bimetallic system including Fe/Si and Ni/Si and addition of surfactant on the dechlorination kinetics and mechanisms were also introduced and reviewed. The dechlorination efficiency and rate of chlorinated hydrocarbons by zerovalent silicon increased upon increasing pH from 7.2 to 9.5. The dechlorination followed the pseudo-first-order kinetics and the rate constant (k_{obs}) for carbon tetrachloride dechlorination increased from 0.5 h^{-1} at pH 7.2 to 2.11 h^{-1} at pH 9.5. In addition, tetrachloroethylene (PCE) can be effectively dechlorinated by zerovalent silicon. Addition of Ni^{2+} and Fe^{2+} significantly enhanced the dechlorination efficiency and rate of PCE as well as changed the distribution of products. However, the roles of Ni^{2+} and Fe^{0} were different. Addition of Fe^{0} to Si could form the buffer solution to maintain the pH at 7.8 when the ratio of Si and Fe is optimized, and subsequently enhanced the dechlorination efficiency and rate of carbon tetrachloride and PCE, while the adsorbed Ni(II) ions were converted to Ni(0) homogeneously onto the surface of zerovalent silicon and exhibited the synergistic effect on the dechlorination of PCE by Si under anoxic conditions. In addition, the introduction of surfactant can also enhance the dechlorination efficiency, and the dechlorination rate of PCE increased 62 and 106 times higher than silicon alone when CTAB and polyethylene glycol were introduced, respectively. X-ray photoelectron spectroscopy results confirmed that the addition of PEG could inhibit the production of silicon dioxide, resulting in the acceleration of dechlorination rate of PCE by Si. These processes would significantly accelerate the dechlorination rate of chlorinated hydrocarbons under anoxic conditions and would be helpful in facilitating the development of processes that could be useful for the enhanced degradation of co-contaminants for long-term performance.

Introduction

Chlorinated hydrocarbons such as carbon tetrachloride (CT), tetrachloroethylene (PCE), and trichloroethylene (TCE) are frequently used chemicals in industrial product generation and cleaning processes. For example, TCE is utilized as an organic solvent for anesthetic and analgesics purposes (*1*, *2*), while PCE is mainly used for laundry cleaning purposes (*3*). Although these artificial organic solvents play an important role when applied in each field, these chemicals are also suspected as carcinogens when released into the environment (*4*). Because of the relatively low water solubility and dense than water, these released dense non-aqueous phase liquids (DNAPLs) are accumulated on the bottom of groundwater. Moreover, the accumulated and precipitated DNAPLs would release into aqueous and poison the groundwater slowly and continuously (*5*).

The development of an efficient technique to remediate the chlorinated contaminants in subsurface environments is one of the most complex challenges (*6*, *7*). Because of the lack of oxygen in the subsurface environments, reductive dechlorination by chemical-based techniques is becoming one of the most favorable methods to decompose chlorinated hydrocarbons (*8*, *9*). In addition, the chemical method is accomplished by the addition of reductive materials to the contaminated sites artificially. The utilization of zerovalent iron as the reductive materials to dechlorinate chlorinated solvents was first patented by Sweeny and Fischer in 1981. This technique makes the *in situ* reductive dechlorination in ground-water reliable, and the permeable reactive barrier (PRB) filled with zerovalent iron as the reductive metal has been developed for the remediation of chlorinated compound as well as inorganics effectively.

The reductive materials used as the filling material in PRB should fulfill 3 major requirements and these are also the reasons why zerovalent iron is frequently used in groundwater remediation. First of all, the dechlorination products after the reaction should be non-toxic or environmental friendliness.

Secondly, the suitable redox potential (E_H^0) is needed. Thirdly, the economic cost of the selected material should be cost-effective. Although zerovalent iron can effectively dechlorinate most chlorinated hydrocarbons and has the advantages described above, several problems including the formation of passivated iron oxide layer, the production of hydrogen, and increase in pH value also occur. To solve these problems, some alternative metals including silicon (*10*, *11*), aluminum (*12*, *13*), magnesium (*14*) and zinc (*15*) have been developed and applied to the groundwater remediation. Zerovalent silicon (Si^0) is the second most abundant element following oxygen in the earth crust. Silicon compounds including silicon dioxide (SiO_2) are the most often found minerals in soil and groundwater environments. The most application of silicon element is in the semiconductor industry and is the subject of numerous investigations such as self-assembled monolayer (*16*, *17*). However, the high reactivity of silicon is still remarkable. In addition, zerovalent silicon has a lower redox potential (E_H^0 = -0.807 V) compared with iron (E_H^0 = -0.44 V). This indicates that zerovalent silicon can be used as a strong reductant for effectively reduction of contaminants. Due to the high reactivity of zerovalent silicon, zerovalent silicon can undergo oxidative dissolution reaction to rapidly form native SiO_2 on the surface of silicon during the exposure to humid air, and, therefore, the commercialized silicon surface are often treated with hydrogen fluoride (HF) and with 40% ammonium fluoride (NH_4F) to form hydrophobic and hydrogen-terminated silicon (Si-H) for preventing the further oxidation before using (*18*).

The basic properties of silicon make it as the good candidate which can apply in groundwater remediation. The first utilization of silicon for the reductive remediation was introduced by Doong et al. (*10*). Results showed that Si(0) is an efficiency reductive material for the dechlorination of carbon tetrachloride and PCE under anoxic conditions. Up to know, zerovalent silicon has been demonstrated to be an excellent reductive material for dechlorination of chlorinated hydrocarbons in subsurface environments. However, little is known about the reaction kinetics and mechanism of chlorinated hydrocarbons by zerovalent silicon under anoxic conditions. In this review, parameters controlling the reactivity of silicon in dechlorination were discussed. In addition, the enhanced effects of bimetallic system including Fe/Si and Ni/Si and addition of surfactant on the dechlorination kinetics and mechanisms were also introduced and reviewed.

The Basic Properties of Silicon in Dechlorination

Reductive dechlorination by zerovalent metal is highly dependent on the electron transfer from metal surface to the target compounds. Both the reduction potentials of zerovalent silicon and zerovalent iron are suitable for reductive dechlorination of chlorinated hydrocarbons. Therefore, the chlorination of chlorinated hydrocarbons by zerovalent silicon as well as zerovalent iron can be described by the pseudo-first-order kinetics with respect to the concentrations of chlrorinated hydrocarbons shown in equation:

$$\ln\left(\frac{[CHCl]_t}{[CHCl]_0}\right) = -k_{obs}t = -k_{sa}\rho_a t \quad (1)$$

where $(CHCl)_t$ and $(CHCl)_0$ represents to the concentration of chlorinated hydrocarbons at time t and 0 during the dechlorination, respectively, k_{sa} is the normalized surface reaction rate constant (L m^{-2} h^{-1}), ρ_a is the surface concentration of zerovalent metal (m^2 L^{-1}), and k_{obs} is the pseudo-first-order rate constant (h^{-1}). Moreover, the electron transfer from metal surface to the pollutant is depended on surface area of metal (*10*, *12*, *19–21*).

The utilization of silicon in the dechlorination of chlorinated hydrocarbons can be described as the equation shown below:

$$Si^0 + 2RX + 3H_2O \rightarrow H_2SiO_3 + 2RH + 2H^+ + 2X^- \quad (2)$$

where RX and RH represent chlorinated hydrocarbons before and after the dechlorination processes, respectively, and X^- is the chloride ions. The equation (2) is spontaneously favorable for zerovalent silicon to treat chlorinated pollutant and will produce 2 mol of protons. Therefore, the pH value of the dechlorination system would be the major dominant parameter should controlling the dechlorination efficiency and rate of chlorinated hydrocarbons (*10*).

Figure 1 shows the dechlorination of carbon tetrachloride by zerovalent silicon under various pH value conditions. From the degradation plot, the dechlorination rates (k_{obs}) increased from 0.5 h^{-1} to 2.1 h^{-1} when the pH value increased from 7.2 to 9.5. The results give the strong support that the reactivity of silicon in the dechlorination is pH dependent and is also fulfilled the eq (2). Zhou et al. (*22*) also reported that the increased pH value could enhance the reaction between silicon and water. Moreover, the mechanism on the enhanced dechlorination of carbon tetrachloride at high pH value is not only because of the Le Chatelier's principle but due also to the solubilization of produced intrinsic silicon dioxide by the hydroxide ions provided at high pH. Figure 2 shows the change in silicon species at different time intervals in bidistilled water system. It is clearly that the intensity of XPS peak at 103-105 eV, representing the silicon dioxide, increased with the incubation time, and the ratio of SiO_2 to Si_{total} increased from 1% at initial to 50% after 12 h of incubation, clearly indicating that silicon dioxide increased with time. The increase in pH value of can provide hydroxide ions for removal of SiO_2 on the surface of Si(0), resulting in the enhancement of the reactivity of silicon.

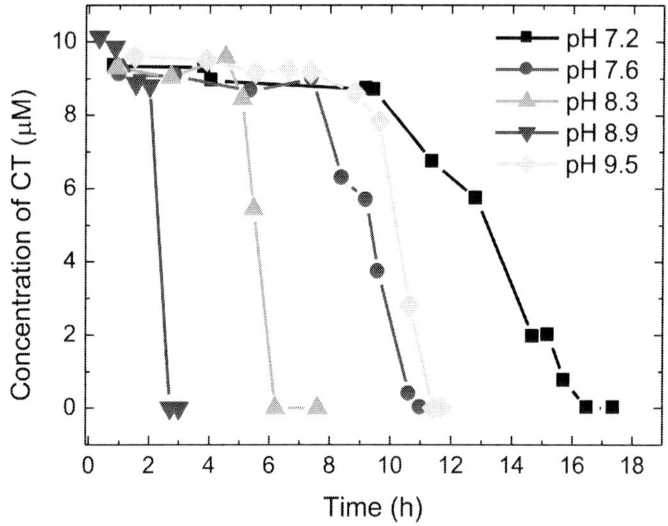

Figure 1. Dechlorination of carbon tetrachloride by 0.6 g zerovalent silicon under various pH values range from 7.2 to 9.5. Adapted from Water Sci. Technol. 2004, 50, 89-96. Copyright 2004 International Water Association.

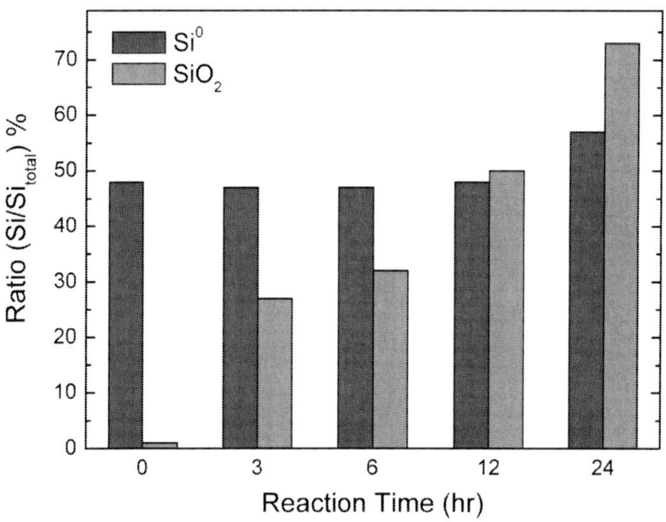

Figure 2. The change in silicon species as a function of incubation time in bidistilled water system under anoxic conditions.

Interestingly, the dechlorination plot shown in Figure 1 is much different from zerovalent iron system. The lag phase is observed when zerovalent silicon is applied to the reductive dechlorination. An increase in pH shortened the lag time, and the lag period decreased from 9.5 to 0.9 h upon increasing the pH from 7.2 to 9.5. Doong et al. (*10*) indicated that the lag period was mainly caused by the hydrogenated silicon and the intrinsic silicon dioxide on the surface of silicon particles. It is also found that hydroxide ion can catalyze the reaction of the hydrogenated silicon with H_2O as well as the removal of the intrinsic silicon oxide. Eq (3) shown below is the scheme to describe the behavior.

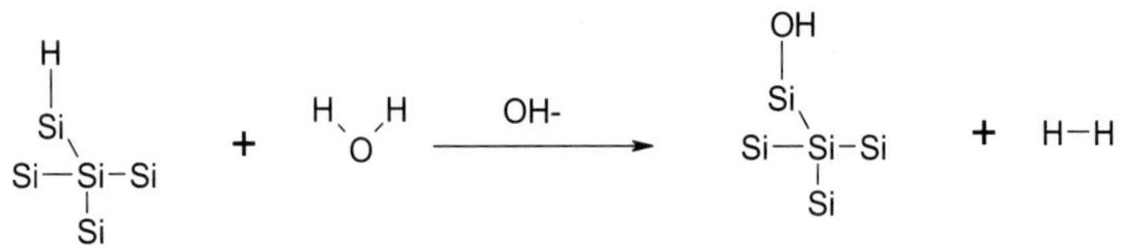

(3)

To prove the enhanced reactivity of zerovalent silicon after removal of silicon dioxide, carbon tetrachloride was re-injected into the silicon system. Figure 3 shows the dechlorination of 23 μM carbon tetrachloride by zerovalent silicon after the re-injection of same concentration of carbon tetrachloride at pH 8.3. It can be seen that a lag phase of 5 h was needed for the activation of silicon surface during the first injection process of carbon tetrachloride. After the activation, a nearly complete dechlorination of carbon tetrachloride was observed in 5 h. Same concentration of carbon tetrachloride was then re-injected into the silicon system after 24 h of incubation. A direct dechlorination without lag phase was clearly observed in the 2nd injection of target compound, and only 90 min was needed to completely dechlorinate carbon tetrachloride. In addition, the k_{obs} values for carbon tetrachloride dechlorination of the 1st and 2nd injection were 0.506 h^{-1} (r^2 = 0.952) and 2.404 h^{-1} (r^2 = 0.988). This result provides evidence that the H-terminated surface can be removed under alkaline conditions and the activated surface has excellent reactivity to dechlorinate the chlorinated hydrocarbons under anoxic conditions.

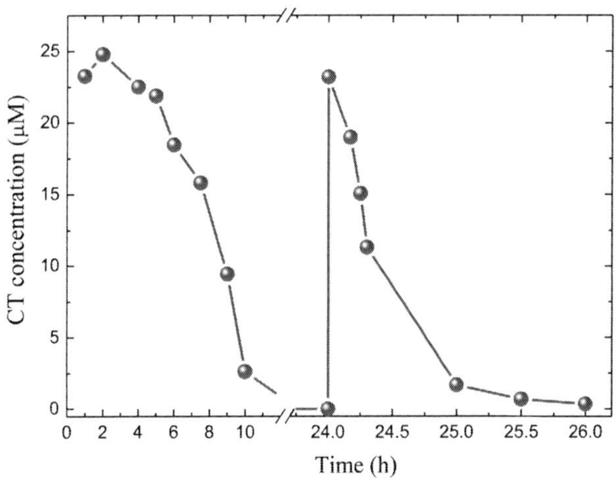

Figure 3. The dechlorination of carbon tetrachloride by zerovalent silicon after the re-injection of same concentration of carbon tetrachloride at pH 8.3. Adapted from Water Sci. Technol. 2004, 50, 89-96. Copyright 2004 International Water Association.

Although the redox potential of zerovalent silicon is lower than that of iron, the conversion of chlorinated hydrocarbons into non-toxic end products is not easy to achieve. Doong et al. (*10*) have found that chloroform (CF) and methylene chloride (DCM) were the major products, which corresponded to around 80% and 20%, respectively, of the total carbon mass. In addition, trichloroethylene (TCE) was the major intermediate of the dechlorination of tetrachloroethylene (PCE) by zerovalent silicon (*11*), showing that reductive dechlorination is the major reaction mechanism for the dechlorination of chlorinated hydrocarbons. Several studies have reported that the general reactions including hydrolysis, β-elimination, reductive elimination and hydrogenolysis are responsible for the degradation of chlorinated hydrocarbons by reductive metals (*20*, *23*). Matheson and Tratnyek (*24*) have pointed out that hydrogenolysis is the predominant pathway in describing the degradation of chlorinated hydrocarbons in zerovalent iron system. Unfortunately, it is noted that without changing the dechlorination pathway, the complete degradation of chlorinated hydrocarbons to non toxic hydrocarbons cannot be reached.

Combination of Silicon and Iron

Several strategies including the addition of second catalytic metal ions and the use of nanoscale zerovalent metals have been employed to change the dechlorination pathways of chlorinated hydrocarbons. Addition of second catalytic metal ions and the converted to nanosized zerovalent metals electrochemically has been proven to serve as catalysts to change the dechlorination pathway (*25*). This gives a great impetus to combine silicon and iron because silicon has a lower reduction potential than that of iron and can reduce iron ions to nanoscale zerovalent iron. It is noteworthy that the reaction of silicon with water would produce proton ions, while the anaerobic corrosion of zerovalent iron would increase the pH value. It would form the naturally-occurring buffer system when combining these two metal systems:

$$Si^0 + 2H_2O \rightarrow H_2SiO_4^{2-} + 2H^+ \tag{4}$$

$$Fe^0 + 2H_2O \rightarrow Fe^{2+} + 2OH^- \tag{5}$$

The equations shown in eq. (4) and (5) represent the reaction of zerovalent silicon and iron with water, respectively. From eq. (5), the pH value can be up to 9-10 in zerovalent iron system in the absence of buffer system, which could deactivate the zerovalent iron. In contrast, proton will be produced after the reaction between silicon and water under anoxic conditions. The decreased pH value in the silicon system will, therefore, compensate the increase in pH value in the iron-containing solution, and subsequently maintains the pH value at a relatively stable value. Figure 4(a) clearly shows the change in pH value to the reaction time in the presence of zerovalent iron and silicon. The result shows that the pH value decreased from 8.5 to 7.5 in first 50 h in the presence of silicon. Moreover, the pH value in iron-containing system increased rapidly from 7.5 to almost 9.1 in the first 3 h. However, the combination of 1.08 g iron with 0.06 g silicon constituted a buffer system and the pH value maintained at near neutral (7.5-7.8) without the addition of buffer system.

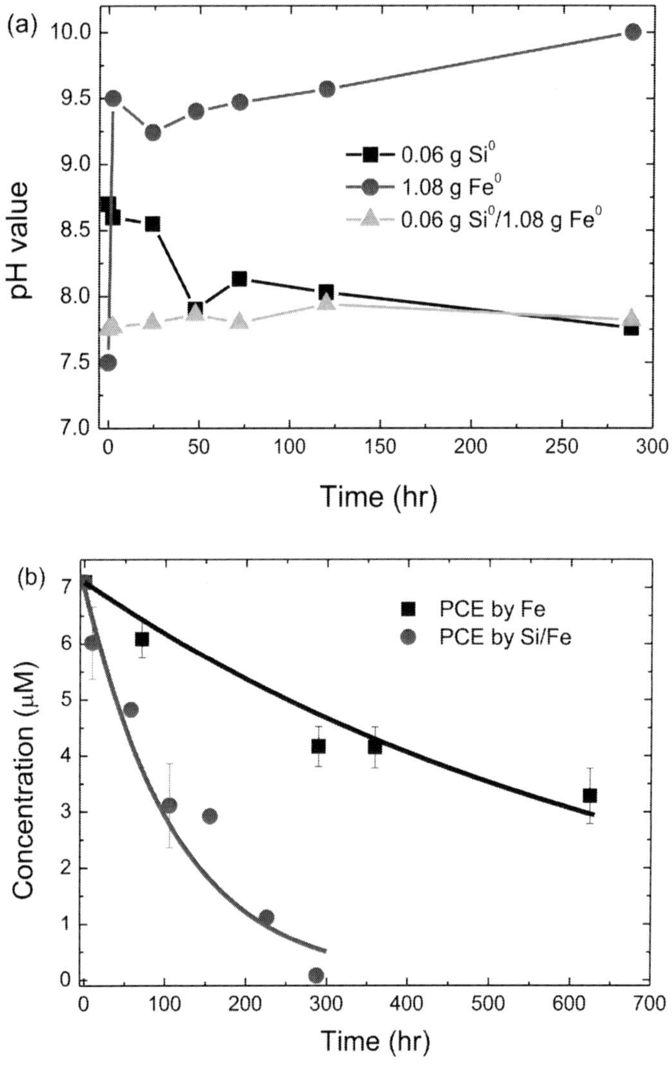

Figure 4. (a) Changes in pH and (b) dechlorination of PCE in Si, Fe and Si/Fe system under un-buffered conditions. The experiments were conducted in serum bottles containing 0.06 g of Si and 1.08 g of Fe without pH control. Adapted from reference (10). Copyright 2003 ACS.

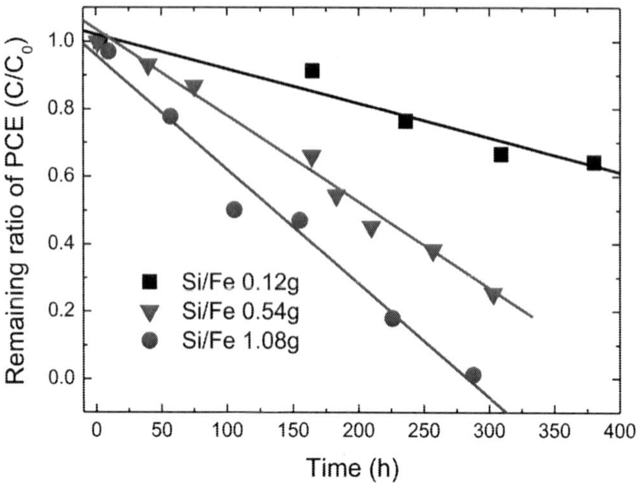

Figure 5. Effect of Fe amount on the dechlorination of PCE in the unbuffered Si/Fe system (10). Adapted from reference (10). Copyright 2003 ACS.

The maintenance of pH at a relatively stable value by bimetallic Si/Fe significantly enhanced the dechlorination of PCE. As shown in Figure 4(b), less than 60 % of PCE was degraded by un-buffered iron system within 620 h. TCE was found as the major product of PCE dechlorination by Fe. In contrast, a nearly complete dechlorination of PCE was observed by 0.06 g of silicon and 1.08 g of iron within 300 h. In addition, ethane, ethylene and trace amount of TCE were found as the major products after the dechlorination of PCE by 0.06 g Si/1.08 g Fe. These results show that the reductive dechlorination was the main reaction pathway for dechlorination of PCE in the presence of zerovalent Fe, while β-elimination and reductive dechlorination were the dominant reaction pathways for PCE dechlorination in the combined system of Si and Fe. The results also indicate that the combination of silicon and iron can control the pH value of the system and further maintains the dechlorination reactivity of zerovalent metals under anoxic conditions.

The dechlorination of PCE by Fe, Si and Si/Fe followed the pseudo-first-order kinetics. The k_{obs} for PCE dechlorination was 0.0013 h^{-1} (r^2 = 0.972) for zerovalent Fe alone and 0.0088 h^{-1} (r^2 = 0.950) for Si/Fe, clearly showing that the combination of Si and Fe significantly enhances the dechlorination efficiency and rate of PCE as well as change the dechlorination pathway. It is noteworthy that the correlation coefficient for dechlorination of PCE in the bimetallic Si/Fe system was not high and seemed to follow zero-order kinetics. To verify this hypothesis, various loadings of iron ranging from 0.12 to 1.08 g in Si/Fe system were used to dechlorinate PCE. As shown in Figure 5, the dechlorination of PCE by various Fe loadings in Fe/Si systems at pH 7.2 followed zero-order kinetics and the correlation coefficients (r^2) were in the range 0.983-0.997. Usually zero-order rate equation is used to describe the reaction behaviors of high concentration of pollutants on the limited reaction sites. In this study, the dechlorination of PCE followed the zero-order kinetics upon increasing the total amounts of zerovalent iron ranging from 0.12 to 1.08 g in Si/Fe system. In the Si/Fe system, the solution pH at 7.2 may favor the reactivity of iron but would inhibit the dechlorination rate of PCE by silicon, which might lead to the change in kinetics order because of the limited amounts of Fe used in the system. Additional study will be needed to evaluate the concentration effect of silicon and iron on the dechlorination kinetics in Si/Fe system.

Combination of Silicon and Nickel

Zerovalent nickel has been proven to be an effective catalytic metal which can enhance the dechlorination rate through the hydrodechlorination as well as change the distribution of final products. In silicon system, the dechlorination efficiency of PCE is enhanced by simply mixing the zerovalent silicon and nickel ions. The nickel ions would convert to zerovalent nickel by Si(0), and the k_{obs} for PCE dechlorination by Ni/Si increased from 0.0047 h^{-1} to 0.019 h^{-1}, which is 4 times higher than that in the zerovalent silicon alone (*11*). In addition, ethane, ethene and trace amounts of *n*-butane were observed as the final non-toxic products. The catalytic effect caused by the combination of silicon and zerovalent nickel is mainly attributed to the electron transfer between these two metals (*26*). Figure 6 shows the XPS spectra of Ni species onto the Si surface under anoxic conditions at pH 8.3. The Ni2p spectrum after Ar sputtering showed two peaks at binding energies of 853.0 and 870.4 eV, which can be assigned as the $Ni^0 2p_{3/2}$ and $Ni^0 2p_{1/2}$, respectively. In addition, the ratio of peak areas of $Ni^0(2p_{1/2})$ to $Ni^0(2p_{3/2})$ is calculated to be 1:2.3, clearly indicating the formation of zerovalent nickel onto the surface of Si. In addition, the Si2p peak at 99.5 eV originates mainly from a single chemical form of $Si2p_{3/2}$ which is the binding energy of unreacted bulk silicon. The broad peak at 103–105 eV is attributed to silicon atoms in more positive oxidation states than those in the bulk. Silicon oxidized by oxygen and/or water gives rise to peaks at 103–105 eV due to SiO_2 and SiO groups. It is noted that the peak intensity at 103–105 eV increased as the Ni(II) concentration increased, and the ratio of Si/SiO_2 decreased from 5.0% to 0.5% when Ni(II) concentration increased from 1 to 4 mM, clearly depicting that the existence of nickel ion accelerates the oxidation of zerovalent silicon.

The dechlorination efficiency and rate of PCE by Si is greatly enhanced by the addition of nickel ions. Lee and Doong (*11*) depicted that the dechlorination efficiency of PCE by Si increased from 40% in the absence of Ni(II) to 99.9% at 0.5 mM Ni(II) after 150 h of incubation, and then deceased to 73% when the Ni(II) concentration further increased to 1 mM. Figure 7 show the PCE dechlorination by zerovalent silicon with various loading of nickel ions. The k_{obs} values increased linearly from 0.0038 ± 0.0005 h^{-1} in the absence of Ni(II) to 0.0516 ± 0.0041 h^{-1} at 0.5 mM Ni(II) and then decreased to 0.0085 ± 0.0004 h^{-1} when further increase the Ni(II) loading to 1.0 mM. This result reflects the fact that addition of 0.5 mM Ni(II) has the maximum enhancement effect on PCE dechlorination by Si.

Several studies have addressed the effect of additive loadings toward the dechlorination efficiency of chlorinated hydrocarbons by zerovalent iron (*27–29*).

An optimal mass loading exists for a wide variety of bimetallic catalysts. Tee et al. (*28*) investigated the role of bimetallic Ni/Fe nanoparticles on the dechlorination of TCE. A higher degradation rate of TCE was observed upon increasing the Ni loading from 2 to 25 wt%, and then the k_{obs} decreased when greater Ni loading was added. Moreover, Lin et al. (*29*) reported that the dechlorination rate of TCE by bimetallic Ru/Fe increased as the Ru loading increased from 0.25 to 1.5 wt%. A decrease in k_{obs} was also observed when Ru loading increased to 2.0 wt%. Several mechanisms have been proposed to explain the concentration effect of catalytic metal ions in bimetallic system. The formation of galvanic cell was used to explain the observed phenomenon (*30*). In the hypothesis, the increased reaction rate was cathodically controlled by the addition of catalytic metal loading. However, the reaction was dominated by the aniodically control when the loading of catalytic metal was higher than its optimized value. In addition, surface coverage of catalytic metal onto the reductive metal was also proposed to explain the phenomenon (*31–33*). This hypothesis suggests that the degradation efficiency would be terminated when the surface of the reductive metal is completely covered by the catalytic metal.

In the silicon system, an optimal mass loading of nickel at 0.5 mM was also obtained for PCE dechlorination (*11*). The possible explanation is that the increased number of Ni nanoparticles caused by higher addition of nickel ion concentration on the surface of silicon may enhance the catalytic activity of hydrodechlorination. However, higher loading of second dopant metal leads to the aggregation of fine catalytic nanoparticles into large ones, and subsequently decreases the reaction rate of chlorinated hydrocarbons. Different form the hypothesis, the combination of nickel as the second catalytic metal in silicon system seems to have another hypothesis to explain the phenomena. Figure 8 shows the SEM image and EPMA elemental maps for Ni species onto the Si surface. the SEM images clearly show that the reduced nickel can be deposited onto the surface of Si. The particles sizes were measured in the range between 20 and 30 nm when 0.5 mM Ni(II) was added, and the particle size increased to 40-80 nm at 1mM Ni(II). In addition, EPMA elemental maps showed that the Ni nanoparticles aggregated at 1mM Ni(II). These results clearly support the hypothesis that addition of low concentration of second metal ions increases the catalytic activity of Si, while the nanoparticles aggregate into large particles at high loading of Ni(II).

The products distribution is also correlated to the dechlorination rates. Figure 9 shows the product distribution of PCE dechlorination by Si in the presence of various concentrations of Ni(II) after 150 h of incubation (*11*). In the absence of Ni(II), only TCE was found to be the major product obtained after PCE dechlorination. The increase in Ni(II) concentration not only increased the extent and rate of PCE dechlorination but also change the end product from TCE to ethane. Since Ni(II) can be reduced to its metallic form onto the Si surface, and PCE is transformed quantitatively to ethane without formation of detectable chlorinated intermediate compounds when Ni(II) concentration is higher than 0.1 mM, hydrodechlorination is supposed as the major reaction pathway for PCE dechlorination by Ni/Si.

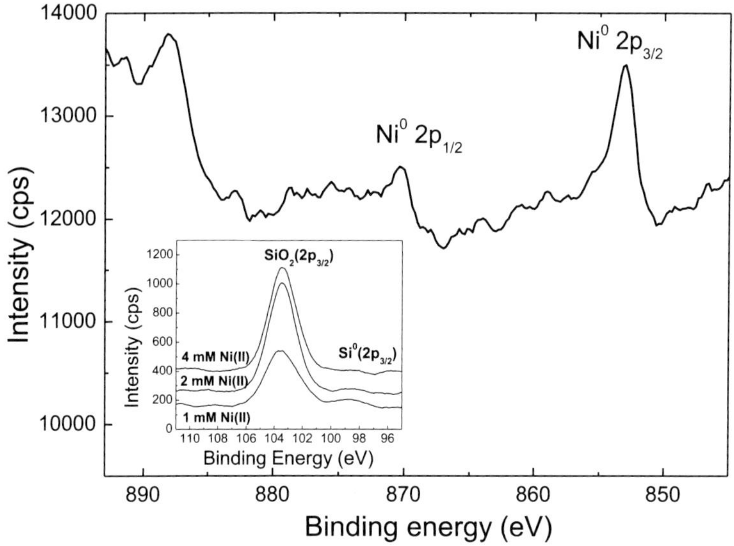

Figure 6. The XPS spectra of Ni(2p) species on the surface of Si at pH 8.3 under anoxic conditions. The inserted figure is the XPS spectra of Si species as a function of Ni(II) concentration. Adapted from reference (11). Copyright 2008 ACS.

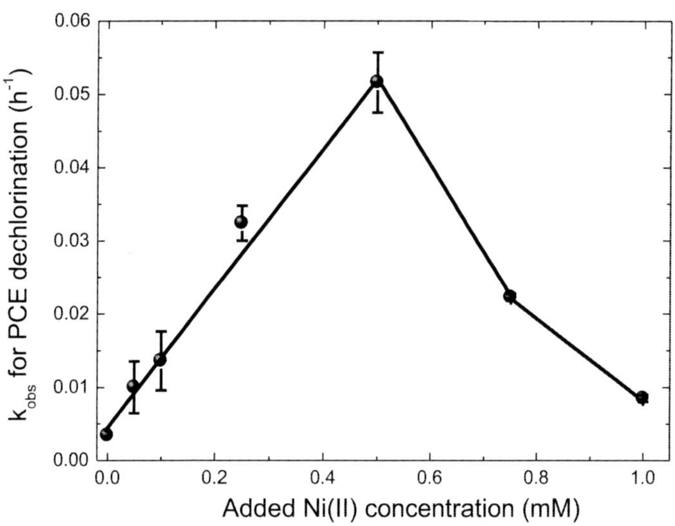

Figure 7. The pseudo-first-order rate constants (k_{obs}) for PCE dechlorination as a function of Ni(II) concentration under anoxic conditions. Adapted from reference (11). Copyright 2008 ACS.

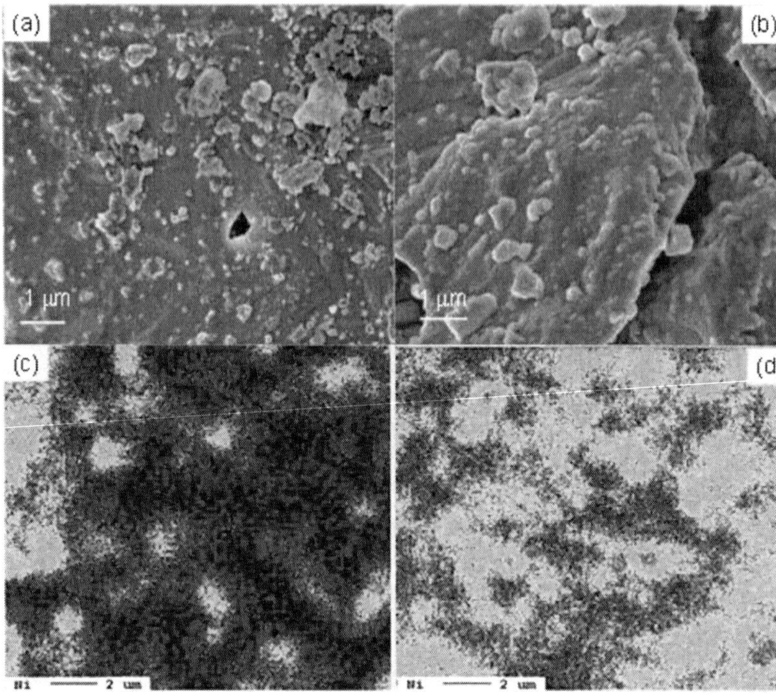

Figure 8. The SEM images and EPMA analysis of Ni-modified Si in the presence of (a, c) 0.5 and (b, d) 1 mM of Ni(II) at pH 8.3. Adapted from reference (11). Copyright 2008 ACS.

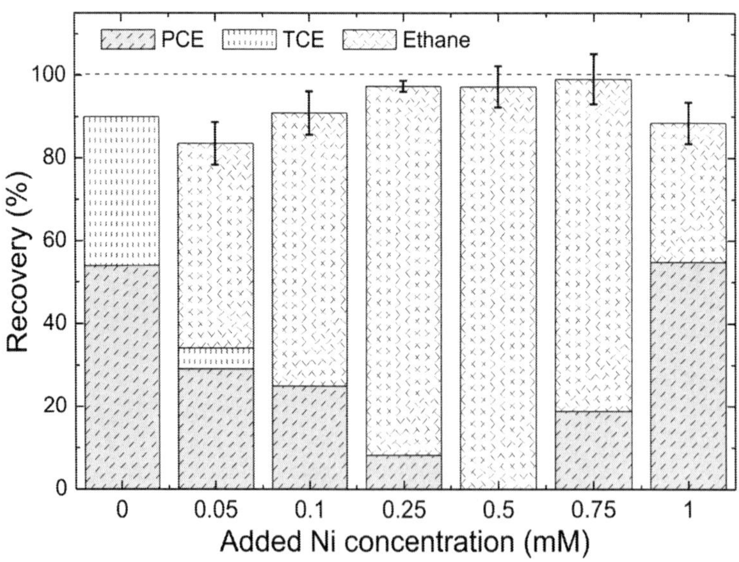

Figure 9. The product distribution of PCE dechlorination by Ni/Si after 150 h of incubation. The concentrations of Ni(II) were in the range 0-1 mM. Adapted from reference (11). Copyright 2008 ACS.

The Enhancement Method by the Addition of Surfactants

The utilization of surfactant is another effective strategy to enhance the dechlorination efficiency ant rate of chlorinated hydrocarbons (*34–36*). It can not only enhance the solubility of aliphatic chlorinated compounds into the bulk solution, but also change the surface of zerovalent metals from hydrophilic into hydrophobic property (*34*). Surfactants such as cetyl trimethylammonium bromide (CTAB), sodium dodecyl sulfate (SDS), Tween 80 and its related compounds are commonly used in iron system (*34*, *36*, *37*). Daniel and the colleague reported that the rate constant for PCE dechlorination increased by a factor of 4-19 with the addition of cationic surfactant when applied to zerovalent iron system (*34*). Sayles group also reported that the degradation rates of DDT, DDD and DDE by zerovalent iron in the presence of surfactant increased at least 2 times when compared with that in the absent of surfactant (*36*). In Pd-Fe system, the enhanced dechlorination of trichlorobenzene was accelerated by 1.5-2.5 times in the presence of CTAB (*37*).

The addition of surfactant in silicon system has the similar effect as in iron system. Lee and Doong (*38*) used several surfactants including SDS, CTAB, Tween 80, and PEG to understand the enhanced reactivity of PCE by silicon. Results shown in Figure 10 clearly indicate that dechlorination efficiency of PCE was enhanced only in the CTAB- and PEG-amended solutions. Little enhancement in PCE dechlorination was obtained when 1 CMC Tween 80 or SDS was added. Several studies have indicated that the sorbed surfactant onto iron surface could enhance the mass transfer of chlorinated hydrocarbons from solution to iron surface (*34*, *37*). However, the addition of Tween 80 and SDS to the silicon system cannot enhance the mass transfer of chlorinated hydrocarbons from solution to the surface, and no further enhancement was observed for PCE dechlorination. Different from SDS and Tween 80, the k_{obs} for PCE dechlorination increased from 0.0035 to 0.21 h^{-1} when CTAB is added. The enhanced aqueous solubility of PCE in the presence of CTAB was found to be the major reason for enhancement of the dechlorination efficiency and rate in silicon system. However, the toxicity and the formation of foam limited the application of surfactants in the environment (*39*, *40*). Addition of PEG significantly enhanced the dechlorination efficiency and rate of PCE by Si, and the k_{obs} for PCE dechlorination was 0.36 h^{-1}. PEG is a bi-functional amphiphilic polymer with reactive OH groups on both ends and has been widely used in biological and food chemistries. PEG has several advantages including high water solubility, non-toxicity, less expensive, biodegradable, and environmentally friendly, and has also been employed as a cross-linker to immobilize the synthesized nano-sized zerovalent iron onto the support (*41*, *42*). Interestingly, results displayed in Figure 10 show that the k_{obs} for PCE dechlorination is 1.5 times higher than that of CTAB and more than 100 times higher than that by Si alone, clearly indicating that the addition of PEG can effectively enhance the dechlorination efficiency and rate of chlorinated hydrocarbons.

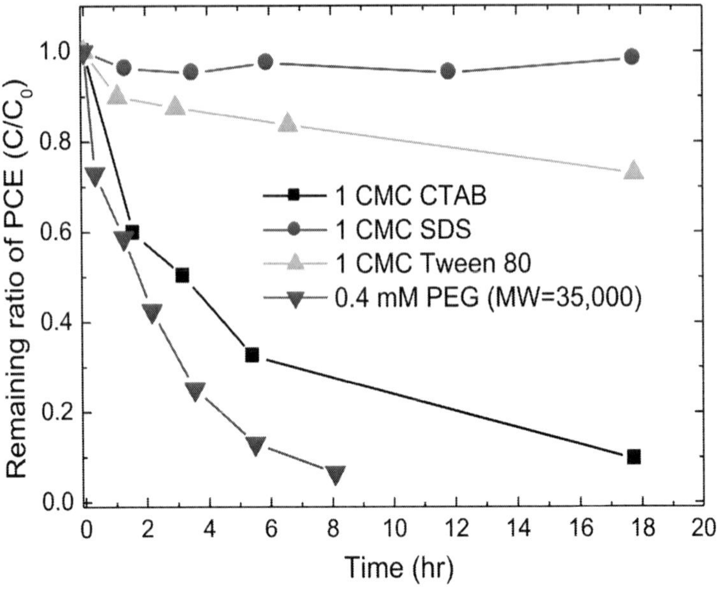

Figure 10. The dechlorination of PCE (62 μM) by zerovalent silicon (2 g L^{-1}) in the presence of different kinds of surfactants. Adapted from reference (38). Copyright 2011 ACS.

Various loading amounts of PEG were added to the silicon system to figure out the concentration effect of PEG on the change in dechlorination rate of PCE. Figure 11 shows the dechlorination of PCE by Si in the presence of various loadings of PEG ranging from 1 nM to 400 μM. The dechlorination efficiency and rate of PCE increased upon increasing PEG concentrations and a nearly complete dechlorination of PCE was observed within 15 h when PEG concentration was higher than 0.2 μM. In addition, the k_{obs} for PCE dechlorination increased from 0.049 h^{-1} at 0.001 μM PEG to 0.36 h^{-1} at 0.2 μM PEG, and then leveled off to a plateau (0.35–0.39 h^{-1}) when higher PEG concentration was introduced. The appearance of plateau is mainly attributed to the fact that the adsorbed polymer can reach a maximum coverage at high concentration of PEG, and thereby forms a protection layer to retain the activity of most reactive sites on Si surface, leading to the occurrence of saturation of reactivity.

The reason for the improvement of the dechlorination rate after the addition of PEG has not been well-understood. Zerovalent silicon has a low redox potential and can easily react with water to produce hydrogen gas and silicon dioxide. It is known that the silicon dioxide can serve as the insulator layer to reduce the electron transfer rate and reactivity of Si. Although the production of large amount of H_2 can be utilized in catalytic bimetallic system under anoxic conditions, the consumption of the reductive metals do existed. Liu et al. (*43*) proposed an idea to preserve the reductive ability by terminating the formation of H_2 in zerovalent iron system, which could increase the dechlorination rate of chlorinated hydrocarbons and keep the benefit for long-term usage. Interestingly, addition of PEG also can prevent the formation of SiO_2 onto the surface of zerovalent iron, leading to the acceleration of dechlorination efficiency and rate of PCE (*38*). To further examine the effect of PEG on the formation of silicon dioxide, various loadings of PEG ranging between 0.01 and 0.6 μM were added into Si-H_2O system. After incubation for 3 h, the released gases in all the tested bottles containing PEG were all around 4-5 mL, which were lower than that in the absence of PEG (10-15 mL). However, the ratio of the produced SiO_2 ($SiO_2/(Si+SiO_2)$, SiO_2/Si_{total}), determined by the ratio of peak areas of SiO_2 and total Si species, decreased upon increasing PEG concentration. As depicted in Figure 12a, the ratio of SiO_2/Si_{total} decreased from 0.77 in the absence of PEG to 0.11 when the PEG loading was 0.6 μM. A good relationship between the SiO_2/Si_{total} ratio and the k_{obs} for PCE dechlorination was also established (Figure 12b). The interaction between PEG and metal particles may involve several interaction mechanisms including ion exchange, chelation, electrostatic interaction, and hydrogen bond. Addition of PEG can adsorb onto the Si surface quickly to form a protection layer, and subsequently decreases the reaction rate of Si with water to inhibit the production of silicon dioxide onto the surface of Si. Therefore, electrons produced from the surface of Si can be easily transferred to PCE, resulting in the enhancement of dechlorination efficiency and rate of PCE.

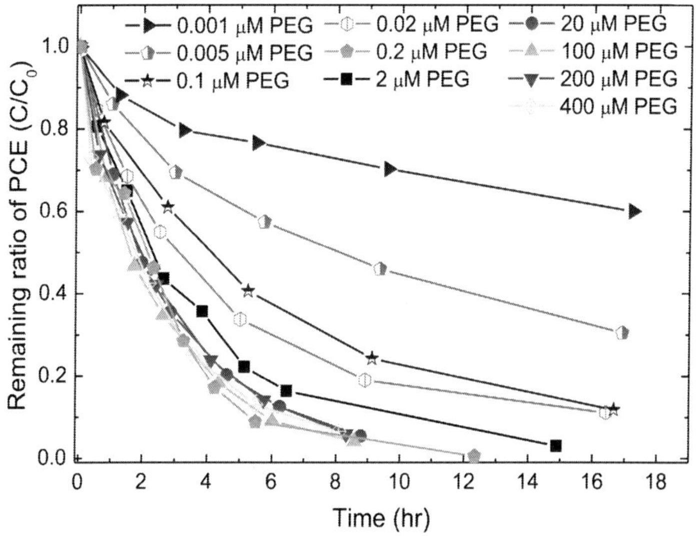

Figure 11. PCE dechlorination by zerovalent silicon (2g L^{-1}) in the presence of various loadings of PEG ranging from 1 nM to 400 μM. Adapted from reference (38). Copyright 2011 ACS.

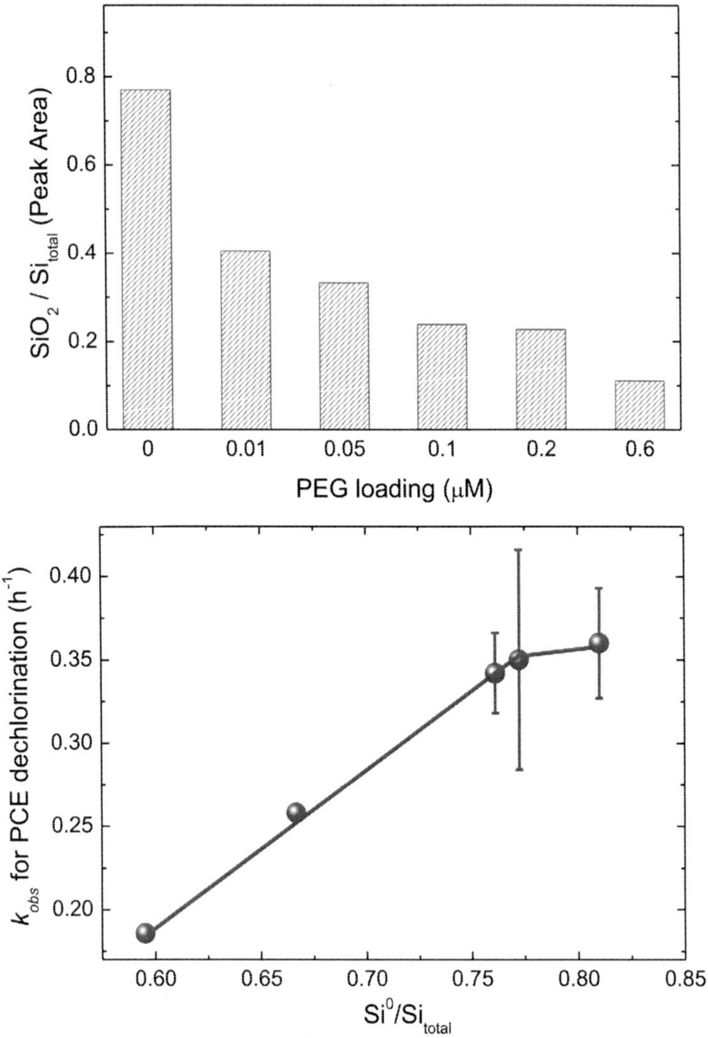

Figure 12. (a) The ratio of SiO_2 to Si total amount and (b) k_{obs} for PCE dechlorination as the function under anoxic conditions. Adapted from reference (38). Copyright 2011 ACS.

Conclusions

In this study, we have reviewed the effective removal of chlorinated hydrocarbons by zerovalent silicon and bimetallic silicon systems under various conditions. The catalytic metal ions including Ni^{2+} and Fe^{2+} can be reduced to nanoscale zerovalent metal by silicon. The XPS results indicated the formation of zerovalent heavy metals through the reactivity of zerovalent silicon. However, the roles of Ni and Fe were different. Addition of Fe^{2+} to Si can form the buffer solution to maintain the pH at a relatively stable value, and subsequently enhance the dechlorination efficiency and rate of carbon tetrachloride and PCE. In contrast, the adsorbed Ni(II) ion exhibited the synergistic effect on the dechlorination of PCE by zerovalent silicon under anoxic conditions. The k_{obs} for PCE dechlorination by zerovalent silicon increased with the increase in nickel ion concentration from 0 to 1.5 wt%, and then decreased when higher loading of nickel ion was added. The change in k_{obs} for PCE dechlorination at various nickel ion concentrations can be explained by the physical change in surface morphology. Moreover, The modification of zerovalent silicon surface with PEG is another strategy to dramatically enhance the dechlorination efficiency and rate of PCE by zerovalent silicon. The degradation rate of PCE by zerovalent silicon in the presence of PEG was 106 times higher than that in the absence of PEG. Similar to the effect of nickel ion on PCE dechlorination, the dechlorination efficiency of PCE increased upon increasing PEG concentration up to 0.2 μM, and then leveled off to a plateau at high loading of PEG, which can be up to 400 μM. The proposed mechanism for enhanced dechlorination by zerovalent silicon is mainly due to the prevention of the formation of passivated layer of SiO_2. These processes would significantly accelerate the dechlorination rate of chlorinated hydrocarbons under anoxic conditions and would be helpful in facilitating the development of processes that could be useful for the enhanced degradation of co-contaminants for long-term performance.

Acknowledgments

The authors thank National Science Council, Taiwan for financial support under contract No. NSC101-2221-E-007-084-MY3.

References

1. Forkert, P. G.; Sylvestre, P. L.; Poland, J. S. *Toxicology* **1985**, *35*, 143–160.
2. Sujatha, T. V.; Hegde, M. J. *Mutat. Res., Genet. Toxicol. Environ. Mutagen.* **1998**, *413*, 151–158.
3. Lash, L. H.; Qian, W.; Putt, D. A.; Desai, K.; Elfarra, A. A.; Sicuri, A. R.; Parker, J. C. *Toxicol. Appl. Pharm.* **1998**, *150*, 49–57.
4. Roberts, A. L.; Gschwend, P. M. *J. Contam. Hydrol.* **1994**, *16*, 157–174.
5. Kueper, B. H.; Redman, D.; Starr, R. C.; Reitsma, S.; Mah, M. *Ground Water* **1993**, *31*, 756–766.

6. Stroo, H. F.; Unger, M.; Ward, C. H.; Kavanaugh, M. C.; Vogel, C.; Leeson, A.; Marqusee, J. A.; Smith, B. P. *Environ. Sci. Technol.* **2003**, *37*, 224a–230a.
7. Henry, S. M.; Hardcastle, C. H.; Warner, S. D. *Chlorinated Solvent and DNAPL Remediation*; ACS Symp. Ser. 837; American Chemical Society: Washington, DC, 2002, pp 1−20.
8. Bigg, T.; Judd, S. J. *Environ. Technol.* **2000**, *21*, 661–670.
9. Vogel, T. M.; Criddle, C. S.; Mccarty, P. L. *Environ. Sci. Technol.* **1987**, *21*, 722–736.
10. Doong, R. A.; Chen, K. T.; Tsai, H. C. *Environ. Sci. Technol.* **2003**, *37*, 2575–2581.
11. Lee, C. C.; Doong, R. A. *Environ. Sci. Technol.* **2008**, *42*, 4752–4757.
12. Lien, H. L.; Zhang, W. X. *Chemosphere* **2002**, *49*, 371–378.
13. Birnbaum, H. K.; Buckley, C.; Zeides, F.; Sirois, E.; Rozenak, P.; Spooner, S.; Lin, J. S. *J. Alloys Compd.* **1997**, *253*, 260–264.
14. Morales, J.; Hutcheson, R.; Cheng, I. F. *J. Hazard. Mater.* **2002**, *90*, 97–108.
15. Feng, J.; Lim, T. T. *Chemosphere* **2005**, *59*, 1267–1277.
16. Oura, K.; Lifshits, V. G.; Saranin, A. A.; Zotov, A. V.; Katayama, M. *Surf. Sci. Rep.* **1999**, *35*, 1–69.
17. Struck, L. M.; Eng, J.; Bent, B. E.; Flynn, G. W.; Chabal, Y. J.; Christman, S. B.; Chaban, E. E.; Raghavachari, K.; Williams, G. P.; Radermacher, K.; Mantl, S. *Surf. Sci.* **1997**, *380*, 444–454.
18. Lee, C. C.; Doong, R. A. *Appl. Catal., B* **2014**, *144*, 182–188.
19. Doong, R. A.; Chiang, H. C. *Environ. Sci. Technol.* **2005**, *39*, 7460–7468.
20. Arnold, W. A.; Roberts, A. L. *Environ. Sci. Technol.* **2000**, *34*, 1794–1805.
21. Doong, R. A.; Lee, C. C.; Chen, K. T.; Wu, S. F. *Water Sci. Technol.* **2004**, *50*, 89–96.
22. Zhou, X. W.; Ishida, M.; Imanishi, A.; Nakato, Y. *J. Phys. Chem. B* **2001**, *105*, 156–163.
23. Lowry, G. V.; Reinhard, M. *Environ. Sci. Technol.* **1999**, *33*, 1905–1910.
24. Matheson, L. J.; Tratnyek, P. G. *Environ. Sci. Technol.* **1994**, *28*, 2045–2053.
25. Liu, Y. Q.; Majetich, S. A.; Tilton, R. D.; Sholl, D. S.; Lowry, G. V. *Environ. Sci. Technol.* **2005**, *39*, 1338–1345.
26. Li, X. Q.; Zhang, W. X. *Langmuir* **2006**, *22*, 4638–4642.
27. Schrick, B.; Blough, J. L.; Jones, A. D.; Mallouk, T. E. *Chem. Mater.* **2002**, *14*, 5140–5147.
28. Tee, Y. H.; Grulke, E.; Bhattacharyya, D. *Ind. Eng. Chem. Res.* **2005**, *44*, 7062–7070.
29. Lin, C. J.; Lo, S. L.; Liou, Y. H. *J. Hazard. Mater.* **2004**, *116*, 219–228.
30. Liou, Y. H.; Lo, S. L.; Lin, C. J.; Hu, C. Y.; Kuan, W. H.; Weng, S. C. *Environ. Sci. Technol.* **2005**, *39*, 9643–9648.
31. Nutt, M. O.; Heck, K. N.; Alvarez, P.; Wong, M. S. *Appl. Catal., B* **2006**, *69*, 115–125.
32. Nutt, M. O.; Hughes, J. B.; Wong, M. S. *Environ. Sci. Technol.* **2005**, *39*, 1346–1353.
33. Cwiertny, D. M.; Bransfield, S. J.; Livi, K. J. T.; Fairbrother, D. H.; Roberts, A. L. *Environ. Sci. Technol.* **2006**, *40*, 6837–6843.

34. Alessi, D. S.; Li, Z. H. *Environ. Sci. Technol.* **2001**, *35*, 3713–3717.
35. Chatterjee, S.; Lim, S. R.; Woo, S. H. *Chem. Eng. J.* **2010**, *160*, 27–32.
36. Sayles, G. D.; You, G. R.; Wang, M. X.; Kupferle, M. J. *Environ. Sci. Technol.* **1997**, *31*, 3448–3454.
37. Zhu, B. W.; Lim, T. T.; Feng, J. *Environ. Sci. Technol.* **2008**, *42*, 4513–4519.
38. Lee, C. C.; Doong, R. A. *Environ. Sci. Technol.* **2011**, *45*, 2301–2307.
39. Isomaa, B.; Reuter, J.; Djupsund, B. M. *Arch. Toxicol.* **1976**, *35*, 91–96.
40. Rosety, M.; Ribelles, A.; Rosety-Rodriguez, M.; Carrasco, C.; Ordonez, F. J.; Rosety, J. M. *Histol. Histopathol.* **2000**, *15*, 1137–1143.
41. Parshetti, G. K.; Doong, R. A. *Water. Res.* **2009**, *43*, 3086–3094.
42. Xu, J.; Bhattacharyya, D. *Environ. Prog.* **2005**, *24*, 358–366.
43. Liu, Y.; Phenrat, T.; Lowry, G. V. *Environ. Sci. Technol.* **2007**, *41*, 7881–7887.

Editors' Biographies

Ruey-an Doong

Ruey-an Doong received his Ph.D. in Environmental Engineering at National Taiwan University in 1992. He is a full professor in the Department of Biomedical Engineering and Environmental Sciences at National Tsing Hua University in Taiwan. Currently, Prof. Doong is the Dean of the College of Nuclear Science at National Tsing Hua University. He is also serving as an editorial member of several reputed journals such as *Sustainable Environmental Research*, *Journal of Biosensors and Bioelectronics*, and *Journal of Environmental Chemical Engineering*. He has authored more than 200 journal articles, book chapters, and proceedings. He was honored as a fellow of the Alexander von Humboldt Foundation of Germany in 2000.

Virender K. Sharma

Virender K. Sharma received his Ph.D. in Marine and Atmospheric Chemistry at the University of Miami after graduating from the Indian Institute of Technology in New Delhi, India with a Master's degree in Technology. He is currently a director of the Center of Ferrate Excellence and a professor of chemistry at Florida Institute of Technology. He was a visiting research scholar at Stanford University under the advisory of Professor Ed Solomon and won both the Florida Tech Faculty of the Year Award and the ACS Orlando Section Outstanding Chemist Award in 2008. He is highly active in pursuing chemistry and environmental applications of high-valent iron species. His other interests include nanoparticles in aquatic environments and in remediation of contaminated water. He has published more than 270 journal articles, book chapters, and proceedings. The books that he has authored and edited include *Ferrates: Synthesis, Properties, and Applications in Water and Wastewater Treatment and Sustainable Nanotechnology and the Environment: Advances and Achievements* (published by Oxford University Press).

Hyunook Kim

Hyunook Kim is a Professor of Environmental Engineering at the University of Seoul in Korea and the Director of the R&D Center of Core Technologies for Water Treatment. Professor Kim earned his B.S. degree in Environmental Science from Yonsei University in Korea in 1994, an M.S. degree in Environmental Engineering from Johns Hopkins University in 1997, and a Ph.D. from The University of Maryland at College Park in 2000. Before he joined the faculty at the University of Seoul in 2002, he worked as Environmental Engineer for the U.S. Dept. of Agriculture (MD, USA).

Professor Kim's research in the area of water pollution control includes a number of projects on process control and the operation of water and wastewater treatment plants. He is especially interested in monitoring and control of contaminants of emerging concern. He has written numerous journal papers and conference presentations. He has served as Associate Editor of *Chemosphere* since 2007 and as an Editorial Board member of *Critical Review in Environmental Science and Technology* since 2012.

Indexes

Author Index

Subject Index

E

F

G

H

I

L

M

N

Z